GROUND MOVEMENTS
—AND THEIR—
EFFECTS
—ON—
STRUCTURES

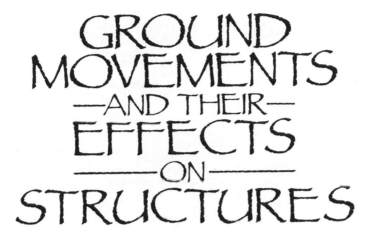

GROUND MOVEMENTS
—AND THEIR—
EFFECTS
—ON—
STRUCTURES

edited by
P.B. Attewell, PhD, DEng, CEng, MICE
Professor of Engineering Geology
University of Durham
and
R.K. Taylor, MSc, PhD, CEng, FIMM, FGS
Reader in Engineering Geology
University of Durham

Surrey University Press

Distributed in the USA by
Chapman and Hall
New York

Published by Surrey University Press
A member of the Blackie Group
Bishopbriggs, Glasgow G64 2NZ and
Furnival House, 14–18 High Holborn, London WCIV 6BX

Distributed in the USA by
Chapman and Hall
in association with Methuen, Inc.
733 Third Avenue, New York, N.Y. 10017

British Library Cataloguing in Publication Data
Ground movements and their effects on
 structures.
 I. Attewell, P.B.   II. Taylor, R.K.
 624.1'762   TH1094

ISBN 0-903384-36-1
For the USA, International Standard Book Number is
0-412-00391-0

Photosetting by Thomson Press (India) Limited, New Delhi
Printed in Great Britain by Bell & Bain Ltd., Glasgow

List of contributors

N.N. Ambraseys, Dip. Ing., PhD, DIC, DSc, C.Eng, FICE, FGS
Professor of Engineering Seismology
Department of Civil Engineering
Imperial College of Science & Technology
London

P.B. Attewell, BSc, PhD, D.Eng, C.Eng, MICE
Professor of Engineering Geology
Department of Geological Sciences
University of Durham

E.N. Bromhead, BSc, MSc, DIC, PhD, C.Eng, MICE, FGS
Senior Lecturer in Civil Engineering
Kingston Polytechnic
Kingston upon Thames

J.A. Charles, MSc(Eng), PhD, C.Eng, MICE
Section Head
Dams and Earthworks Section
Geotechnics Division
Building Research Establishment
Garston
Watford

A.E. Cobb, BSc, PhD, C.Eng, MIMM, FGS
Senior Engineer,
Geoffrey Walton Consulting Mining &
Engineering Geologists
Oxford

I. Cooper, BSc, MSc, C.Eng, MIMM
Assistant Director
WRC Engineering
Swindon

J.C. Cripps, BSc, MSc, PhD, FGS
Lecturer in Geology
University of Sheffield
Sheffield

R. Driscoll, BSc, MSc, C.Eng, MICE
Head of Foundation Section
Geotechnics Division
Building Research Establishment
Garston
Watford

R. Foott, BSc, ScD, C.Eng, MICE, MASCE
Managing Principal
Dames & Moore,
San Francisco
California

J.D. Geddes, BSc, PhD, C.Eng, FICE, FASCE, MIHT, FGS
Professor of Civil Engineering
Head of Department of Civil Engineering
& Building Technology
UWIST
Cardiff

M.R. Hurrell, BSc,
Demonstrator
Department of Geological Sciences

J.A. Jackson, MA, PhD
University of Durham
Research Fellow in Seismology
Bullard Laboratories
Madingley Rise
Cambridge.

D. Koutsoftas
Principal
Dames & Moore
San Francisco
California

R.T. Murray, BSc, PhD, MIHE
Senior Principal Scientific Officer
Transport and Road Research laboratory
Crowthorne
Berkshire

P. Rumsey, BSc(Eng), MSc(Eng), DIC, C.Eng, MICE
Manager
Applied Technology Group
WRC Engineering
Swindon

B.O. Skipp, BSc, PhD, C.Eng, MICE, FGS
Consultant
Soil Mechanics Limited
Bracknell
Berkshire

I.F. Symons, MSc, C.Eng, MICE, MIHE
Principal Scientific Officer
Transport and Road Research laboratory
Crowthorne
Berkshire

R.K. Taylor, BSc, MSc, PhD, C.Eng, FIMM, FGS
Reader in Engineering Geology
Department of Geological Sciences
University of Durham

G. Walton, BSc, C.Eng, MIMinE, MIMM, FGS
Principal
Geoffrey Walton Consulting Mining & Engineering Geologists
Oxford

J. Yeates, BSc, C.Eng, MICE
Design Engineer
Northumbrian Water Authority
Gosforth
Newcastle upon Tyne

Contents

6   Tunnelling in soil   132
P.B. Attewell and J. Yeates

7   Mining subsidence   216
G. Walton and A.E. Cobb

Editorial introduction

The ground movements discussed in this book are mainly those that need to be considered by the practising geotechnical engineer. Some of these movements are quite small, as in tunnelling and trenching, but they can create problems for in-ground and above-ground structures in urban areas. Others, such as the ground lowering associated with coal-mining, can be relatively large. Considerations of ground movements are not restricted to vertical displacements since with some ground engineering operations horizontal movements may be dominant. Nor should one think only of quasi-static—or long-term—movements. Man-made vibration and natural seismicity impose engineering and environmental constraints on construction. The common theme running through the various chapters is that the processes and phenomena described are amenable to a reasonable degree of prediction and analysis. Specialist authors are drawing upon their own experience in an attempt to provide the necessary data for a design rationale which will accommodate the movements involved.

Seismicity is included in the book because of the worldwide problems it poses, and also because of the growing appreciation of its relevance in areas such as the United Kingdom which is customarily regarded as experiencing low seismic activity (an aseismic area). In contrast, a number of other mega-scale processes have been excluded from this volume. Prominent on a regional scale is frost heave, culminating in permafrost conditions of northern climes. This complex phenomenon is covered extensively in recent texts which present the views and design criteria of both North American and Soviet authors (Tsytovich, 1975; Andersland and Anderson, 1978; Jumikis, 1966).

It is pertinent to comment briefly on the extraction of fluids—water, brine, gas and oil—all of which can lead to 'land subsidence' because of changes in crustal loading. A well-documented review of these largely uncontrollable, widespread processes, which are often categorized as geological hazards (Scheidegger, 1975; De Freitas, 1978), is given by Poland and Davies (1969) and also by Allen (1969).

Large ground settlements have caused major damage in many leading cities of the world as a consequence of water abstraction from both deep and shallow wells. Lowering the groundwater level reduces the ambient pore-pressure and thus increases the effective stress. Freeze and Cherry (1979) make the point that the common feature at these land subsidence (well) sites is a thick sequence of unconsolidated or poorly-consolidated sediments forming an aquifer-aquitard system. Although the drainage from the sand-gravel aquifers causes consolidation and settlement, it is aquitard drainage which produces the greatest degree of settlement, albeit at a slower rate. Jacob (1940) concluded that the *chief* source of water on tapping an elastic artesian aquifer would be from the adjacent and included clay beds (or shales) following a time delay.

In Mexico City the groundwater table has been lowered some 30 m by wells abstracting from aquifers ranging in depth from 30 m to 600 m below ground level. Until the nature of the problem was recognized a settlement rate of 1 mm/day was recorded (see Carrillo, 1948). Settlements totalling about 8 m have been measured in this century. Considerable damage to the city's drainage and sewerage system has resulted, piled foundations have either been exposed or overloaded as a consequence of negative skin friction, and differential settlements have damaged numerous famous and antique structures (Hiriart and Marsal, 1969).

Over 2 million people in Tokyo live in a 78 km area of the city which has undergone a maximum subsidence of about 4 m. The lowest land is now over 2 m below sea level. In Osaka City, water abstraction has produced a maximum ground lowering of about 3 m over an area of about 280 km (Poland and Davis, 1969).

Perhaps one of the most currently emotive issues concerns the celebrated city of Venice, where in 1858 the wells were typically artesian with a 4 m head above sea level By the year 1900 the head had fallen to 3 m and by 1970 it was about 6 m below sea level (Penman, 1978). The fall in piezometric head was found to correlate closely with ground lowering (Ricceri and Butterfield, 1974). It has been postulated that the increase in settlement rate in the post-1952 period can be attributed to the effective pressure being in excess of previous overburden pressure and thus acting on the virgin part of the consolidation curve of the compressible clays. The settlement induced by over-abstraction, largely as a result of encroaching industry, has combined with a positive rise in sea level to produce periodic flooding of the historic city and the islands. Land subsidence in the region close to Venice has also arisen from another source. Methane extraction from the Po Delta had by 1966 resulted in a 300 mm/year maximum rate of ground lowering.

The examples cited above concern compressible, poorly-consolidated sediments. It should not be overlooked, however, that underdrainage of overconsolidated clays can also result from excessive water abstraction. An analysis by Wilson and Grace (1942) suggests that between 1865 and 1913 the decline in artesian head in the Chalk gave rise to a maximum settlement of about 213 mm in the overlying London Clay of the London Basin.

An *increase* in ground loading by water impounded in reservoir basins is believed to induce seismicity in many parts of the world. (Perhaps the classic case is the seismicity induced by the impounding behind Kremasta Dam on the Acheloos River, near Agrinion in Greece.) Indeed, Howells (1973) concluded that an increase in local seismicity on filling occurs in about one in a hundred cases.

Historically, wild brine-pumping in Cheshire, England resulted in serious subsidence at distances up to 8 km from pumping centres (see Bell, 1975, 1978; also Calvert, 1915). Brine abstraction from natural springs by the natives in the Northwich and Nantwich areas of Cheshire was first recorded by Caesar's *salinators* in pre-Christian times. Salt subsidence is important regionally, and information on the subject from other parts of the world is now being reported (Ege, 1979; Walters, 1977; Jeremic, 1975; Grube, 1973; cf. Allen's 1969 comments).

A well-known example of land subsidence as a consequence of oil extraction is the cracking of the lining of the Baldwin Hills reservoir in California, and partial washing out of the embankment, during pumping from the Inglewood anticline in the years 1943–57. In this case the pore-fluid pressure was reduced by over $3000 \, kN/m^2$ but the pressure drop was not transmitted uniformly through the field (De Freitas, 1978).

In a number of small oilfields recharging by fluid injection has been modestly successful in reducing subsidence, although it should be recalled that deep injection of waste fluids in the Denver well in 1962 has caused local earth tremors.

The aforementioned land subsidence examples concern increasing crustal loads. It is important to consider the alternative, namely, the removal of overburden, which leads to rebound phenomena. Rebound must be considered by engineers whenever the equilibrium of crustal materials is disturbed by large-scale excavations or by natural processes. A useful review is given by Nichols (1980).

In the near-surface, undisturbed ground zone, horizontal stresses are commonly equal to, or greater than, the vertical stresses. Hoek and Brown (1980) have plotted the ratio, K, of average horizontal to vertical stress against depth below surface (z) for 116 results elicited from the literature. Values of K lie within the following limits:

$$(100/z) + 0.3 < K < (1500/z) + 0.5$$

where z is in metres.

For depths of less than 500 m the horizontal stresses are significantly greater than the vertical stresses, whereas at 1 km depth or more, vertical and average horizontal stresses are more equal. At these greater depths, and over a protracted period of geological time, processes of plastic deformation at the high stress levels prevailing must tend to equalize the all-round state of stress (Heim, 1912). Nevertheless, stress differences must still exist at these depths, as evidenced by the existence of deep-focus earthquakes.

The highest K value recorded by Hoek and Brown (1980) is 5.56, although they recognize the existence of (local) values up to about 10 in (currently tectonic) mountain belts. Blackwood (1970) reported K to be about 7.40 in Australian rocks at very shallow depths. In overconsolidated clays the ratio is usually denoted as K_0 (earth pressure at rest) on an effective stress basis. A maximum value of K_0 equal to 2.8 at 7 m depth is given by Skempton (1961) for London Clay at Bradwell, whilst computations by Bishop et al. (1965) suggest that an even higher maximum value might apply to London Clay. Since the passive earth pressure coefficient, K_p, is of the same order as K_0 in this depth range (that is, the maximum shear stress is equal to the shearing resistance of the clay), Terzaghi (1961) suggested that failure would be occurring at shallow depths. This is not incompatible with Bjerrum's (1967) hypothesis that under natural conditions expansion in clays and mudrocks is restricted less in the vertical than in a horizontal direction, so that the degree of vertical load-shedding is greater than any reduction in a horizontal direction. In clays and mudrocks having strong diagenetic bonds Bjerrum visualized deformation, water uptake, and progressive failure as being functions of the time-dependent release of strain energy stored during compaction and diagenesis. The opening of (additional) fractures and fissures because of high horizontal stresses and relaxation in mudrocks is seen by Cripps and Taylor (1981) as constituting the onset of weathering (the converse of diagenesis).

The question of stress field generation in rocks at shallow depth presents a number of interpretive problems. With the advent of plate tectonics, however, the analysis of plate motions has indicated that continental areas are generally in compression. Apparently contradictory stress field measurements—$K > 1$ in extensional areas such as the mid-Atlantic ridge and the African rift valley (Herget, 1973)—are not necessarily incompatible with the existence of a compressive boundary force helping to push the plates apart at, for example, an ocean ridge (Bott, 1983). The contribution made by present-day, as opposed to past (residual) stress fields is problematic and it is currently more realistic to consider the cumulative effects of both. In this way the cumulative strain energy imprint of tectonic, thermal and gravitational compaction forces is unlocked progressively, either by natural uplift and erosion or by man in digging large-scale excavations.

The earlier references to 'land subsidence' are cautionary examples of uncontrolled ground movements of a hazardous character. In the chapters which follow, the contributing authors have each been encouraged to outline a problem involving ground movements and to present the salient features of prediction, analysis and design relating to those movements with a view to controlling or accommodating them with geotechnical understanding.

PBA

RKT

References

Allen, A.S. (1969). Geologic settings of subsidence. In *Reviews in Engineering Geology*, (eds. D.J. Varnes and G. Kiersch), 2, The Geological Society of America, Boulder, Colorado, pp. 305–342.

Andersland, O.B. and Anderson, D.M. (1978). *Geotechnical Engineering for Cold Regions*. McGraw–Hill Book Co., London.

Bell, F.G. (1975). Salt and subsidence in Cheshire, England. *Eng. Geol.* 9, 237–247.

Bell, F.G. (1978). Subsidence due to mining operations, In *Foundation Engineering in Difficult Ground* (ed. F.G. Bell). Newnes–Butterworths, London, pp. 322–362.

Bishop, A.W., Webb, D.L. and Lewin, P.I. (1965). Undisturbed samples of London Clay from the Ashford Common shaft: strength–effective stress relationships. *Géotechnique* 15, 1–31.

Bjerrum, L. (1967). Progressive failure in slopes of over-consolidated clay and clay shales. *Proc. J. Soil Mech. Fdns. Div. ASCE*, 93, (SM5), 1–49.

Blackwood, R.L. (1977). The tectonophysical significance of large lateral in situ stresses in the Australian continent. In *Int. Symp. on Geodynamics in South-West Pacific*, Techip, Paris, pp. 395–404.

Bott, M.H.P. (1983). Personal communication.

Calvert, A.F. (1915). *Salt in Cheshire*. E. and F.N. Spon Ltd., London.

Carrillo, N. (1948). Influence of artesian wells on the sinking of Mexico City. *Proc. 2nd. Int. Conf. Soil Mech. & Fdn. Engng.* 7, 156–159.

Cripps, J.C. and Taylor, R.K. (1981). The engineering properties of mudrocks. *Q. J. eng. Geol.* 14, 325–346.

De Freitas, M.H. (1978). Geological hazards. In *Industrial Geology* (ed. J.L. Knill), Oxford Univ. Press, pp. 287–309.

Ege, J.R. (1979). Selected bibliography on ground subsidence caused by dissolution and removal of salt and other soluble evaporites. *Open file report* U.S. Geological Survey (Washington D.C.) 79/1133.

Freeze, R.A. and Cherry, J.A. (1979). *Groundwater*. Prentice-Hall, Inc., New Jersey.

Grube, F. (1973). Experience of engineering geology on the top of salt dome Othmarschen–Langenfelde (Hamburg) *Proc. Symp. on Sink-holes and Subsidence, Hanover*. Deutsche Gesellschaft fur-und Grundbau (Essen).

Heim, A. (1912). Zur Frage der Gebirgs-und Gesteinsfestigkeit. *Schweiz. Bauztg.* 50, February, 1912.

Herget, G. (1973). Variation of rock stresses with depth at a Canadian iron mine. *Int. J. Rock Mech. and Min. Sci.* 10, 37–51.

Hiriart, F. and Marsal, R.J. (1969). *The Subsidence of Mexico City*, Nabor Carrillo Commemorative Volume. Mexican Geotechnical Soc., 109–147.

Hoek, E. and Brown, E.T. (1980). *Underground Excavations in Rock*. The Institution of Mining and Metallurgy, London.

Howells, D.E. (1973). Man-made earthquakes; the reservoir designers problem. *Proc. Percolation Through Fissured Rock*. Symp. Int. Soc. Rock Mechanics and Int. Assoc. Eng. Geol., Stuttgart. Paper T2-E.

Jacob, C.E. (1940). On the flow of water in an elastic artesian aquifer. *Am. Geophys. Union Trans.* 2, 574–586.

Jeremic, M. (1975). Subsidence problems caused by solution mining of the rock salt and potassium deposits. *Proc. 10th Can. Rock Mech. Symp. (Kingston)*.

Jumikis, A.R. (1966). *Thermal Soil Mechanics*. Rutgers University Press, New Brunswick.

Nichols, T.C. (1980). Rebound, its nature and effect on engineering works. *Q. J. eng. Geol. London* 13, 133–152.

Penman, A.D.M. (1978). Ground water and foundations. In *Foundation Engineering in Difficult Ground* (ed. F.G. Bell), Newnes–Butterworths, London, pp. 204–225.

Poland, J.F. and Davis, G.H. (1969). Land subsidence due to withdrawal of fluid. In *Reviews in Engineering Geology* (eds. D.J. Varnes and G. Kiersch), **2**, The Geological Society of America, Boulder, Colorado, pp. 187–269.

Ricceri, G. and Butterfield, R. (1974). An analysis of compressibility data from a deep borehole in Venice. *Géotechnique* **24**, 175–192.

Scheidegger, A.E. (1975). *Physical Aspects of Natural Catastrophes.* Elsevier, Amsterdam.

Skempton, A.W. (1961). Horizontal stresses in an over-consolidated Eocene Clay. *Proc. 5th Int. Conf. Soil Mech. and Fdn. Engng. Paris*, **1**, 351–7.

Terzaghi, K. (1961). In discussion of A.W. Skempton, 'Horizontal stresses in an over-consolidated Eocene clay'. *Proc. 5th Int. Conf. Soil Mech. and Fdn. Engng. Paris*, **3**, 144–5.

Tsytovich, N.A. (1975). *The Mechanics of Frozen Ground.* McGraw–Hill Book Co., London.

Walters, R.F. (1977). Land subsidence in central Kansas related to salt dissolution. *Bull. State Geol. Surv. Kansas* (Lawrence), **214**.

Wilson, G. and Grace, H. (1942). The settlement of London due to underdrainage of the London Clay. *J. Instn. Civ. Engrs.*, **19**, Paper No. 5294, 100–127.

Acknowledgements

The Editors of this volume are indebted to the individual authors for agreeing in the first place to contribute, and for weathering the pressures in attempting to write to deadlines. Each author has been given freedom to develop his own chapter in the manner thought best by himself. The Editors may not always share the views expressed by the contributors.

Particular thanks are due to Mrs Wendy Lister for her secretarial assistance throughout this project.

1 Settlement of natural ground under static loadings

ROGER FOOTT and DEMETRIOUS KOUTSOFTAS

1.1 Introduction: settlement and its prediction

Application of a vertical load to a soil stratum causes deformations within the soil which result in settlement. When the load is applied over a finite area of the stratum, the soil deformations are multi-dimensional, involving both vertical compression of the soil beneath the load and lateral deformation of the soil out from under the loaded area.

> *Granular soils* are usually free-draining and of relatively low compressibility. Subject to normal static engineering loads, they generate settlements which typically occur rapidly and are usually quite small. The engineering practice which has developed for predicting these settlements makes little attempt to model in detail the complex behaviour of granular soils (that is, variations with relative density, method of deposition, stress history, strain history, and so on; see Ladd *et al.*, 1977). Instead, predictions are generally made using either direct empirical correlations or elastic theory with an empirically selected modulus.
>
> Settlements in *cohesive soils* are typically much greater than in granular materials. They also typically take much longer to occur, due to the slow drainage characteristics and the time-dependent stress-strain behaviour of clay materials. The engineering practice which has developed for cohesive soils therefore considers settlement in terms of a number of discrete components which have different relative importances.

The first settlement component for a clay soil is an *initial, undrained* and essentially instantaneous response to the placement of load. With the common assumptions that the clay is fully saturated and relatively little consolidation occurs during load application, the clay is considered substantially incompressible during this initial settlement phase. Initial, undrained settlement therefore results from lateral displacement of the clay from beneath the loaded area. This component of settlement is usually relatively small except in the case of highly plastic, organic clays (see Foott and Ladd, 1981). It is typically predicted using elastic theory or finite element modelling, with empirical selection of modulus.

The second settlement component in clay soil is *'undrained' creep*, whereby lateral displacement of soil from beneath the loaded area continues on a long-term basis during the early stages of consolidation. Again, these movements are generally very small and/or insignificant, except with the same highly plastic, organic clay soils noted above. Undrained creep movements are extremely difficult to predict. Routine engineering practice can only attempt to screen out potentially troublesome situations and improve them by modifying the loading and drainage conditions (Foott and Ladd, 1981).

1

The third settlement component for clay soils is due to *consolidation*, as pore water is squeezed out of the clay fabric. This component is the best known and the one for which prediction techniques are most highly developed. Consolidation generally yields the largest component of settlement on clay soils, and settlement prediction using the one-dimensional Terzaghi consolidation model frequently gives a reasonable practical estimate of total settlement due to all causes (Seed, 1965).

The fourth and final settlement component in clay soils occurs after completion of primary consolidation. It is broadly referred to as *drained creep* and, for practical purposes, it can usefully be considered as the long-term equilibrium of the clay's fabric. It is commonly predicted using the one-dimensional case of *secondary compression* as measured in a consolidation test.

1.2 Prediction of stress increments within a soil stratum

Application of load to a soil stratum causes stress increments which vary with the relative size and geometry of the loaded area and the stratum, with the stress-strain properties of the soil, and with the boundary conditions.

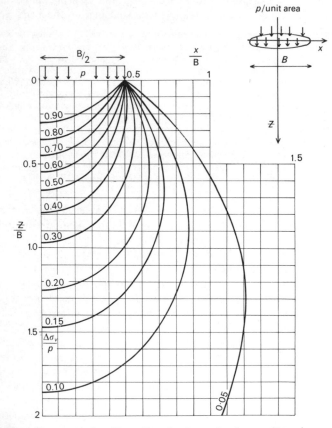

Figure 1.1 Vertical stresses induced by uniform load on a circular area (Boussinesq solution after Lambe and Whitman, 1969).

Hand computation of stress increments is usually performed via solutions based on elastic theory and often presented in chart form. The most basic solutions are for uniformly-distributed loads applied to the surface of an elastic layer of infinite depth and width and having uniform and isotropic properties throughout. Examples are shown in Fig. 1.1, which indicates the distribution of vertical stress increments $(\Delta\sigma_v)$ beneath a uniform circular loading, and Fig. 1.2, which presents charts for estimating the vertical stress increment beneath the corner of a uniformly-loaded rectangular area on the surface of a similar 'semi-infinite' layer. These solutions can be superimposed to yield stress increment data at any point within the soil mass and for loadings of less regular geometry. Thus, for example, the stress increments beneath point X due to uniform loading p on area ABCD in Fig. 1.2c can be computed by using the charts of Fig.1.2 and summing solutions for a downward loading p on area AB'XC', and upward loading p on areas BB'XD" and CD'XC', and a downward loading p on area DD"XD'.

The particular charts of Figures 1.1 and 1.2 are independent of Poisson's ratio (v) and therefore apply to all types of 'elastic' soils, provided only that the layer is of sufficient extent and can be considered as having uniform and isotropic properties. However, solutions for other than vertical stress increments (for example, for principal stresses, horizontal stresses, shear stresses) and for layers of limited extent, vary considerably with change in v, which in turn varies considerably with soil type. A fully saturated clay loaded without drainage would typically have $v = 0.5$ (that is, incompressible material), but with drainage this might become 0.3 to 0.4 for normally consolidated or slightly

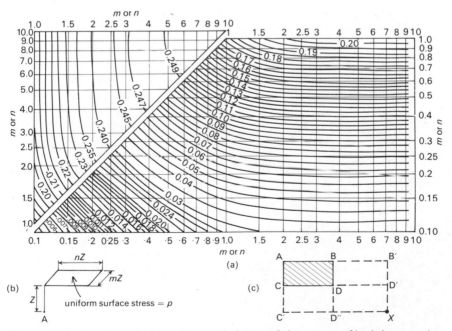

Figure 1.2 (a) Chart for use in determining vertical stresses below corners of loaded rectangular surface areas on elastic, isotropic material. Chart gives $f(m,n)$. (b) At point A, $\Delta\sigma_v = p \times f(m,n)$. (c) Use of superposition. (Data from Newmark, 1942; after Lambe and Whitman, 1969).

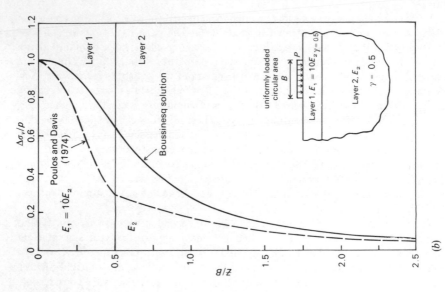

(b)

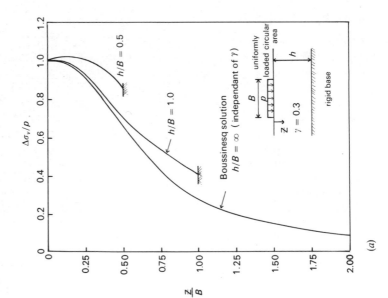

(a)

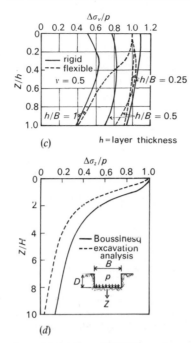

Figure 1.3 (a) Elastic layer over rough rigid base. (b) Stiff layer over soft layer. (c) Effects of footing rigidity and depth of rigid base on vertical stress increments for a uniformly loaded circular area (after Poulos and Davis, 1974). (d) Vertical stresses beneath strip load at base of excavation ($B/D = \frac{1}{2}$) (after Burland, et al., 1977).

overconsolidated material and as low as 0.15 to 0.3 for heavily overconsolidated clays. Values of v for free-draining sand might similarly range from zero (collapsing sand) to 0.5 or higher (very dense sand which dilates on shearing), although values would more typically be in the range 0.3 to 0.4 at stress levels well below failure. The aforementioned values are very approximate and only represent a starting point for using chart solutions. If the engineering problem is found to be very sensitive to the v value used, then detailed evaluation may be appropriate.

A wide range of standard solutions similar to the charts of Figs. 1.1 and 1.2 are available for estimating stress increments within soil layers under more complicated conditions of loading or soil characteristics. Some of the typical trends revealed by these solutions are described below and in Fig. 1.3, in which the vertical stress increments below the centre of a circular loaded area are compared with values from the Boussinesq solution of Fig. 1.1.

The following conditions should be carefully considered.

(a) *The effect of a limited thickness of soil layer, with underlying incompressible material.* Tends to inhibit the spreading of load with depth and to increase the vertical stress increments directly beneath the loaded area compared to the Boussinesq case, particularly just above the base of the layer (Fig. 1.3a).

(b) *The effect of increasing soil stiffness with depth.* Tends to inhibit load spreading with depth, the lower and stiffer material beneath the loaded area being apt to experience increased vertical stress increments compared to the Boussinesq case (Gibson and Sills, 1971).

(c) *The effect of a stiff layer overlying a softer layer.* Tends to spread the vertical load faster and reduce vertical stress increment in the soft layer immediately below the loaded area (Fig. 1.3b).

(d) *The effect of the loaded area being rigid.* Produces a non-uniform load application at the soil surface and a decrease in centreline vertical stress increments compared to a uniform loading, as shown in Figure 1.3c for various finite layer thicknesses (h). Note that the effective rigidity of the loaded area can have major impact on the stress distribution, which in turn affects the resulting settlement.

(e) *The effect of embedment of the loaded area within the soil stratum.* Tends to reduce vertical stress increments below the load, compared to the Boussinesq values for the same depth below the loaded area (Fig. 1.3d).

For the most comprehensive documentation of elastic solutions available, the reader is referred to Poulos and Davis (1974).

In addition to chart solutions based on elastic theory, several alternative techniques exist for estimating stress increments. These range from simply assuming a constant rate of load-spreading with depth (often 1 horizontal to 2 vertical) and distributing the total applied load over the increased area at the depth under consideration, to the use of finite element analyses with detailed modelling of soil profiles, variations in v, differences in stress-strain relationships, and so on. The simple assumption of constant rate of load spreading with depth has application in routine design procedures (for example, local Codes of Practice) for which an empirical justification has been developed. The computer models have application where high precision is needed, where the soil profile is very complex, or (particularly) where very stiff or soft inclusions in the profile may cause stress concentrations of practical significance. It should be noted that the use of finite element solutions for stress increment predictions is not as expensive or time-consuming as one might assume. Continuous improvement in electronic computers will eventually make the use of such programs routine.

An important aspect of the prediction of vertical stress increments at depth is the interaction of adjacent footings. Several relatively simple computer programs are available which can compute stress increments at any given point caused by a number of surface loads applied simultaneously. Alternatively, elastic solutions can be superimposed to provide an indication of interaction effects. Similarly, the complex question of soil–structure interaction, whereby the relative stiffnesses of the loaded area and the foundation soils influence the stress distribution and settlement profiles, can be addressed through an iterative stress distribution/settlement analysis (see Burland, 1977, for a discussion of the practical importance of soil–structure interaction, and Focht *et al.*, 1978, for an example of soil–structure interaction analysis).

Good engineering practice in predicting vertical stress increments for settlement prediction is to choose the most appropriate method having regard to the complexity of the soil profile and loading, the sensitivity of the design, the cost impact of design refinement, and the availability of evaluated experience with various methods and design procedures. Whatever method is used, considerable variation between predicted and actual stress increments should be expected and considered in design. Lambe and Whitman (1969) suggest that these variations may be ± 25 per cent or more.

1.3 Prediction of settlement of granular soils

Techniques for predicting settlement in granular soils have developed primarily around the need to limit differential settlements within buildings to avoid structural or cosmetic damage. A common building requirement with spread footings on granular soils is to keep the maximum total settlement to one inch (25.4 mm)*, which has been correlated to maximum angular distortions in the building of below 1/300 (Terzaghi and Peck, 1948, 1968; Skempton and McDonald, 1956). This requirement is frequently neither very difficult nor very costly to achieve, so rather simple prediction techniques have usually been adequate for engineering practice.

In general, the predictive approach to be used is either a direct empirical correlation of settlement with *in situ* testing data, or the development of empirical correlations of modulus for use with elastic theory for settlement prediction. This section briefly reviews the techniques available and comments when possible on the accuracy which might be expected from them.

1.3.1. Standard penetration test methods

The Standard Penetration Test (SPT) developed from the use of a conventional split spoon sampler to obtain disturbed samples of the soil being penetrated. By recording the number of standard blows (a 140-lb hammer falling 30 inches; 65 kg: 760 mm; BS 1377, (1975)) required for the sampler to penetrate 12 inches (300 mm) into the soil, the *SPT blow-count* (N) is obtained and this has been correlated with soil characteristics. Techniques for settlement prediction using blow-count data have also been developed.

The physical similarity between driving a sampler into the ground and settlement due to a static load applied slowly to the ground surface is clearly limited, and a correspondingly limited accuracy of settlement predictions from blow-count data might therefore be presumed. The sensitivity of the blow-count data itself to the operator's technique is frequently cited as a further limitation of SPT settlement predictions. Nevertheless, SPT procedures are used world-wide for settlement prediction and they probably have the greatest body of data available for correlation purposes.

The first technique for predicting settlement using SPT blow-counts was proposed by Terzaghi and Peck (1948, 1968) as a conservative, envelope design aid for sizing footings on sand. They suggested that settlement varies as a function of blow-count, footing width, depth of water table and embedment of the footing, and presented approximate and conservative procedures for making settlement estimates.

Several attempts have been made to refine the Terzaghi and Peck procedure. Meyerhof (1965) noted its conservatism and suggested that the predicted settlements be reduced by 33 per cent and that no correction for high-water table was needed, since the effect of water table would already be implicit in the blow-counts. Correction of blow-counts to reflect variations in overburden stress using the relationship of Gibbs and Holtz (1957) has also been suggested and discussed (see, for example, Sutherland, 1963; Thornburn, 1963; Alpan, 1964; D'Appolonia et al., 1968; Holtz and Gibbs, 1969). Peck

---

*Grant et al. (1974) suggested that the allowable settlement could be increased to 1.5 inches (38 mm).

Table 1.1 Computation of settlement from STP blow-counts (after D'Appolonia *et al.*, 1970)

Factor considered (1)	Terzaghi and Peck (2)	Meyerhof (3)	Peck and Bazaraa (4)
Load-settlement relationship*	$S = C_W C_D$ $\times \dfrac{3P}{N}\left(\dfrac{2B}{B+1}\right)^2$	$S = C_W C_D$ $\times \dfrac{2P}{N}\left(\dfrac{2B}{B+1}\right)^2$	$S = C_W C_D$ $\times \dfrac{2P}{N_B}\left(\dfrac{2B}{B+1}\right)^2$
SPT	Use N as measured	Use N as measured	$N_B = \dfrac{4N}{1+4\bar{\sigma}_V}\bar{\sigma}_V$ $\leqslant 0.75$ tons per sq ft $N_B = \dfrac{4N}{3.25+\bar{\sigma}_V}\bar{\sigma}_V$ > 0.75 tons per sq ft
Ground water correction	$C_W = 1.0$ for $D_W \geqslant 2B$ $C_W = 2.0$ for $D_W \leqslant B$	$C_W = 1.0$	$C_W = \dfrac{\sigma_V}{\bar{\sigma}_V}$
Depth correction	$C_D = 1.0 - \dfrac{D}{4B}$	$C_D = 1.0 - \dfrac{D}{4B}$	$C_D = 1.0 - 0.4\left(\dfrac{\gamma D}{P}\right)^{1/2}$

* S = settlement, in inches; P = bearing pressure, in tons per square foot; and B = footing width, in feet. N = average blow-count in depth B below footing

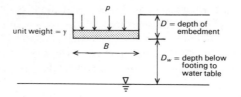

and Bazaraa (1969) then proposed a modified relationship which included Meyerhof's 33 per cent reduction in settlement estimate and incorporated explicit blow-count and water table corrections.

For a discussion of the relative merits of the various procedures, the reader is referred to D'Appolonia *et al.* (1968), to discussion of that paper by Gibbs and Holtz (1969), Peck and Bazaraa (1969) and Bolognezi (1969), and to the closure by D'Appolonia *et al.* (1970). Table 1.1 is reproduced from the concluding comments and it presents the Terzaghi and Peck, Meyerhof, and Peck and Bazaraa methods as used by D'Appolonia *et al.* Further comprehensive reviews of SPT techniques are also presented by Sutherland (1974) and Oweis (1979).

It seems clear from the various discussions that the Terzaghi and Peck method is very conservative and should not generally be used. The Meyerhof and Peck and Bazaraa methods are more suitable for practical use and they represent appropriate techniques for estimating the upper limit of expected settlement in conventional design. When 'best estimate' instead of 'upper limit' predictions of settlement on granular soil are required, D'Appolonia *et al.* found that the use of Meyerhof's method with blow-counts corrected using the Gibbs and Holtz (1957) method gave best overall agreement, but noted a considerable variation in accuracy for individual footings and that this experience related to only one site.

1.3.2 *Static cone penetration test methods*

Static penetration tests, most commonly using the Dutch Cone, have been widely utilized in Europe for several decades and are now finding increasing application in the United States and elsewhere in the world. They provide alternative *in situ* resistance data to the SPT blow-count and it can be argued that the increased standardization involved in Dutch Cone testing represents an advantage over the operator-sensitive SPT data.

The Dutch Cone test involves pushing a standard cone through the soil at a constant rate (normally 20 mm/s) and measuring cone penetration resistance at frequent intervals (almost continuously in the case of the best equipment). The penetration resistance of the cone (q_c, commonly measured in kg/cm^2 on a 10 cm^2 tip) can then be used via empirical correlations to predict settlement.

A simple means of making these predictions is by empirical correlation between N and q_c, whereby an equivalent blow-count is estimated and used with the SPT prediction procedure described above. A correlation of the form

$$q_c(\text{kg/cm}^2) = KN(\text{blows/ft})$$

has been proposed, with recommended K values in the range 5 to 10 being cited by Terzaghi and Peck (1968). Subsequent recommended conversion factors by Schmertman (1970, 1975) and Sutherland (1974) extend this range to between about 3 and 15 for granular soils, as a function of soil type. DeMello (1971) also presents a detailed discussion of these conversion factors. Schmertman (1970) produced a more direct approach to settlement prediction based on cone data, by integrating strains through the soil stratum. The method uses the relation

$$\varepsilon(Z) = \frac{pI(Z)}{E_s(Z)}$$

where $\varepsilon(Z)$, $I(Z)$ and $E_s(Z)$ are, respectively, strain, strain influence factor and deformation modulus, at depth Z, and p is the loading pressure at foundation level. E_s in this equation is empirically determined as

$$E_s = \alpha q_c$$

Schmertman initially (in 1970) suggested an α value of 2 but subsequently (in 1978) modified this recommendation to 2.5 for square footings and 3.5 for strip footings. He also proposed (in 1970) a strain influence factor distribution based on elastic theory, finite element studies and model tests, which was later (in 1978) revised to account for the shape of the loaded area. Figure 1.4 presents Schmertman's method in its revised form.

The modifications in Schmertman's recommendation between 1970 and 1978 reflect the development of a new empirical procedure which no doubt will continue to be

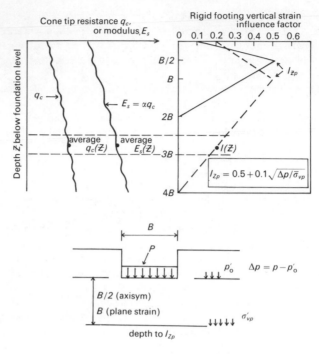

Figure 1.4 Prediction of settlement using static cone penetration tests (after Schmertman, 1970, 1978).

Computation of settlement as follows:

$$S = C_1 C_2 \Delta p \sum_{i=1}^{n} \frac{I(Z_i) \Delta Z_i}{E_s(Z_i)}$$

n = number of layers into which the depth of influence is subdivided

$C_1 = 1 - 0.5 (p'_o / \Delta p)$ embedment correction

$C_2 = 1 + 0.2 \log_{10} \left(\frac{t}{0.1} \right)$ creep correction t in years

refined as experience is gained with its use. Mitchell and Gardner (1975), for example, present an extensive list of $E_s - q_c$ correlations with most of the recent values falling within the range 1.5 to 3.5 (E_s and q_c being in the same units, say kg/cm²), and it should be expected that this correlation will be sensitive to the field situation (particularly the overconsolidation ratio of the granular deposit). However, this prediction technique is promising, incorporating both the large amount of data obtained during cone testing and variations in compressibility with depth. It is a method which is directed towards a best estimate of settlement, rather than a safe upper bound, and it should be used accordingly.

1.3.3 Plate load tests

The plate load test is generally performed using a 1 ft × 1 ft (0.3 m × 0.3 m) square plate placed at site formation level and loaded to yield an *in situ* load/settlement relationship.

This relationship is then extrapolated to the footing size under consideration using the Terzaghi and Peck (1948, 1968) relation

$$S = S_1 \left(\frac{2B}{B+1} \right)^2$$

where S_1 is the settlement of the 1-ft plate and S is the settlement of a footing of width B (ft).

Although offering apparent advantages in terms of simulating the actual field loading conditions, the plate load test has considerable practical disadvantages. The test procedure requires very careful attention to detail (see Terzaghi and Peck, 1968) and a large number of (expensive) tests will probably to be required to provide a

1   Settlement at centre of uniformly loaded circular area

$$S = \frac{pB}{2E} \; 2(1-v^2)$$

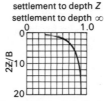

settlement to depth Z
settlement to depth ∞

Effect of considering only limited depth of strains

2(a)   Settlement at corner of uniformly loaded rectangular area

$$S = pB \frac{(1-v^2)}{E} \; I_p$$

(b)   Settlement at centre of uniformly loaded square area

$$S = p\frac{B}{E} \; 1.12(1-v^2)$$

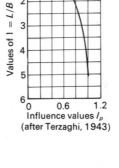

Values of $I = L/B$

Influence values I_p
(after Terzaghi, 1943)

3   Settlement of circular rigid loaded area

$$S = p\frac{B}{E}\frac{\pi}{4}(1-v^2)$$

4   Settlement of rectangular rigid loaded area

$$S = \frac{p(1-v^2)}{\beta_2 \sqrt{BLE}}$$

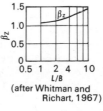

L/B
(after Whitman and Richart, 1967)

S = settlement L = length
p = uniform or average loading pressure B = width or diameter
     E = Young's modulus
v = Poisson's ratio

Figure 1.5 Elastic settlement solutions for surface loading on a uniform, isotropic, linear elastic medium of infinite depth.

statistically meaningful data base for prediction purposes. Serious problems in extrapolating to larger footing sizes can also occur if the soil characteristics vary significantly with depth, or in view of the questionable accuracy of the above extrapolation relationship (Bjerrum and Eggestad, 1963). D'Appolonia *et al.* (1968) found the plate load test to give relatively poor estimates of settlement and it is doubtful if this procedure has wide practical application.

1.3.4 *Techniques based on elastic theory*

A number of settlement prediction techniques based on elastic theory are available, generally utilizing either empirical or laboratory evaluation of modulus. The techniques can be broadly classified as those involving a direct elastic model of settlement, and those which use an elastic model to predict stress increments from which local strains are computed and integrated to yield settlement.

The elastic model for settlement has the general form

$$S = \frac{pBI}{M}$$

where S is the settlement, p is the foundation loading pressure, B is the foundation width, I is a settlement influence factor which varies with geometry and elastic parameter values, and M is a modulus of compressibility for which Young's modulus, E, is often used.

Elastic solutions are available for a variety of geometries and soil conditions, and the reader is referred to Poulos and Davis (1974) for a summary. Lambe and Whitman (1969, Section 14.8), also present a useful review of basic elastic settlement solutions, from which the summary information of Fig. 1.5 has been largely abstracted. Janbu *et al.* (1956) presented approximate influence factors for the average settlement of a flexible, embedded foundation on an elastic layer underlain by a rough rigid base. Christian and Carrier (1978) revised this solution to provide the information of Fig. 1.6, in which I is the product of the two terms μ_0 and μ_1.

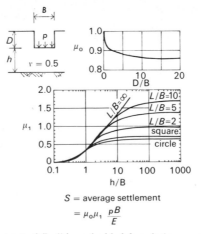

Figure 1.6 Average settlement of flexible embedded foundation on a uniform isotropic linear elastic layer on a rigid base (after Christian and Carrier, 1978)

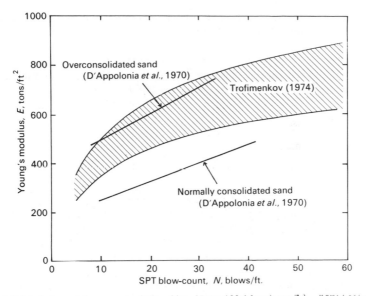

Figure 1.7 Modulus v. blow-count relationships. *Note*: 100 (short) ton/ft$^2 = 9576\,\text{kN/m}^2$

Application of these elastic solutions requires selection of a modulus, usually E for drained conditions in the case of granular soils. Empirical correlations are generally employed for this purpose and Mitchell and Gardner (1975) summarize numerous relationships including the correlations with cone penetration data that have already been discussed. Correlations with SPT blow-count data are also available. Fig. 1.7 presents USSR practice (Trofimenkov, 1974) and also two tentative relationships presented by D'Appolonia *et al.* (1970). The average blow-count in a depth at least equal to the loading width should be used in applying these relationships. A further means of observing compressibility *in situ* and developing estimates of deformation modulus is presented by the screw plate test (Janbu and Senneset, 1973).

Deformation modulus can also be measured in laboratory tests, either directly as E in triaxial test samples or as the constrained modulus, D, in oedometer tests. Young's modulus can then be evaluated from D as a function of v, using the expression

$$D = \frac{E(1 - v)}{(1 + v)(1 - 2v)}$$

However, the sensitivity of this expression to the value of v is obvious and measurement of *in situ* modulus via laboratory tests usually requires empirical calibration in view of the extreme difficulty in obtaining or preparing truly representative samples.

The second general approach for applying elastic theory to the prediction of settlement on granular soils involves

(a) evaluation of the initial stress *in situ*
(b) estimation of the applied stress increment at various locations beneath the loaded area (from elastic theory)
(c) evaluation of the corresponding strains at the locations of (b) and
(d) integration of those strains through the soil profile to yield a settlement estimate.

These are the basic steps of the Lambe (1967) stress path method and the approach constitutes a very valuable conceptual model for the soil behaviour beneath the applied load. However, practical application of the method is difficult, particularly when laboratory tests are used to evaluate strains through the profile (see Ladd *et al.*, 1977, for discussion of the high stress-strain sensitivity of granular soils to minor changes in sample preparation and both stress and strain history). Thus, routine application of elastic theory for evaluation of settlement would probably best use the direct elastic models discussed above, with evaluation by strain profile limited to complex soil geometries or the more sophisticated investigations.

1.4 Prediction of settlement in cohesive soils

Settlement of cohesive soils can usefully be considered as having separate 'initial-undrained', 'undrained creep', 'consolidation', and 'drained creep' components. The most significant of these components is usually consolidation, and routine practice often only considers consolidation settlement in design prediction. However, the undrained movements can be important in special situations as with plastic and organic soils, and drained creep can be significant when consolidation rates are fast. Prediction of each settlement component in turn will therefore be reviewed.

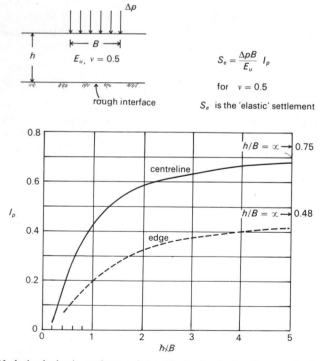

Figure 1.8 Undrained elastic settlement for uniform circular loading on elastic layer (after Section 5.2 of Poulos and Davis, 1974)

1.4.1 Initial undrained settlement

Initial undrained settlement predictions are generally performed using elastic theory and the assumption of undrained conditions (that is, $v = 0.5$). Figure 1.8 indicates the typical computational procedures, in which the undrained modulus, E_u, is usually selected empirically and frequently via a relationship with the undrained shear strength, that is, E_u/c_u. Numerous values of this relationship are presented in the literature, Bjerrum (1972) for example, suggesting values in the range 500 to 1500, with the lower end of the range applying to heavy loading of very plastic clays and the high end applying to lean clays with small loadings. Figure 1.8 represents circular loading but would be reasonably applicable to relatively equidimensional loadings of other shapes. The data of Fig. 1.5 could also be used for predicting elastic undrained settlements with the assumption of $v = 0.5$.

A more sophisticated model, in which the elastic prediction of Fig. 1.8 is corrected for the effect of local yielding in the foundation soils, was proposed by D'Appolonia et al. (1971) and is presented in Fig. 1.9. D'Appolonia et al. also evaluated E_u/c_u for use in the analysis, based on data from 10 case studies. They concluded that values of 1000 to 1500 were appropriate for 'lean inorganic clays of moderate to high sensitivity' but that 'considerably lower' values apply to 'highly plastic clays and for organic clays'.

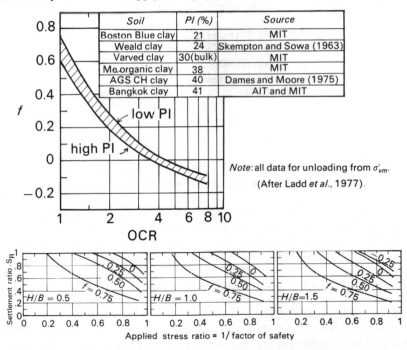

Soil	PI (%)	Source
Boston Blue clay	21	MIT
Weald clay	24	Skempton and Sowa (1963)
Varved clay	30(bulk)	MIT
Me.organic clay	38	MIT
AGS CH clay	40	Dames and Moore (1975)
Bangkok clay	41	AIT and MIT

Note: all data for unloading from σ'_{vm}.

(After Ladd et al., 1977).

Figure 1.9 D'Appolonia et al. (1971) method for computing initial settlement.
1. Compute the elastic settlement, S_e, using the procedure of Figure 1.8.
2. Determine initial shear stress ratio, f, as function of overconsolidation ratio (OCR) of foundation soil—use average value of OCR in depth B beneath loaded area.
3. Determine settlement ratio, S_R, as a function of layer thickness, H, and loading width, B. (Charts apply to strip loadings but are also reasonably applicable to circular loads).
4. Compute initial settlement = S_e/S_R.

B

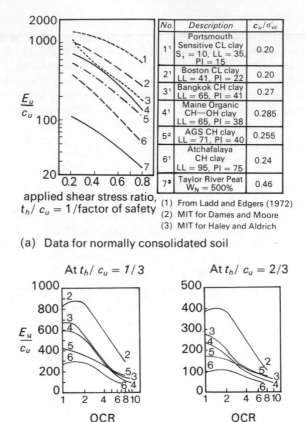

No.	Description	c_u/σ_{vc}
1[1]	Portsmouth Sensitive CL clay $S_t = 10$, LL $= 35$, PI $= 15$	0.20
2[1]	Boston CL clay LL $= 41$, PI $= 22$	0.20
3[1]	Bangkok CH clay LL $= 65$, PI $= 41$	0.27
4[1]	Maine Organic CH—OH clay LL $= 65$, PI $= 38$	0.285
5[2]	AGS CH clay LL $= 71$, PI $= 40$	0.255
6[1]	Atchafalaya CH clay LL $= 95$, PI $= 75$	0.24
7[3]	Taylor River Peat $W_N = 500\%$	0.46

applied shear stress ratio,
$t_h/c_u = 1/$factor of safety

(1) From Ladd and Edgers (1972)
(2) MIT for Dames and Moore
(3) MIT for Haley and Aldrich

(a) Data for normally consolidated soil

(b) Data for overconsolidated soil

Figure 1.10 Normalized modulus data from direct simple shear tests.

Foott and Ladd (1981) extended this latter observation, presenting three case studies wherein highly plastic and organic soils had experienced very large initial settlements which were much greater than would conventionally be expected. They suggested use of direct simple shear modulus data for approximate prediction of undrained deformations using the D'Appolonia *et al.* procedure, and presented the data of Fig. 1.10 to allow preliminary selection of modulus values based on general soil classification. The data are secant moduli which therefore vary as a function of stress level, so analytical values should be selected corresponding to the foundation stability factor of safety. Whenever E_u/c_u values below about 100 are indicated, the possibility of larger initial settlements should be anticipated and specific testing is appropriate to establish the deformation characteristics of the soil. Note that low modulus values are again generally associated with highly plastic and/or organic soils.

1.4.2 *Undrained creep*

Undrained creep settlements can be investigated using time-dependent mathematical models (see, for example, Singh and Mitchell, 1968, 1969; Edgers *et al.*, 1972; Mitchell,

1976) to produce approximate predictions of field behaviour. However, these methods tend to be very complex and can be considered as primarily research tools usually giving results of limited precision. For general engineering practice, the qualitative observation that undrained creep movements can become significant when initial settlements are large and consolidation is slow, is of more significance. The initial settlement predictions discussed above then become an important screening test for potential undrained creep problems, which again are most likely in the case of plastic and organic soils.

1.4.3 *Consolidation settlement*

Consolidation settlements are generally predicted from laboratory consolidation tests using Terzaghi's one-dimensional consolidation theory. The laboratory test data can conveniently be presented as vertical strain v. log stress, and an unload/reload cycle from the virgin stress range can be used to minimize disturbance effects in the recompression zone. Figure 1.11 shows this type of laboratory data and demonstrates its use for prediction of one-dimensional consolidation settlement. It should be noted that typical values of the virgin compression ratio, CR, are 0.30 ± 0.15, while the recompression ratio, RR, is typically only 10 to 20 per cent of the CR value. Thus,

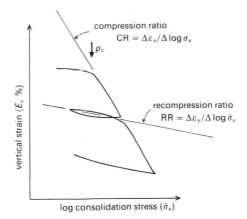

Figure 1.11 Computation of consolidation settlement using one-dimensional model.

1. Divide soil profile into layers and for each layer:
   —evaluate initial vertical effective stress due to overburden, $\bar{\sigma}_{v0}$
   —evaluate vertical stress increment ($\Delta\sigma_v$) due to applied loading
   —evaluate final vertical effective stress, $\bar{\sigma}_{vf} = \bar{\sigma}_{v0} + \Delta\sigma_v$
   —evaluate prestress, P_c
   —compute layer settlement, S_l

If $\bar{\sigma}_{vf} > P_c$, S_l = layer thickness $\times \left(\text{RR} \log \dfrac{P_c}{\bar{\sigma}_{v0}} + \text{CR} \log \dfrac{\bar{\sigma}_{vf}}{P_c} \right)$

If $\bar{\sigma}_{vf} < P_c$, S_l = layer thickness $\times \left(\text{RR} \log \dfrac{\bar{\sigma}_{vf}}{\bar{\sigma}_{v0}} \right)$

2. Total consolidation settlement = sum of layer settlements.

evaluation of the preconsolidation pressure, P_c, at which the soil loading changes from recompression to virgin compression is of particular importance in consolidation settlement prediction.

As indicated in Figure 1.11, consolidation settlement estimates are frequently performed by dividing the foundation soil into layers and using the one-dimensional model to predict settlement in each layer based on the profile of vertical stress increment predicted by elastic theory. By estimating stress increment profiles at different locations beneath the applied loading, an indication of differential settlements can also be obtained. It should be noted, however, that both vertical stress increments and differential settlements will be influenced by the stiffness of the applied loading, so this approach to predicting differential settlements requires realistic modelling of the loading conditions and soil strata. It should also be recognized that the results are inevitably approximate. Skempton and Bjerrum (1957) presented a technique for correcting consolidation settlement predictions for the effect of undrained deformations; further details are given in Chapter 11. This correction can have a substantial proportional effect with overconsolidated clays (when consolidation settlements are usually small), but it tends to be less significant with normally consolidated and lightly overconsolidated materials.

Terzaghi's consolidation theory also allows prediction of the rate of consolidation as a function of the coefficient of consolidation, c_v, of the clay and the drainage conditions. The curve at the right of Fig. 1.12 presents the classical solution for the one-dimensional Terzaghi case, with the average per cent consolidation (U_v) being plotted against a time factor, T, defined in the Figure. When the loading is over a limited area, two- or three-dimensional consolidation occurs and the rate of consolidation is increased due to lateral drainage. The additional curves of Fig. 1.12 indicate this effect for a uniform

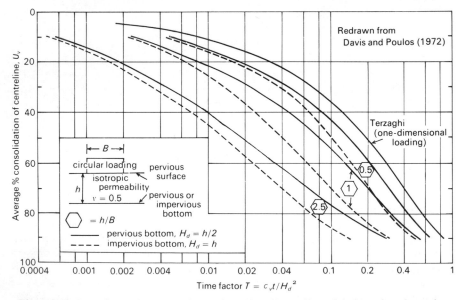

Figure 1.12 Rate of consolidation settlement including effects of lateral drainage from beneath a circular loading. c_v = coefficient of consolidation, t = time after load application, H_d = length of drainage path. (Redrawn from Davis and Poulos, 1972).

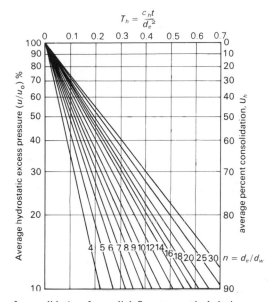

Figure 1.13 Rate of consolidation for radial flow to vertical drains. c_h = coefficient of consolidation with horizontal drainage; d_w = effective diameter of vertical drain; d_e = equivalent diameter of drainage area per drain ($d_e = 1.05s$ for triangular spacing, $d_e = 1.14s$ for square spacing); s = drain spacing. (After Johnson, 1970).

circular loading on a clay of isotropic permeability, as a function of the geometry of the loading. If the horizontal permeability is greater than the vertical permeability, the consolidation rate increases still further due to the increased ability for excess pore pressures to dissipate laterally. The reader is referred to Poulos and Davis (1974) for further information on this subject.

When vertical drains are used to accelerate consolidation, the resulting average consolidation (U_h) can be estimated from the chart of Fig. 1.13. This consolidation is due only to radial flow to the drains and it can be combined with consolidation due to vertical drainage (Fig. 1.12) to yield a composite consolidation rate (U) as follows:

$$(1 - U) = (1 - U_v)(1 - U_h)$$

1.4.4 Drained creep settlement

Long-term drained creep movements can be predicted using drained parameters in the same complex models as discussed for undrained creep. However, it is probably rare that drained creep movements would be of such a magnitude as to require such modelling and the more customary approach is to use the one-dimensional drained creep case known as secondary compression, for which the characteristic behaviour is measured in consolidation tests. Controversy exists regarding the precise mechanisms involved in secondary compression (see Ladd et al., 1977), but the prediction technique presented in Fig. 1.14, and which assumes that secondary compression commences at the end of primary consolidation, has considerable practical application and merit.

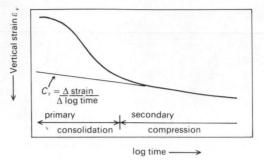

Figure 1.14 Computation of secondary compression settlement.
1. Evaluate c_α from consolidation test for stress range applicable to field situation
2. Compute secondary compression, S_s, using

$$S_s = hc_\alpha \log(t/t_p)$$

h = initial layer thickness
t = time since middle of loading period in field
t_p = time at end of primary consolidation since middle of loading period in field.

1.5 Reduction of settlement

Numerous techniques exist for reducing settlement under structural loadings, many of them involving procedures to expedite the movements which will inevitably occur. Thus, for example, dynamic techniques are frequently used to improve granular soils before constructing a building, the resulting pre-construction settlement reducing the important post-construction movements. Settlement of granular soils can also be reduced by grouting. Techniques for dynamic improvement range from vibratory rolling, through vibroflotation of large bodies of soil, to dynamic compaction (by dropping large weights on the soil—see Chapter 2) and even blasting. Terzaghi and Peck (1968) recommend that granular soils be improved in some way before construction, whenever their SPT blow-counts are less than five.

Modest changes in post-construction settlement can be achieved by simply varying the load dimensions and the applied bearing pressure. This approach is a standard element of design and is applicable to both granular and cohesive soils. More advanced approaches to reducing settlement in cohesive soils consist primarily of techniques for stageloading, surcharging, and expediting consolidation. A temporary load (such as an earth fill) can also be employed to achieve most of the settlement before construction of the final structure. This approach is termed preloading or precompression (Johnson, 1970).

Stage loading consists of applying part of the load and waiting for consolidation to strengthen the soil before application of additional load stages. It is used when full loading will take the foundation soils close to or beyond failure, in order to strengthen the soil before full load applications. It is also likely to be somewhat effective in reducing large initial and undrained creep movements, although the degree of improvement would be hard to assess.

Surcharging is used to expedite settlement and also to reduce secondary compression movements, for which purpose the surcharge must be left in place throughout primary consolidation. As an approximate rule of thumb, a 25 per cent increase in effective stress increment via surcharging will reduce secondary compression rates by 50 per cent or more.

Expediting consolidation, most frequently using vertical drains, will increase consolidation settlement rates and can reduce undrained creep. By reducing the primary consolidation period it may also increase secondary compression, although this effect may or may not be significant. Drains can be used very effectively with both stage construction and surcharging, and all these approaches can be used with a preload.

Acknowledgement

A review of this chapter by Professor Charles C. Ladd is gratefully acknowledged, as is his teaching in 1969/70 which formed the framework for much of the chapter's contents

Notation

B width of loaded area; also diameter of circular area

C_D settlement correction factor to account for the depth of foundation embedment

CR   compression ratio, compressibility parameter obtained from 1-dimensional consolidation test $= \Delta\varepsilon/\Delta\log\bar{\sigma}_v$

C_h coefficient of horizontal consolidation

C_w settlement correction factor to account for the depth of water level below foundation level

C_α coefficient of secondary compression

C_1 empirical settlement coefficient $= 1 - 0.5(\bar{p}_0/\Delta p)$ used in Schmertmann's method of analysis

C_2 empirical settlement coefficient used in Schmertmann's method to account for secondary compression $= 1 + 0.2\log_{10}(t/0.1)$, where t is in years

c_u undrained shear strength

c_v coefficient of vertical consolidation

D depth of foundation embedment; also constraint modulus obtained from 1-dimensional consolidation tests $= \bar{\sigma}/\varepsilon$

D_w depth of water table below foundation level

d_e equivalent diameter of drained area around a single drain

d_w effective equivalent diameter of vertical drain

E Young's soil modulus

E_s soil modulus estimated from cone point resistance

E_u undrained soil Young's modulus

f	initial shear stress ratio $= \frac{1}{2}(1 - K_0)\bar{\sigma}_v$
H	thickness of soil layer
H_d	vertical drainage path
h	thickness of soil layer
I	settlement influence factor
I_p	settlement coefficient
I_z	strain influence factor at depth z
I_{zp}	maximum value of strain influence factor for Schmertmann's method of analysis
K	an empirical constant relating cone tip resistance to SPT blow count values
K_0	coefficient of earth pressure at rest
L	length of foundation
M	soil modulus of compressibility
N	number of blows from the standard penetration test
n	d_e/d_w
OCR	overconsolidation ratio $= \bar{\sigma}_{v\ max}/\bar{\sigma}_{v0}$
PI	Soil Plasticity Index
p	uniform load applied to the foundation
p_c	preconsolidation stress
$\bar{p}_0$	in-situ vertical effective stress at foundation level
q_c	cone penetration tip resistance
RR	recompression ratio, compressibility parameter obtained from 1-dimensional consolidation test $= \Delta\varepsilon/\Delta\log\bar{\sigma}_v$
S	settlement
SPT	Standard Penetration Test
S_R	settlement ratio
S_e	elastic settlement
S_s	settlement resulting from consolidation
S_1	settlement of a square plate 1 ft in size
T	dimensionless time factor used in consolidation analysis
T_h	dimensionless time factor for consolidation due to horizontal drainage $= C_h t/d_e 2$

T_v dimensionless time factor for consolidation due to vertical drainage $= C_v t / H_d 2$

t time

t_p time at the end of primary consolidation since the midpoint of the loading period

U average degree of consolidation

U_h average degree of consolidation for horizontal drainage

U_v average degree of consolidation for vertical drainage

u average excess pore-water pressure

u_0 average initial excess pore pressure

α empirical coefficient relating soil modulus to cone tip resistance

β_z settlement coefficient

γ soil unit weight

ε strain

$\varepsilon_{(z)}$ strain at depth z

Δ this symbol when proceeding any variable indicates incremental value

$\Delta\sigma_z = \Delta\sigma_v$ incremental vertical stress

Δ_p incremental vertical stress

μ_0 settlement coefficient

μ_1 settlement coefficient

ν Poisson's ratio

$\bar{\sigma}_v$ vertical effective stress

$\bar{\sigma}_{v0}$ in-situ vertical effective stress due to overburden

$\bar{\sigma}_{v\,max}$ maximum past pressure from consolidation test

$\bar{\sigma}_{vf}$ final vertical effective stress

τ_h horizontal shear stress

References

Alpan, I. (1964) Estimating the settlements of foundations on sands. *Civil Engineering and Public Works Review*, 1415.

Bjerrum, L. (1972) Embankments on soft ground. *Proc. ASCE Speciality Conference on Performance of Earth and Earth-Supported Structures, Lafayette, Ind.*, Vol. 1, pp. 1–54.

Bjerrum, L. and Eggestad, A. (1963) Interpretation of loading test on sand. *Proceedings, European Conference on Soil Mechanics and Foundation Engineering, Wiesbaden*, Vol. 1, p. 199.

Bolognesi, A.J.L. (1969) Discussion of 'Settlements of spread footings on sand'. *J. Soil Mechanics and Foundations Division, ASCE*, Vol. 95, SM3.

British Standard 1377 (1975) Methods of test for soils for civil engineering purposes. British Standards Institution, London, 143 pp.

Burland, J.B., Broms, B.B., Legrand, J. and de Mello. V.F.B. (1977) Behavior of foundations and structures, State-of-the-Art Report. *Proc 9th International Conference on Soil Mechanics and Foundation Engineering, Tokyo*, Vol. 2, pp. 495–546.

Christian, J.T. and Carrier, W.D., III. (1978) Janbu, Bjerrum and Kjaernsli's chart reinterpreted. *Canadian Geotechnical J.* **15**, 123–128.

D'Appolonia, D.J., Poulos, H.G. and Ladd, C.C. (1971) Initial settlement of structures on clay. *J. Soil Mechanics and Foundation Division, ASCE,* **97** (SM10) 1359–177.

D'Appolonia, D.J., D'Appolonia, E. and Brissette, R.F. (1968) Settlement of spread footings on sand. *J. Soil Mechanics and Foundations Division, ASCE,* (SM3) 735–759.

D'Appolonia, D.J., D'Appolonia, E. and Brissette, R.F. (1970) Closure to 'Settlement of spread footings on sand,' *J. Soil Mechanics and Foundations Division, ASCE* (SM2) 754–761.

Davis, E.H. and Poulos H.G. (1972) Rate of settlement under two- and three-dimensional conditions. *Géotechnique* **22** (1) 95–114.

de Mello, V.F.B. (1971) The standard penetration test. State-of-the-Art Paper, *Proc. 4th Panamerican Conf. on Soil Mechanics and Foundation Engineering, San Juan, Puerto Rico* 1971, Vol. I, pp. 1–86.

Edgers, L., Ladd, C.C. and Christian, J.T. (1973) *Undrained Creep of Atchafalaya Levee Foundation Clays.* MIT, Dept. Civil Eng. Res. Rep. R73–16.

Focht, J.A., Jr., Khan, F.R. and Gemeinhardt, J. (1978) Performance of One Shell Plaza deep mat foundation. *J. Geotechnical Engineering Division, ASCE,* **104** (GT5) 593–608.

Foott, R. and Ladd, C.C. (1981) Undrained settlement of plastic and organic clays. *J. Geotechnical Engineering Division, ASCE,* **107** (GT8) 1079–1094.

Gibbs, H.J. and Holtz, W.G. (1957) Research on determining the density of sands by spoon penetration testing. *Proceedings, 4th International Conference on Soil Mechanics and Foundation Engineering, London*, Vol. 1, 35.

Gibson, R.E. and Sills, G.C. Some results concerning the plane deformation of a non-homogeneous elastic half-space. *Proc. Roscoe Memorial Symposium, Cambridge*, Foulis, pp. 564–572.

Grant, B., Christian J.T. and Vanmarke, E.H. (1974) Differential settlement of buildings. *J. Geotechnical Engineering Division, ASCE,* **100** (GT9) 973–991.

Holtz, W.G. and Gibbs, H.J. (1969) Discussion of 'Settlement of spread footings on sand', *J. Soil Mechanics and Foundations Division, ASCE,* (SM3).

Janbu, N., Bjerrum, L. and Kjaernsli, B. (1956) *Veiledning ved Lasning au Fundamentering Soppgauer.* Norwegian Geotechnial Institute, Publication No. 16.

Janbu, N. and Senneset, K. (1973) Field compressometer—principles and applications. *Proceedings 8th International Conference on Soil Mechanics and Foundation Engineering, Moscow*, 1973, Vol. 1–1, pp. 191–198.

Janbu, N. (1967) *Settlement Calculations Based on the Tangent Modulus Concept.* (3 guest lectures at Moscow State University). Bulletin No. 2, Soil Mechanics, NTH, Trondheim, Norway.

Johnson, S.J. (1970) Precompression for improving foundation soils. *J. Soil Mechanics and Foundations Division, ASCE,* **96** (SM1) pp. 111–144.

Johnson, S.J. (1970) Precompression with vertical sand drains. *J. Soil Mechanics and Foundations Division, ASCE,* **96** (SM1) pp. 145–175.

Ladd, C.C., Foott, R., Ishihara, K., Schlosser, F. and Poulos, H.G. (1977) Stress-deformation and strength characteristics. *Proceedings 9th International Conference on Soil Mechanics and Foundation Engineering, Tokyo*, 1977, Vol. 2, pp. 421–494.

Lambe, T.W. (1967) Stress path method. *J. Soil Mechanics and Foundations Divisions, ASCE,* **93** (SM6) 309–331.

Lambe, T.W. and Whitman, R.V. (1969) *Soil Mechanics.* John Wiley & Sons, New York.

Meyerhoff, G.G. Shallow foundations. *J. Soil Mechanics and Foundation Division, ASCE,* II, (SM2) 21–31.

Mitchell, J.K. (1976) *Fundamentals of Soil Behavior.* John Wiley & Sons, New York.

Mitchell, J.K. and Gardner, W.S. (1975) In situ measurement of volume change characteristics. *Proceedings, ASCE Geotechnical Engineering Division Specialty Conference on In-situ Measurement of Soil Properties*, Vol. 2, pp. 279–345.

Oweis, I.S. (1979) Equivalent linear model for predicting settlements of sand bases. *J. Geotechnical Engineering Division, ASCE,* **105** (GT12) 1525–1544.

Peck, R.B. and Bazaraa, A.R.S. (1969) Discussion of 'Settlement of spread footings on sand'. *J. Soil Mechanics and Foundation Division, ASCE*, (SM3).

Poulos H.G. and Davis, E.H. (1974) *Elastic Solutions for Soil and Rock Mechanics*. John Wiley & Sons, New York.

Schmertmann, J.H. (1970) Static cone to compute static settlement over sand, *J. Soil Mechanics and Foundation Engineering Division, ASCE*, (SM3) 1011–1043.

Schmertmann, J.S. Measurement of in situ shear strength, State-of-the-Art Report. *Proc. ASCE Specialty Conference on In Situ Measurement of Soil Properties, Raleigh*, 1975, Vol. 2, pp. 57–138.

Schmertmann, J.H. (1978) *Guidelines for Cone Penetration Test Performance and Design*. Manual FHWA-TS-78-209, U.S. Dept. of Transportation Federal Highway Administration.

Schmertmann, J.H., Hartman, J.P. and Brown, P.R. (1978) Improved strain influence factor diagrams. *J. Geotechnical Engineering Division, ASCE*, **104** (GT8) 1131–1135.

Seed, H.B. (1965) Settlement analyses, a review of theory and testing procedures, *J. Soil Mechanics and Foundation Division, ASCE*, **91** (SM2) 39–48.

Singh, A. and Mitchell, J.K. (1968) General stress-strain time function for soils. *J. Soil Mechanics and Foundations Division, ASCE*, **94** (SM1) 21–46.

Singh, A. and Mitchell, J.K. (1969) Creep potential and creep rupture of soils. *Proc. 7th International Conference on Soil Mechanics and Foundation Engineering, Mexico*, 1969, Vol. 1, pp. 379–384.

Skempton, A.W. and Bjerrum, L. (1957) A contribution to the settlement analysis of foundations on clay. *Géotechnique*, **7**, 168–178.

Skempton, A.W. and MacDonald, D.H. (1957) Allowable settlement of buildings. *Proc. Institution of Civil Engineers* **3** (5) 727–768.

Sutherland, H.B. (1963) The use of in-situ tests to estimate the allowable bearing pressure of cohesionless soils. *The Structural Engineer* **41** (3).

Sutherland, H.B. (1974) Granular materials—state of the art report. *Proc. Conference on Settlement of Structures, Cambridge*, Pentech Press, London, pp. 473–499.

Terzaghi, K. (1943) *Theoretical Soil Mechanics*. John Wiley & Sons, New York

Terzaghi, K. and Peck, R.B. (1948, 1968) *Soil Mechanics in Engineering Practice*. John Wiley & Sons, New York.

Thornburn, S. (1963) Tentative correction chart for the SPT in non-cohesive soils. *Civil Engineering and Public Works Review* **58** (683).

Trofimenkov, J.B. Penetration testing in USSR. State-of-the-Art Report. *Proc. of the European Symposium on Penetration Testing, Stockholm*, Vol. 1, pp. 147–154.

2 Settlement of fill[*]

J.A. CHARLES

2.1 Introduction

There can be serious problems where construction takes place on 'made ground' or 'fill'. Nevertheless the scarcity of good building land and the desirability of redeveloping inner city areas and reclaiming industrial waste lands have increasingly led to the development of filled sites.

In considering the geotechnical problems which may arise when building on fill, it is important to make a basic distinction between sites where filling has yet to take place and earthmoving can be controlled and supervised, and sites which have already been filled with little or no control being exercised. In the former situation selection of suitable material and supervision of placement and compaction should ensure an adequate foundation material. It is the latter situation in which problems are much more likely and with which this chapter is principally concerned.

Urban areas are likely to have large areas of filled ground. Urban fills may comprise soil, rubble and refuse, and although extensive in area, are usually only of shallow depth. Disused docks, quarries and clay pits have often been infilled with similar materials.

Large areas of land have been used for the deposition of industrial, mining and domestic wastes. The overburden from opencast mining is often replaced without systematic compaction leaving loose fills.

The development of these various types of filled sites can present a wide variety of problems. These could include chemical attack on foundations, the risk of explosion or fire within the waste deposit, and hazards to health from contaminated land (Smith and Charles, 1983). This chapter will be concerned only with the evaluation of the load-carrying characteristics of the filled ground. Several types of fill commonly encountered are listed in Table 2.1, with a brief account of some of their more important characteristics.

2.2 Settlement of fills

Building on filled ground can involve a number of geotechnical problems. Where fill has been placed above natural ground level the stability of the slopes at the edge of the fill and the stability of the underlying natural ground may need to be examined. If structures are built on piles which pass through the fill into underlying natural strata, then negative skin friction caused by the fill settling under its own weight

---

[*] Crown Copyright 1982

Table 2.1 Commonly encountered fills

Fill type	Comments	Reference
Opencast mining backfill	Generally these fills have not been compacted during the backfilling operation. The method of opencast working (whether by dragline, face shovel or scraper) will have had a significant effect on the condition of the fill. Features of the opencast mining operation such as the position of a lagoon or an overburden heap can also be important.	Charles *et al.* (1977) Charles *et al.* (1978) Kilkenny (1968) Knipe (1979) Leigh and Rainbow (1979)
Domestic refuse	The age of the refuse is of major significance. Older refuse has had longer for decay and decomposition of organic matter to occur and was often an inherently better material with a much higher ash content than recent refuse. Generation of methane can be a major problem. For many years controlled tipping methods (in which the refuse is placed and compacted in shallow layers, each layer being covered with soil) have been advocated for domestic refuse.	Bratley (1977) BRE Digest 168 (1974) Refuse disposal (1971) Sowers (1968) Sowers (1973)
Colliery spoil	This is the waste product of mining coal and is derived from the rocks, (mainly siltstones and mudstones with seat-earths) adjacent to the coal seams. Older spoil tips often ignited and materials in these are generally a mixture of unburnt, partially burnt and well-burnt spoil. Variability can be a problem. Following the Aberfan disaster, coarse waste material has been placed in tips built in layers with compaction. When colliery spoil is placed in this way the possibility of spontaneous combustion appears to be remote. Weathering would seem to affect only the surface layer of the fill. Fine waste is usually pumped to lagoons.	Bishop (1973) Isaac and Troughton (1971) Isaac (1972) Sherwood (1975a) Taylor (1975) Taylor and Spears (1972) Thomson and Rodin (1972)
Pulverized fuel ash	This is the waste from power stations burning pulverized coal and consists mainly of minute glass spheres in the silt size range. The ash may be discharged as a slurry in lagoons. The particle size will vary within the lagoon, the finer material being furthest from the outfall. This variability can be a major problem. Alternatively it may be mixed with just sufficient water to make it suitable for placement and compaction as a fill. The fill has a low specific gravity and a corresponding low maximum dry density.	Ballisager and Sorenson (1981) Knight (1979) Leonards and Bailey (1982) PFA data book (1967) Sherwood (1975b) Swain (1979) Weatherley (1979)

Table 2.1 (*continued*)

Fill type	Comments	Reference
Demolition wastes	Concrete and brick rubble from the demolition of buildings may contain other poorer materials such as wood, glass and plaster.	Gray and Thomson (1979) Nixon (1976) Wilde and Crook (1979)
Mineral waste tailings and slurries	It has already been mentioned that colliery spoil fine waste and PFA are often discarded by allowing them to settle in large ponds. Many other wastes are disposed of in this manner, e.g. tailings from various mining industries, the micaceous residue of china clay waste, red mud (the waste from the production of alumina from bauxite by the Bayer process). A firm crust may form over the surface of the lagoon deposit but this may be thin and overlie very soft material. Variability of particle size within the deposit may also be a problem.	Hughes and Windle (1976)

around the piles may be a major consideration in the foundation design. The bearing capacity of the fill (and if the fill is only shallow, also of the underlying natural ground) should be investigated. However, in most situations the major geotechnical problem will be associated with long-term settlement of the fill. This settlement may be attributable to a number of different causes, such as the self-weight of the fill, the decay of organic matter, structural loads or a change in groundwater level.

2.2.1 *Creep settlement*

Where the fill is deep, self-weight will often be the principal cause of long term settlement. With granular fills and indeed poorly compacted unsaturated fills of all types, primary compression occurs almost immediately a load is applied, and as a consequence the major part of settlement due to self-weight occurs as the fill is placed. Nevertheless some significant further movements do occur under conditions of constant effective stress and moisture content, and can be termed 'creep' settlement. With many fills it is found that creep compression shows an approximately linear relationship when plotted against the logarithm of the time that has elapsed since the deposit was formed (Fig. 2.1). A parameter, α, can be defined as the percentage vertical compression of the fill that occurs during a log cycle of time, say between one year and ten years after the fill was placed (Sowers *et al.*, 1965). Some typical values of α are given in Table 2.2. The magnitude of α can depend on the depth of the deposit as well as on the nature and degree of compaction of the fill. It should be remembered that the use of an α parameter to predict the settlement of a fill can only be valid where conditions in the fill remain unaltered. An increase in stress due to an applied load or a change in moisture content could cause much greater movements.

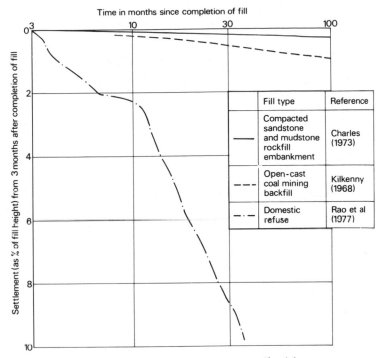

Figure 2.1 Creep settlement due to self-weight.

Table 2.2 Creep settlement rate parameter α

Fill type	Typical values of α	References
Well compacted sandstone rockfill	0.2%	Charles (1973)
Uncompacted opencast mining backfill	0.5%–1%	Charles *et al.* (1977) Kilkenny (1968) Knipe (1979) Leigh and Rainbow (1979)
Domestic refuse	2%–10%	Merz and Stone (1962) Rao *et al.* (1977) Sowers (1973) Yen and Scanlon (1975)

2.2.2 *Compression due to structural loads*

The fill will be compressed by the loads applied to it by the building constructed on it. The compressibility of fill materials shows wide variation depending on the nature of the fill, its particle size distribution, compactness, the existing stress level, the stress increment and the moisture content. Assuming that the stress increments due to the

Table 2.3 Compressibility of fills

Fill type	Compressi-bility	Typical value of constrained modulus kN/m^2	References
Dense well-graded sand and gravel	very low	40 000	BRE data
Dense well-graded sandstone rockfill	low	15 000	Penman et al. (1982)
Loose well-graded sand and gravel	medium	4 000	BRE data
Old urban fill	medium	4 000	Charles and Driscoll (1981)
Uncompacted stiff clay fill above water table	medium	4 000	Charles et al. (1978)
Loose well-graded sandstone rockfill	high	2 000	BRE data
Poorly compacted colliery spoil	high	2 000	BRE data
Old domestic refuse	high	1 000–2 000	BRE data
Recent domestic refuse	very high		

structural loads do not bring the fill to a condition close to a bearing capacity failure, then the settlements can be most simply predicted using a compressibility parameter related to one-dimensional compression. If in one-dimensional compression an increment of vertical stress $\Delta\sigma_v$ produces an increase in vertical strain $\Delta\varepsilon_v$, then the constrained modulus is defined as $D = \Delta\sigma_v/\Delta\varepsilon_v$ (Lambe and Whitman, 1979). Some typical values of D for a number of different fill types are quoted in Table 2.3. They are applicable to small increments in vertical stress, say $\Delta\sigma_v = 100\,kN/m^2$, where the initial vertical stress was about $30\,kN/m^2$. The modulus of the uncompacted stiff clay fill quoted in Table 2.3 might seem relatively high compared with the other types of poorly-compacted fill. It should be recognized that such a fill may be particularly prone to collapse settlement on wetting.

With the exception of fine-grained saturated fills, most of the compression of a fill will occur almost immediately the load is applied (as in the case of settlement due to self-weight). However, movements that occur during construction of a building are likely to be much less of a problem than those which occur after completion of the structure. Consequently, the long-term creep component of the settlement is of particular significance. Values of the parameter α have been listed in Table 2.2 for creep settlement under self-weight in various fills. In general these values are equally applicable to settlement produced by structural loads although zero time now corresponds to the application of the load, not the placement of the fill. However, the α values for domestic refuse are different under the two different loading situations. The values of α for domestic refuse quoted in Table 2.2 are largely governed by the decay and decomposition of organic matter. Clearly, when old domestic refuse is loaded by the weight of a structure the creep rate due to the increase in stress will not be principally a function of this organic decay. A value of 1% would probably be more appropriate in this situation.

2.2.3 Compression due to inundation

Loose unsaturated fill materials are usually liable to collapse settlement on inundation with water. If inundation occurs after construction on the fill, a serious settlement problem may arise. It is believed that this is often a major factor where settlement problems have occurred in building development on restored opencast mining sites. Charles et al. (1977) have presented a detailed case history of the effect of a rising ground-water table on the settlement of a 70-m deep opencast mining backfill. Problems can also be caused by water penetrating into the backfill from the surface through deep trench excavation for drains associated with the building development. The percentage compression caused by the inundation of two opencast mining backfills is given in Table 2.4.

2.2.4 Consolidation

Most of the foregoing discussion of the causes of settlement has been related principally to loose uncompacted fills formed by tipping. When fine material is placed under water as in many tailings lagoons, a soft cohesive fill is formed which is characterized by low permeability. Consequently settlement due both to self-weight and to applied loads is controlled by a consolidation process in which excess pore-water pressures dissipate slowly as water is squeezed out of the voids of the fill. Both the magnitude and the rate of settlement can be estimated from one-dimensional consolidation theory using values of compressibility measured in laboratory tests and field measurements of permeability. Geotechnical investigations of a number of such waste fills have been carried out (Ball, 1979; Krizek and Salem, 1977; Somogyi and Gray, 1977).

2.2.5 Differential settlement

It has to be remembered that it is differential settlement rather than total settlement that leads to distortion and damage to buildings. However, the poorer types of fill, which contain organic matter or were placed without control, are not only liable to large total settlements but also because of their variability are liable to large differential

Table 2.4 Field measurements of collapse compression in uncompacted opencast mining backfills

Fill type	Cause of inundation	Maximum measured compression	Reference
Sandstone and mudstone fragments	Rising ground-water table	2%	Charles et al. (1977)
Stiff clay lumps	Through surface trenches	6%	Charles et al. (1978)

movements. Differential movements should also be expected at the edges of filled areas and in places where the depth of fill changes rapidly.

2.3   Investigation of fills

The investigation of a filled site to assess its suitability for development generally should include two complementary approaches.

2.3.1 *Historical review*

A thorough review should be undertaken of all the available historical evidence relating to previous use of the site and in particular the placement of the fill (Dumbleton, 1979). This could include oral testimony, business records, old plans and maps, and air photographs. Sources of information might include previous occupiers of the site, local authorities and libraries.

2.3.2 *Ground investigation*

A ground investigation should be carried out and would generally include trial pits and boreholes. Trial pits are particularly useful as they enable large quantities of the fill to be inspected and reveal the nature of the fill, its composition and variability. Standpipe piezometers can be sealed into boreholes and provide valuable information about water levels within the fill. With a variable fill material, small-scale laboratory tests may be of limited use whereas a programme of field tests may yield much important information. Some simple field tests are now described.

2.3.2.1 *Precise levelling.* The most useful field test on deep fills may be simply to monitor the current rate of settlement of the fill by precise levelling. The levelling stations need not be elaborate. Knipe (1979) found that 1.2 m long × 20 mm diameter steel rods, driven into the ground so that less than 100 mm protruded, recorded almost identical settlements to those given by more sophisticated levelling stations formed from sleeved rods concreted into the bases of shallow holes. Stable datums (bench marks) need to be established away from the filled ground. It is useful to have a number of levelling stations close to each other so that differential movements likely to occur over the area of a building can be estimated. Settlement at different depths within the fill can be measured by installing magnet extensometers in boreholes (Charles *et al.*, 1977). Settlement measurements of this type coupled with observations of piezometric level in standpipe piezometers can give a real understanding of the performance of a fill.

2.3.2.2 *Field loading tests.* Simple field loading tests can prove very useful in some situations since they provide direct evidence of performance. Lightweight structures with strip footings stress the ground significantly only to depths of 1.5 metres to 2.5 metres. Consequently, it is relatively simple to test-load the fill to reproduce the actual stress level and distribution with depth. Charles and Driscoll (1981) have described a simple test using a weighted rubbish skip (see Fig.2.2). The tests were carried out

Figure 2.2 Simple field loading test using a skip filled with sand.

over a period of a month, which is probably the minimum period required. Settlements were plotted against the logarithm of time and extrapolated to predict the likely settlement during the life of the structure (Fig. 2.3).

2.3.2.3 *Trench water test.* If collapse settlement is considered to be a potential problem, an appropriate field test may prove useful. A shallow trench can be dug through the surface crust of the fill and filled with water. The movement of adjacent settlement stations should be monitored by precise levelling and the rate of fall of the water level in the trench should also be recorded.

2.3.3 *Non-geotechnical considerations*

It is clearly desirable, both in the review of historical evidence and in the ground investigation, that the geotechnical aspects should be integrated with the investigation of any other relevant factors which could, in some situations, include chemical attack,

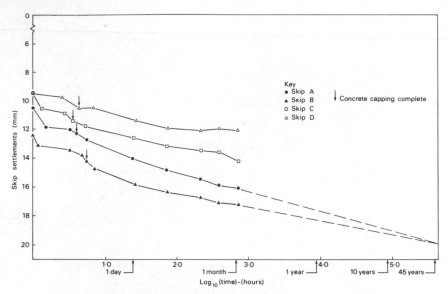

Figure 2.3 Settlement records of field loading tests on 2m depth of urban fill.

gas generation, combustibility and toxicity (Smith and Charles, 1983). On some sites special precautions may be necessary to ensure the safety of personnel carrying out the investigation. The British Standards Institution is currently preparing a code of practice for identification and investigation of contaminated sites. This will complement the existing code of practice for site investigation (BS 5930: 1981).

2.4 Classification of fills

Fill materials show quite as wide a range of engineering properties as do natural soils and rocks. Properly engineered fills such as those typically used in highway embankment construction may have excellent engineering properties. Recently placed domestic refuse would be at the opposite end of the range. It is important, therefore, to classify the fill material initially to gain some insight into whereabouts within this wide range of behaviour a particular fill lies. The investigation that has been carried out should make it possible to classify the fill in a qualitative manner under the headings listed in Table 2.5.

From a classification of this type important deductions can be made about the causes and possible magnitude of settlement of the fill subsequent to construction on it. Whenever this is possible, field tests should be carried out to make quantitative estimates of settlement with more confidence. It is useful then to distinguish between three different situations on the basis of these estimated movements (Charles and Burland, 1982).

(a) *Very small movements*—vertical compression of the fill subsequent to construction everywhere smaller than 0.5 per cent. This is likely to be the case with a granular fill that has been placed under controlled conditions and received

Table 2.5 Qualitative classification of fills

Classification	Description

Nature of material	Chemical composition;
	    organic content;
	    combustibility;
	    homogeneity.

Particle size distribution	Coarse soils, less than 35% finer than 0.06 mm; fine soils, more than 35% finer than 0.06 mm.

Degree of compaction	Largely a function of method of placement;
	    thin layers and heavy compaction—high relative density;
	    high lifts and no compaction—low relative density;
	    end tipped into water—particularly loose condition;
	    fine-grained material transported in suspension
	    and left to settle out produces fill with high
	    moisture content and low undrained shear strength,
	    e.g. silted-up abandoned dock or tailings lagoon.

Depth	Boundary of filled area; changes in depth

Age	Time that has elapsed since placement;
	    particularly significant when organic matter
	    is present.

Water table	Does one exist within the fill? Do fluctuations in level occur?

adequate compaction. Such a material forms a good foundation material and there should be few problems.

(b) *Significant movements*—vertical compression of the fill subsequent to building estimated to have a maximum value between 0.5 per cent and 2 per cent. A granular fill that has been placed without compaction, but has little organic matter within it and which has already been in place for some years, could come into this category. In this situation special attention needs to be given to foundation design. If piling is considered to be uneconomic, the basic alternatives are either to use some ground treatment technique to improve the load-carrying properties of the fill and effectively produce a category (a) situation, or to design the foundations to withstand the differential movements caused by settlement of the fill. Reinforced concrete rafts with edge beams have commonly been used for two-storey dwellings. It should be realized, however, that where large differential settlements can occur very substantial foundations may be needed and building units should be kept small and simple in plan.

(c) *Very large movements*—vertical compression of the fill subsequent to building estimated to exceed 2 per cent. This might include both recently-placed domestic refuse with high organic content liable to decay and decomposition, and fine-grained materials which have been transported in suspension and discharged into lagoons forming highly-compressible cohesive fill which might be liable to liquefaction. Problems due to settlement will be very severe and ground improvement techniques may be quite limited in what they can achieve. With recently-placed domestic refuse, methane emission may be a major problem, and such sites may be prohibitively expensive to develop.

2.5 Improvement of fills

Where buildings are to be founded on fill and the site investigation has indicated that significant differential movement may occur over the area of a building, improving the load-carrying characteristics of the fill by the use of a ground improvement technique prior to development of the site may be an attractive solution. Many treatment techniques are essentially methods of increasing the density of the fill. (Other methods such as grouting may be applicable in some situations.) This should make the fill less compressible, and consequently movements in the fill subsequent to construction will be reduced. In considering the applicability of ground improvement techniques, two different types of fill should be clearly distinguished.

Type A. Loose uncompacted fills placed by tipping. Such fills are characterized by high permeability and therefore tend to be granular in character. (This applies even to a fill composed of lumps of stiff clay unless the fill is subsequently inundated and the lumps then soften into a relatively impermeable cohesive mass.) Ground improvement techniques are used to compact this type of fill.

Type B. Soft saturated fine-grained fills formed by sedimentation under water (Hughes and Windle, 1976). These are usually characterized by a relatively low permeability and often tend to be cohesive in character. Ground improvement techniques are used to consolidate this type of fill.

In Table 2.6, several ground improvement techniques are listed. Some brief notes are included on their applicability to different types of fill. Various ground improvement methods are shown in Fig. 2.4.

When considering which ground improvement technique to use in a particular situation both the effectiveness of the different methods and their costs need to be examined. Where the fill is deep, the costs of all the methods will be closely related to the depth to which the fill is to be treated (Cross, 1977). Refilling and pre-loading with a surcharge are purely earthmoving operations and it should be a simple matter to estimate costs. The mobilization costs of dynamic consolidation are high, as a very large crane is needed to lift the weight to the required height. This means that this method is unlikely to be economic on small sites; vibro techniques may be much cheaper. In built-up areas dynamic consolidation may be prohibited by the possibility of damage being caused to existing buildings by vibrations produced by impacts. Menard and Broise (1975) quoted 30 metres as a safe distance from the point of impact to a building. Vibro methods could be used much closer to existing structures. Charles *et al.* (1978) have shown that pre-loading with a 9 metre high surcharge was a more effective method of improving a 24 m deep clay fill than was dynamic consolidation, the costs of the two methods on that site being quite similar.

Any ground improvement technique that is considered for use on a filled site should be examined in the light of the full range of problems existing at the site. In some cases a particular improvement method may be beneficial in more than one respect. For example, compaction of a colliery spoil will not only improve its load-carrying characteristics but may also eliminate the risk of combustion within the fill. In other situations the various problems associated with a waste fill may lead to conflicting requirements, for example, where gases are being generated during organic decay, solutions involving either piling or vibro techniques could lead to the formation of paths through which methane gas could enter the foundations of a building. Also, situations can arise where excavation and refilling with adequate compaction might improve the load-carrying properties of a waste fill but could also lead to mixing and reactivation of chemicals, with associated ground movements.

Table 2.6 Ground improvement methods for fills

Method	Description	Applicability	Design	References
Refilling	The fill is excavated and then recompacted in thin layers with adequate compaction.	Applicable to type A fills provided that most of the material is suitable (e.g. does not contain much organic matter) and is above the water table. Not usually applicable to type B fills.	Guidance on layer thickness and type of compaction suitable for different types of fill is given in DTp *Specification for Road and Bridge Works*. Adequate supervision of fill placement is essential. The minimum depth of excavation and refilling should be related to likely magnitude and cause of settlement, size of structure and ability of structure to withstand some differential movement. Depth is unlikely to be less than 4 to 5 metres and could be much deeper in some situations.	Department of Transport (1976)
Pre-loading	The fill is pre-loaded with a temporary surcharge of fill. A temporary lowering of the water table can also sometimes be used to pre-load the ground.	Applicable to both type A and type B fills. The method is only likely to be economic where there is a local supply of surcharge material.	With type A fills, compression of the fill will usually occur almost immediately the surcharge is applied. It is therefore not necessary to leave the surcharge in position for an extended period. A small area can be surcharged initially and then the surcharge can be moved around the site in a continuous earthmoving operation.	Charles *et al.* (1978) Skopek (1979) Tomlinson and Wilson (1973)

Table 2.6 (*Contd.*)

Method	Description	Applicability	Design	References
(Pre-loading continued)			With type B fills, consolidation of the fill under the surcharge loading may be a lengthy process due to low permeability of the fill and the surcharge may need to be left in position for a long period. It may be possible to speed up the consolidation process by the installation of vertical drains in the fill. Surcharges are usually designed to apply stresses larger than those which will be subsequently applied by buildings. However if other types of settlement (e.g. due to self-weight of the fill) are likely to be significant, then additional criteria must be used when calculating the required height of surcharge.	
Dynamic consolidation	The fill is compressed by repeated impacts of a heavy weight dropped from a considerable height onto the surface of the fill	Applicable to type A fills where it is essentially a compaction process. Its use on type B fills is much less well established.	In U.K. type A fills have been compacted typically using a 15 tonne weight dropped from 20 metres with an overall average energy input of the order of 2000 $kN.m/m^2$. Menard suggested that $z < \sqrt{WH}$ where z is the depth of effectiveness in metres W is the weight in tonnes H is the drop height in metres There is some field evidence to suggest that $z = 0.35\sqrt{WH}$ for a stiff clay fill $z = 0.40\sqrt{WH}$ for old refuse $z = 0.50\sqrt{WH}$ for sand fill This approach should be used with caution as clearly other factors besides W and H affect the depth of effectiveness of the treatment. For a more realistic analysis see Charles (1979b).	Charles (1979a) Charles *et al.* (1978) Charles *et al.* (1981) Charles and Watts (1982) Leonards *et al.* (1980) Menard and Broise (1975)

Method	Description	Applicability	Comments	References
Vibro methods e.g: vibroflotation vibrocompaction vibroreplacement	The basic tool is a large cylindrical poker vibrator which contains an eccentric weight. Rotation of the weight results in vibrations in a horizontal plane being transmitted to the fill and the poker penetrates into the fill. Either water or air can be used in a jet from the bottom of the vibrator. The long cylindrical hole produced by the vibrator is back-filled with granular material.	Applicable to many type A and possibly some type B fills. Its use is not recommended in fills with a significant organic content e.g. recent domestic refuse. Decay and decomposition of the organic matter reduce the lateral support of the granular columns.	Compaction points are typically spaced at something like 2-metre centres. Often only the upper 5 or 6 metres of deep fills are treated. With type A fills this technique is basically a method of compaction. With type B fills the technique will have little effect on the cohesive fill itself, but the granular columns will reinforce the fill and may improve its drainage characteristics.	Charles and Watts (1983) Gray and Thomson (1979). Greenwood (1970) Thorburn (1975)
Inundation	The fill is saturated either by a rising groundwater table or by water penetrating into it from the surface via trenches	Applicable only to type A fills susceptible to collapse settlement on saturation.	Where the water table rises the fill is saturated in a uniform manner and the method can be quite effective. Attempts to inundate fills via surface trenches may not be very effective due to non-uniformity of the treatment. Where large collapse settlements are expected on inundation, the possibility of carrying out inundation in conjunction with surcharging might be considered.	Charles et al. (1977) Charles et al. (1978)

Figure 2.4 Ground improvement methods. Top, pre-loading with a temporary surcharge of fill; top right, dynamic consolidation: bottom right, vibroflotation.

2.6   Foundation design for fills

Where a relatively large structure is to be built on shallow fill, poor load-carrying characteristics of the fill can be circumvented by using piled foundations and a suspended floor. The piles should be designed for negative skin friction caused by settlement of the fill. The skin friction can be calculated in terms of effective stress (Burland, 1973) as

$$\tau = (\gamma \cdot d - \gamma_w h_w) \cdot K \tan \delta$$

where τ = shaft friction at depth d below ground level,

   γ = bulk density of fill
   γ_w = density of water
   h_w = depth below water table
   K = earth pressure coefficient = $(1 - \sin \phi')$, approximately
   ϕ' = angle of shearing resistance
and δ = angle of friction of the fill on the pile surface.

For a loose granular fill a value of $K \tan \delta = 0.2$ might be appropriate. In some situations it may be considered that negative skin friction is taking too large a proportion of the allowable bearing capacity of the pile. The use of a slip coating can greatly reduce its effect. Special attention is needed in the design of services which span from the filled ground into buildings founded on piles. In a fill in which methane gas is being generated by the decay and decomposition of organic matter, piles could form paths for the escape of the gas.

Where small structures have to be built on deep fill, piling through the fill to an underlying firm stratum is not likely to be an economic solution and the structure may have to be founded directly on the fill. Reinforced concrete rafts with edge beams have commonly been used, but it should be realized that where large differential movements may occur, very substantial foundations may be required. In a previous section three different situations were distinguished on the basis of the percentage vertical compression that could occur in the fill. In the foundation design it is also important to distinguish between settlement due to the weight of the building and settlement due to other causes such as self-weight of the fill. With small structures on deep fills, almost invariably the latter will predominate and consequently 'bearing capacity' can be a misleading concept. Foundation design should be based on an assessment of the magnitude of movements of the fill subsequent to construction on it. Where large movements could occur, the use of a ground improvement technique before construction on the fill should be considered.

The problems associated with low-rise buildings on filled ground can be minimized by avoiding building across the edges of filled areas where the structure would be partly founded on fill and partly on undisturbed ground. It is desirable to restrict construction to small units and to avoid building long terraces of houses on existing filled ground. The relative movements between the building and services entering it and between various sections of the piping merit careful consideration. If the movements are not likely to be large, the use of short lengths of pipes with flexible connections may be sufficient. In more severe cases it may be necessary to use flexible pipes or carry services on piles bearing on a firm stratum beneath the fill. Generous falls should be given to drains to reduce the risk of backfalls.

Burland and Wroth (1974) reviewed the settlement of buildings and the associated damage. Their approach was based on the concept of limiting tensile strain in masonry and finishes and from this deflection ratios were calculated at which damage could be expected in frame structures and structures with loadbearing walls. From the theoretical relationship and field data which they brought together it is evident that loadbearing walls subject to hogging are much more vulnerable to damage than when subjected to sagging. Also frame structures can tolerate more relative movement than loadbearing walls.

Acknowledgement

The work described has been carried out as part of the research programme of the Building Research Establishment of the U.K. Department of the Environment and this chapter is published by permission of the Director.

References

Ball, M.J. (1979) The investigation of a quarry waste silt lagoon on the M42 motorway. *Proc. Symposium on Engineering Behaviour of Industrial and Urban Fill, Birmingham*, Midland Geotechnical Society, pp. C11–C24.

Ballisager, C.C. and Sorensen, J.L. (1981) Flyash as fill material. *Proc. 10th International Conference on Soil Mechanics and Foundation Engineering, Stockholm*, vol. 2, pp. 297–301.

Bishop, A.W. (1973) The stability of tips and spoil heaps. *Q. J. Engineering Geology* 6, 335–376.

Bratley, K.J. (1977) A description of comparative performance tests of mobile plant on a major landfill site. *Solid Wastes*, February, pp. 57–80.

British Standard 5930 (1981). Code of practice for site investigations. British Standards Institution, London.

Building Research Establishment Digest 168 (1974) *Refuse Handling*. HMSO, London.

Burland, J.B. (1973) Shaft friction of piles in clay—a simple fundamental approach. *Ground Engineering* 6 (3), pp. 30, 32, 37, 38, 41, 42.

Burland, J.B. and Wroth, C.P. (1974) Settlement of buildings and associated damage. State-of-the-art review. *Proc. Conference on Settlement of Structures, Cambridge*, Pentech Press, London, pp. 611–654.

Charles, J.A. (1973) Correlation between laboratory behaviour of rockfill and field performance with particular reference to Scammonden dam. Ph.D. thesis, University of London.

Charles, J.A. (1979a) Field observations of a trial of dynamic consolidation on an old refuse tip in the east end of London. *Proc. Symposium on Engineering Behaviour of Industrial and Urban Fill, Birmingham*, Midland Geotechnical Society, pp. E1–E13.

Charles, J.A. (1979b) Contribution to discussion on design parameters for artificially improved soils. *7th European Conference on Soil Mechanics and Foundation Engineering, Brighton*, 4, 307–310.

Charles, J.A. Naismith, W.A. and Burford, D. (1977) Settlement of backfill at Horsley restored opencast coal mining site. *Proc. Conference on Large Ground Movements and Structures, Cardiff*, Pentech Press, pp. 229–251.

Charles, J.A., Earle, E.W. and Burford, D. (1978) Treatment and subsequent performance of cohesive fill left by opencast ironstone mining at Snatchill experimental housing site, Corby. *Proc. Conference on Clay Fills, London*, Institution of Civil Engineers, pp. 63–72.

Charles, J.A., Burford, D., and Watts, K.S. (1981) Field studies of the effectiveness of 'dynamic consolidation'. *Proc. 10th International Conference on Soil Mechanics and Foundation Engineering, Stockholm*, vol. 3, pp. 617–622.

Charles, J.A. and Driscoll, R. (1981) A simple in-situ load test for shallow fill. *Ground Engineering* 14 (1) pp. 31, 32, 34, 36.

Charles, J.A. and Burland, J.B. (1982) Geotechnical considerations in the design of foundations for buildings on deep deposits of waste materials. *The Structural Engineer* 60A (1) 8–14.

Charles, J.A. and Watts, K.S. (1982) A field study of the use of the dynamic consolidation ground treatment technique on soft alluvial soil. *Ground Engineering* 15 (5), 17–22, 25.

Charles, J.A. and Watts K.S. (1983) Compressibility of soft clay reinforced with granular columns: *Proc. 8th European Conference on Soil Mechanics and Foundation Engineering, Helsinki*, vol. 1, pp. 347–352.

CP 2004 (1972) Code of practice for foundations. British Standards Institution, London.

Cross, J.E. (1977) *An Economic Assessment of Methods of Accelerating the Consolidation of Natural Soils.* Transport and Road Research Laboratory, Supplementary Report 203.

Dept. of Transport (1976). *Specification for Road and Bridge Works.* HMSO, London.

Dumbleton, M.J. (1979) Historical investigation of site use. *Proc. Conference on Reclamation of Contaminated Land, Eastbourne, Soc. of Chemical Industry*, pp. B3, 1–13.

Gray, J. and Thomson, G.H. (1979) Some observations on settlements of houses on treated urban fill. *Proc. Symposium on Engineering Behaviour of Industrial and Urban Fill, Birmingham*, Midland Geotechnical Society, pp. E51–E66.

Greenwood, D.A. (1970) Mechanical improvement of soils below ground surface. *Proc. Conference on Ground Engineering, London*, Institution of Civil Engineers, pp. 11–22.

Hughes, J.M. and Windle, D. (1976) Some geotechnical properties of mineral waste tailings lagoons. *Ground Engineering* 9 (1), 23–28.

Isaac, A.S. and Troughton, S.J. (1971) *An Instrumented Demonstration Embankment at Cortonwood Colliery—Interim Report.* Coal Research Establishment, National Coal Board, Stoke Orchard.

Isaac, A.S. (1972) *An Instrumented Demonstration Embankment at Cortonwood Colliery—Final Report.* Coal Research Establishment, National Coal Board, Stoke Orchard.

Kilkenny, W.M. (1968) *A Study of the Settlement of Restored Opencast Coal Sites and their Suitability for Building Development.* Bulletin no. 38, Dept. of Civil Engineering, University of Newcastle-upon-Tyne.

Knight, P.G.K. (1979) The engineering use of PFA. *Proc. Symposium on Engineering Behaviour of Industrial and Urban Fill, Birmingham*, Midland Geotechnical Society, pp. D63–D69.

Knipe, C. (1979) Comparison of settlement rates on backfilled opencast mining sites. *Proc. Symposium on Engineering Behaviour of Industrial and Urban Fill, Birmingham*, Midland Geotechnical Society, pp. E81–E98.

Krizek, R.J. and Salem, A.M. (1977) Field performance of a dredgings disposal area. *Proc. Conference on Geotechnical Practice for Disposal of Solid Waste Materials, University of Michigan*, ASCE, pp. 358–383.

Lambe, T.W. and Whitman, R.V. (1979) *Soil Mechanics* (SI version). Wiley, New York.

Leigh, W.J.P and Rainbow, K.R. (1979) Observations of the settlement of restored backfill of opencast mine sites. *Proc. Symposium on Engineering Behaviour of Industrial and Urban Fill, Birmingham*, Midland Geotechnical Society, pp. E99–E128.

Leonards, G.A. and Bailey, B. (1982) Pulverized coal ash as structural fill *J. Geotechnical Engineering Division, ASCE*, 108 (GT4) 517–531.

Leonards, G.A. Cutter, W.A. and Holtz, R.D. (1980) Dynamic compaction of granular soils. *J. Geotechnical Engineering Division, ASCE*, 106 (GT1) 35–44.

Menard, L. and Broise, Y. (1975) Theoretical and practical aspects of dynamic consolidation. *Géotechnique* 25 (1) 3–18.

Merz, R.C. and Stone, R. (1962) Landfill settlement rates. *Public Works* 93 (9) 103–106, 210, 212.

Nixon, P.J. (1976) The use of materials from demolition in construction. *Resources Policy*, December, pp. 276–283.

Penman, A.D.M, Charles, J.A. and Humphreys, J.D. (1982) Sandstone rockfill in two dams, *Proc. 14th International Congress on Large Dams, Rio de Janeiro*, vol. 4, pp. 279–291.

PFA Data Book (1967) *Loadbearing Fill.* Central Electricity Generating Board.

Rao, S.K., Moulton, L.K. and Seals, R.K. (1977) Settlement of refuse landfills. *Proc. Conference on Geotechnical Practice for Disposal of Solid Waste Materials, University of Michigan*, ASCE, pp. 574–598.

Refuse Disposal (1971) *Report of the Working Party on Refuse Disposal*, Dept. of Environment, HMSO, London.

Sherwood, P.T. (1975a) *The Use of Waste and Low-Grade Materials in Road Construction. 2: Colliery Shale.* Laboratory Report 649, Transport and Road Research Laboratory, Crowthorne.

Sherwood, P.T. (1975b) *The Use of Waste and Low-Grade Materials in Road Construction. 3: Pulverized Fuel Ash.* Laboratory Report 686, Transport and Road Research Laboratory, Crowthorne.

Skopek, J. (1979) Geotechnical problems encountered in moving the church at Most. *Canadian Geotechnical J.* **16** (3) 473–480.

Smith, M.A. and Charles, J.A. (1983) Technical problems associated with the development of derelict and former industrial land. *Proc. International Land Reclamation Conference, Grays, Essex,* pp. 113–123.

Somogyi, F. and Gray, D.H. (1977) Engineering properties affecting disposal of Red Muds. *Proc. Conference on Geotechnical Practice for Disposal of Solid Waste Materials, University of Michigan,* ASCE, pp. 1–22.

Sowers, G.F. (1968) Foundation problems in sanitary landfills. *J. Sanitary Engineering Division, ASCE,* **94** (SA1) 103–116.

Sowers, G.F. (1973) Settlement of waste disposal fills. *Proc. 8th International Conference on Soil Mechanics and Foundation Engineering, Moscow,* Vol. 2.2, pp. 207–210.

Sowers, G.F., Williams, R.C. and Wallace, T.S. (1965) Compressibility of broken rock and the settlement of rockfills. *Proc. 6th International Conference on Soil Mechanics and Foundation Engineering,* vol. 2, pp. 561–565.

Swain, A. (1979) Field studies of PFA in partially submerged conditions. *Proc. Symposium on Engineering Behaviour of Industrial and Urban Fill, Birmingham,* Midland Geotechnical Society, pp. D49–D61.

Taylor, R.K. (1975) English and Welsh colliery spoil heaps—mineralogical and mechanical interrelationships. *Engineering Geology* **9**, 39–52.

Taylor, R.K. and Spears, D.A. (1972) The geotechnical characteristics of a spoil heap at Yorkshire Main colliery. *Quarterly Journal of Engineering Geology* **5**, 243–263.

Thomson, G.McK, and Rodin, S. (1972) *Colliery Spoil Tips—After Aberfan.* Institution of Civil Engineers, London.

Thorburn, S. (1975) Buildings structures supported by stabilized ground. *Géotechnique* **25** (1) 83–94.

Tomlinson, M.J. and Wilson, D.M. (1973) Pre-loading of foundations by surcharge on filled ground. *Géotechnique* **23** (1) 117–120.

Weatherley, N. (1979) Trench filled PFA in colliery waste supporting old peoples' bungalows. *Proc. Symposium on Engineering Behaviour of Industrial and Urban Fill, Birmingham,* Midland Geotechnical Society, pp. D71–D76.

Wilde, P.M. and Crook, J.M. (1979) Problems and solutions in developing large areas of filled ground at Warrington New Town. *Proc. Symposium on Engineering Behaviour of Industrial and Urban Fill, Birmingham,* Midland Geotechnical Society, pp. D39–D47.

Yen B.C and Scanlon, B. (1975) Sanitary landfill settlement rates. *J. Geotechnical Engineering Division, ASCE,* **101** (GT5) 475–487.

3 Slopes and embankments

EDWARD N. BROMHEAD

3.1 Introduction

Movements of soil and rock downslope under the influence of gravity are a widespread problem, and occur in response to a variety of stimuli. This chapter considers the understanding and control of such movements and the behaviour of groundwater, which is a major factor in the stability of slopes.

3.2 Classification of types of slope instability

Many systems of classification for different types of slope instability have been proposed. These include notable schemes by Sharpe (1938), Varnes (1958), and Hutchinson (1968) to which Skempton and Hutchinson (1969) give a comprehensive list of illustrative case records. Unfortunately, the plethora of types of movements and

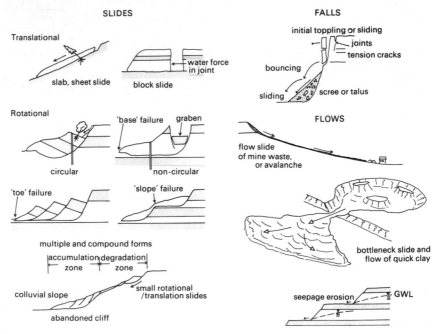

Figure 3.1 Classification of types of slope instability.

46

the limited range of appropriate descriptive words in English give all the schemes a superficial similarity which can lead to confusion.

It is usual to discriminate between three major classes of slope movements, falls, slides and flows, as illustrated in Fig. 3.1. In a *fall*, the moving material becomes separated from the parent mass of soil or rock and contact is infrequent or intermittent during movement. In *slides* the moving material remains largely in contact with underlying rocks and movement takes place on a discrete boundary shear surface. The term *flow* is applied to a movement where the material becomes disaggregated and moves without the concentration of displacement at a boundary shear. Although a flow may remain largely in contact with the surface of the ground it travels over, this is by no means always the case.

Further subdivisions in this classification can be made according to the shape of the moving mass (in section or in plan); according to the nature of its internal structure or mechanics; by virtue of some aspect of the behaviour of the rocks involved; or with reference to the water content of the materials or on the basis of the rate of movement.

Rock and soil falls (Fig. 3.1) may involve *toppling* or *sliding* in their initial movements and may progress with elements of *sliding*, *rolling* or *bouncing* to form a *scree* or *talus* at the foot of the slope. This *accumulation zone* of *colluvium* may be susceptible to further movement by flow or sliding if disturbed. Colluvial slopes formed at the base of abandoned cliffs when the erosion ceases are often a source of continuing trouble from small movements. Such slopes reflect former erosive conditions in the same way that cambered or solifluctcd slopes reveal the former presence of glacial or periglacial conditions.

Slides may be *translational* (that is have linear movement) and involve *blocks*, *slabs*, *sheets*, *lobes* or *elongate* masses of material. Where they have a *rotational* component, this may take place on approximately *circular* surfaces, or on *compound* (*non-circular*) surfaces. The latter will often display *graben* features due to internal deformation of the slide mass.

Movements may be *retrogressive* or *progressive* (working up-slope or down-slope, respectively) or may involve degeneration of the disturbed material into other, secondary, forms of mass movement. Very rapid movements with or without large *run-out* are the result of significant loss of strength. Examples of these include failure by *lateral spreading* in water-bearing silts, *bottleneck* slides in the quick clays of Scandinavia, and *flows* of loose, non-cohesive debris. In every case, the nature of the mass movements that occur is controlled by the geological structures and lithologies present, and even very localized conditions can have a major impact on this (Henkel, 1967; Bromhead, 1978; Zaruba and Mencl, 1968).

In the text that follows, a soil-mechanics rather than a rock-mechanics approach is adopted. Discontinuity patterns in a rock mass dominate its engineering behaviour, and as a rule they must be taken into account explicitly in analysis and design. In contrast, the soil-mechanics approach is to incorporate the effect of discontinuities by suitably modifying the bulk parameters considered in design. This approach is valid for a high proportion of the sediments and weak rocks exposed on the earth's surface. Indeed, in Britain, rocks down to and including the Lower Jurassic have properties which lend themselves to this approach. Furthermore, the concepts of stability analysis are identical between the two once the critical discontinuities in a rock mass have been determined: the major difference is that in a soil mass, failure is less constrained to take place along predefined surfaces.

C

3.3 Limit Equilibrium Design

Most deformation problems with natural and artificial cut slopes arise from movement on the sliding surfaces of landslides, whereas in fill slopes the problems are
equally divided between these causes and settlements due to self-weight. There are
additional problems with pre-failure shear deformations in some fills due to their flat
stress strain curves. Except for settlement under self-weight (see Chapter 1), in all cases
the ground movement is the result of shear deformations usually concentrated into
zones. Many of these shear zones are so thin that the orthodox definition of shear strain
ceases to be meaningful, and we then have to refer to the *displacement* on the shear
surface. Under such conditions most of the presently-available continuum-mechanics
methods break down and the approach of using strains throughout the soil mass to
predict deformations is no longer valid. Current practice is to use the equilibrium
equations in analysis and to control displacements by providing and maintaining an
adequate factor of safety against sliding. This limit equilibrium approach will be
followed here.

3.3.1 *Mechanics of slope stability*

Certain kinds of stability analysis for slides, slow-moving flows and the onset of
movements leading to falls have become conventionally accepted. They are usually
carried out on two-dimensional sections through the slope: three-dimensional effects
are treated by taking appropriately weighted combinations of the results from different
sections. The exception to this is the stability analysis of rock wedges sliding along
intersecting discontinuities where a fully three-dimensional method (Hoek and Bray,
1977) can be employed. In this section the two-dimensional approach is followed.

The Factor of Safety (F) is usually defined as the ratio between the forces available to
resist movement and those forces which cause movement, but a useful alternative
definition is to take F to be the amount by which the shear strength has to be factored to
bring the resisting forces into equilibrium with the destabilizing forces. This equilibrium is obtained by resolving forces and moments for the whole slide or for each of the n
slices or wedges into which the mass can be divided. Take for example the typical slice (i)
shown in Fig. 3.2. All known force and moment components, including pore-water
pressures, body forces due to gravity or seismic loading and structural loads, can be
resolved into the two forces H_i and V_i, and the moment M_i, leaving as unknown the
shear and normal effective forces on the sides and base of the slice, together with the
positions at which the normal forces act. On the base of the slice, the shear (S_i) and
normal effective (N_i') forces can be related through the shear strength parameters for the
soil:

$$S_i = \frac{1}{F}(c_i' l_i + N_i' \tan \phi_i') \qquad (3.1)$$

but even with this simplification, the system of equations that is obtained has too many
unknowns for solution.

Numerous attempts at overcoming this problem have been made. For instance, if the
forces between the slices are ignored, then n trivial problems result, from which the
following average factor of safety is obtained by finding the sums of the forces causing

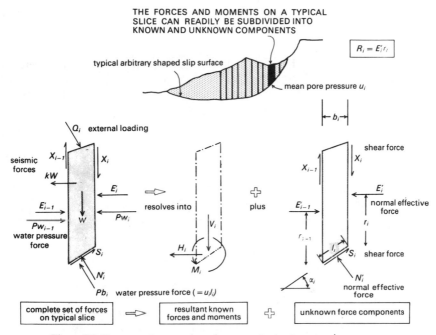

Figure 3.2 Forces and moments acting on a single slice of a sliding mass.

and resisting instability and taking a direct ratio.

$$F = \frac{\sum_{i=1}^{n} c_i' l_i + (W_i \cos \alpha_i - u_i l_i) \tan \phi_i'}{\sum_{i=1}^{n} W_i \sin \alpha_i} \qquad (3.2)$$

Bishop (1955) reports that the use of this simple expression leads to a conservative result, especially where the slip surfaces are deep or the pore-water pressures are high. It can, however, be simplified into the so-called *infinite slope* expression (Skempton and Delory, 1957) for sliding parallel to the slope surface which, with z as the vertical depth to the slip surface, and the ground water table h_w above this, yields:

$$F = \frac{c' + (\gamma z - \gamma_w h_w) \cos^2 \alpha \tan \phi'}{\gamma z \sin \alpha \cos \alpha} \qquad (3.3)$$

Inspection of this equation reveals that the limiting slope angles in non-cohesive soils for no pore pressure or with the groundwater level at the surface are equal to ϕ' and approximately $1/2(\phi')$, respectively. This result is of great use in regional studies of slope instability where a simple measurement of natural slope angle is all the investigation that can be undertaken (Skempton and Delory, 1957).

More refinement is obtained if only the shear component of the interslice force is omitted from consideration (Bishop, 1955), or by assuming, for example, a position for the line of thrust, or a direction for the resultant (see, for example, Morgenstern and Price, 1965; Spencer, 1967; Janbu, 1973; Sarma, 1973).

Certain assumptions lead to internal inconsistencies if all three conditions of equilibrium are to be satisfied, and often one or more of the conditions is violated in order to simplify the equations. These violations may be stated explicitly, or be implicit in the derivation. One major simplification can be achieved if the slip surface is the arc of a circle in section: all forces normal to the base of each slice then act through this point and can be ignored in taking moments. (The normal forces still influence the frictional resistance and have to be taken into account in some way in the solution.) With this approach, the factor of safety is defined in terms of moments.

In Bishop's (1955) method, which falls into this class of 'slip-circle' methods, interslice shear forces are neglected. Resolving vertically so as not to include the other interslice force components explicitly, the following expression for the normal effective force N_i' on the base of a typical slice is obtained:

$$N_i' = \frac{(c_i' b_i + (W_i \cos \alpha_i - u_i b_i) \tan \phi_i')}{\left(\cos \alpha_i + \sin \alpha_i \dfrac{\tan \phi_i'}{F} \right)} \tag{3.4}$$

The ratio of the summed resisting and destabilizing moments about the centre of rotation then gives (after some manipulation!)

$$F = \frac{\sum\limits_{i=1}^{n} \left\{ (c_i' b_i + (W_i - u_i b_i) \tan \phi_i') \dfrac{\sec \alpha_i}{\left[1 + \dfrac{(\tan \phi_i \tan \alpha_i)}{F} \right]} \right\}}{\sum\limits_{i=1}^{n} W_i \sin \alpha_i} \tag{3.5}$$

This equation can be solved from a first estimate for the factor of safety F by evaluating the right-hand side and using the result to re-evaluate this expression until a negligible difference is found. Convergence and other problems with this method are discussed by Whitman and Bailey (1967) among others, but in general good results are achieved. Spencer (1967) attributes this to the reliance on the moment condition of equilibrium. Janbu (1973) uses force, rather than moment equilibrium, to derive a similar, simple equation to that of Bishop for use with slip surfaces of any shape. (To obtain his expression from the above, for 'sec' read 'sec^2', and for 'sin' read 'tan'.) He found that errors in the use of this modified equation were related to the depth of the slip surface, and proposed an empirical correction factor of up to 13 per cent based on a depth-to-length criterion for the slip surface. The results are still prone to some error and his work contains more sophisticated procedures in recognition of this.

To obtain the benefit of a more accurate treatment of the interslice forces the concept of 'summing over all the slices' has to be dropped. Instead we have to work from one end of the slide to the other (say from the toe to the head) using the basic equations of equilibrium to find the out-of-balance forces and moments that are implied by an initial estimate for the factor of safety, and the assumption made regarding the state of the interslice forces. Either or both of these factors can then be modified until equilibrium (negligible out-of-balance) is obtained. For instance, the following pair of simultaneous equations, for force and moments, can be obtained from the forces acting on a typical slice (Fig. 3.2) and used in this process:

$$E_i'(1 - f_i a_i) = E_{i-1}'(1 - f_{i-1} a_i) + \frac{b_i c_i'}{F}(1 - \tan \alpha_i a_i) + a_i V_i - H_i \qquad (3.6)$$

$$R_i = R_{i-1} - \frac{b_i \tan \alpha_i}{2}(E_i' + E_{i-1}') + \frac{b_i}{2}(f_i E_i' + f_{i-1} E_{i-1}') - M_i \qquad (3.7)$$

where E' and R are respectively the effective force and moment at the interface between slices, the subscripts $i - 1$ and i indicate respectively the downslope and upslope sides of slice i, and where the coefficient a_i represents the function

$$a_i = \frac{\dfrac{\tan \phi_i'}{F} - \tan \alpha_i}{1 + \dfrac{\tan \phi_i' \tan \alpha_i}{F}} \qquad (3.8)$$

In the derivation of this equation it has been assumed that the common Mohr-Coulomb assumptions regarding the shear strength (equation 3.1) still hold, and that the shear (X_i) and normal effective (E_i') components of the interslice force are related by

$$X_i = \lambda f_i E_i' \qquad (3.9)$$

In this equation, λ is a variable determined as part of the solution, hence giving the absolute ratio, and f is a second parameter defined *a priori* at each slice interface which allows some control to be exerted over the relative inclination of the interslice force resultant (f = a constant yielding *parallel* interslice forces). From estimated starting values of F and λ, the out-of-balance force and moment at the very head of the slide can be evaluated and systematically reduced to zero in an iterative process relying on the use of successively modified F and λ values. This is also the basis of the Morgenstern and Price (1965, 1967) method, although those authors use two simultaneous partial differential equations rather than the recurrence formulae quoted above which are due to Maksumovic (1970). The Morgenstern and Price method is applicable to somewhat wider slices than the above which assumes that the normal effective force on the base of the slice has a negligible eccentricity from the slice centreline.

A systematic procedure for altering F and λ requires estimates of the derivatives of both E' and R with respect to both F and λ, and the use of a Newton-based iterative formula. These derivatives can be obtained either by differentiating the basic formula, or approximating by trials using slightly different F and λ values.

3.3.2 *Location of critical slip surfaces*

An infinite number of possible slip surfaces may be drawn on the cross-section of an earth slope. Many of these will have a very low likelihood of failure, and inspection of the slope details can immediately reveal the most critical areas for investigation. A fairly systematic approach to the selection of slip surfaces can, however, throw light on problem areas that might otherwise be missed. The use of slip circles helps enforce discipline in this.

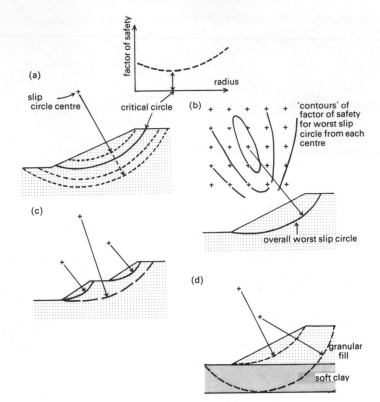

Figure 3.3 Locating a critical slip circle. There are 'critical' slip surfaces associated with surface features (c), or with changes in soil strength (d), or with pore-water pressure concentrations.

Take a simple soil slope (Fig. 3.3a) and a range of slip circles all centred on the same point. One of these will have a lower factor of safety than the rest. Repeating the process for a range of centres on an regular grid produces a set of these minimum safety factors from which the true critical slip circle can be found (Fig. 3.3b). Computer programs which automatically search out this 'minimum of minima' have been developed. Charts have also been prepared which enable factors of safety to be evaluated for simple cases without further calculations. Examples of these are given by Spencer (1967) and by Pilot and Moreau (1973) who follow the earlier, total stress approach of Taylor (1937). Unfortunately, real cases are far too complicated for this simple approach to yield satisfactory results. Take, for instance, a slope with two berms. There are critical slip circles for the failure of each berm as well as for failure of the whole slope. Each of these slip circles has its own centre. In principle, any of the three could be most critical, but an automatic search routine could be 'trapped' in a non-critical local minimum factor of safety zone (Fig. 3.3, c and d). These local minima are associated with the presence of weak zones, or concentrations of pore-water pressure in the slope as well as with irregular surface forms.

It is important to identify each mode of local or overall failure and to ensure that the critical slip circle for each mode has been found. It will often be the case that the 'least safe' slip circle is for an inconsequential mode of failure: an example of this is where non-cohesive fills are used. The local factor of safety for shallow slides parallel to the slope face can be shown (based on equation 3.3) to be given by the ratio

$$F = \frac{\tan \phi'}{\tan (\text{slope angle})} \tag{3.10}$$

a value which can approach very close to unity without necessarily having any ill effect. A cautionary note for users of automatic search computer programs is that the critical centre for a slip circle which approximates to this sliding surface is located an infinite distance away from the slope face and the slip circle has an infinite radius.

3.3.3 Use of computer methods

The calculations for slope stability, especially the repetitive ones involved in searching for a critical slip surface, are extremely tedious and, regardless of computation aids, are inevitably error-prone. They were therefore among the earliest of the civil engineering applications of computers when these became available in the 1950s (Little and Price, 1958). Computer programs for slope stability are now widely used, to the virtual exclusion of hand calculations, but they must be used with caution.

A check must always be on 'correctness' through comparisons with published cases (beware—these are not always correct!) or by relying on a validating body. The U.K. Department of the Environment, for instance, has a register of approved programs kept by the Highway Engineering Computer Branch. It is probably not worth using any method less sophisticated than Bishop's iterative method (equation 3.5); the accuracies will be too suspect.

3.3.4 Part submergence

External water loads can be added directly into an analysis or can be eliminated by a simple artifice due to Bishop (1955). Those parts of each slice below the level of the external water surface are given a submerged unit weight (that is, from c to d in the typical slice of Fig. 3.4), and only the pore-water pressure head in excess of the external water level is used in the analysis proper. Examination of the equilibrium of the slide mass will show that this removes forces and moments exactly equivalent to the external water load. For upstream slip surfaces, however, we may be left with an apparently negative net pore-water pressure to consider.

Slip surfaces in a dam which has external water loads on both upstream and downstream faces (for example slip XY in Fig. 3.4) can be treated by using Bishop's technique on the downslope water load, and adding the additional forces from the upstream water to the slide mass. A simple way of doing this is to take the reservoir as being filled with a strengthless 'soil' with water pressures related to the external water level. Any slip surface extension from point 'X' through this 'soil' will ensure that the correct additions are made.

3.3.5 Seismic techniques

Seismic stability is dealt with in Chapter 12. It is noteworthy at this point, however, while dealing with slope stability in general terms, that the effect of uniform seismic

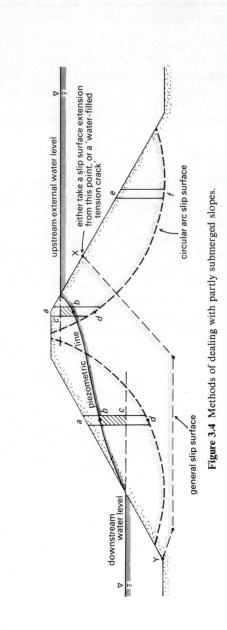

Figure 3.4 Methods of dealing with partly submerged slopes.

accelerations on the factor of safety can be simulated by rotating the slope cross section, usually so as to steepen it. The rotation angle θ is related to the horizontal $(k_h g)$ and additional vertical $(k_v g)$ accelerations by:

$$\theta = \tan^{-1} \frac{k_h g}{(1 + k_v)g} \tag{3.11}$$

All soil unit weights must be multiplied by the factor

$$[k_h^2 + (1 + k_v)^2]^{\frac{1}{2}}$$

to reflect the net acceleration.

Alternatively, non-uniform accelerations can be included by adding in the appropriate force and moment components to the equations of stability (Fig. 3.2 and equations 3.6 and 3.7).

In many cases, the factor of safety under seismic conditions is of lesser importance than the critical seismic acceleration k_{cr} under which the slope just becomes unstable. This can be found by taking $F = 1.0$, and separating out the seismic force elements. A solution scheme based on k_{cr} rather than F can then be developed. Sarma's (1973) method is based on this approach, but with a little added refinement which makes it possible to solve directly for k_{cr} instead of iterating.

3.3.6 Tension cracks

Tension cracks at the head of a slide are sometimes considered in slope stability analysis. Usually, they are taken as full of, or part filled with, water and the resulting destabilizing force is added in to the analysis. However, the appearance of a tension crack at the crest of a soil slope usually indicates failure: stability analysis is therefore redundant since the factor of safety is close to unity. In contrast, tension cracks at the head of a rock slope may open due to stress relief, or as a result of blasting and do not necessarily signify the onset of failure.

Where such cracks have formed, not by slide-induced tension but by desiccation shrinkage during dry weather, or by joint opening in rocks, then extremely high lateral thrusts can be experienced should they fill with water. These thrusts may be of short duration but do pose a severe test of stability.

A second case where the tension crack concept is useful is as a technique for overcoming numerical problems in stability analysis where soil strengths at the head of a slide are high. The assumption of a constant factor of safety around all of the slip surface in this situation will imply tension between the slices. In turn, this may have a great effect on the calculated factor of safety. The effect is least in the *slice summation* methods (where indeed the treatment of the interslice forces is cursory) and greatest in the *integration* methods where the order of the slices and their interaction is of significance. A tension crack inserted at the appropriate place can return realism to the calculations.

3.3.7 Back analysis and internal force distributions

Back analysis is the term used to describe the process of obtaining soil strength parameters from failed slopes by stability analysis. Repeated trial analysis using different parameters in an effort to achieve $F = 1$ is unnecessary, bearing in mind the definition of F. All that is needed to find the mobilized parameters (required for equilibrium) is to factor the parameters c' and $\tan \phi'$ used in analysis by the factor of safety obtained.

It is usual to compare these back-analysed parameters to those obtained from laboratory testing as an important check. To do this properly, the laboratory strength tests must be performed with normal effective stresses comparable to those acting in the field. These can best be found from the individual N' forces, but this step in the analysis can be circumvented by the following method. If the critical slip surface is re-analysed using an arbitrary cohesion and without friction, a 'mobilized cohesion' can be found. This is identical to the average shear stress on the sliding surface. Returning to the original effective stress analysis, and with the use of equation 3.1, the average normal effective stress can be found.

Individual interslice force values are also useful if excavation into a slide mass is proposed, as they give some indication of the likely forces on retaining walls.

3.4 Influence of groundwater on stability

The water pressures in the pores of a soil play a fundamental role in determining the strength and deformation properties of that soil. When left to equilibrate, these pressures may be distributed hydrostatically (where there is no flow) or in accordance with the laws of seepage. However, they may be perturbed from this equilibrium state by changes in the hydraulic or physical boundaries to the seepage regime, or by the response to applied loads. Usually the effects act in conjunction rather than singly, and an example will illustrate this. Imagine a uniform bed of clay with the groundwater table at the surface and into which a cutting is to be excavated. First of all, the relief of stress on the cut faces causes a change (a decrease) in the pore-water pressure in the remaining clay. With time, and the infiltration of water, the depressed pore-water pressures will rise until the seepage pattern is in equilibrium with the physical shape of the cut slope and the mean surface water pressure conditions. The term *undrained* is applied to the initial pore-water pressure response, and *drained* to the final state where the pore-water pressure is in equilibrium. In this context the term 'drained' implies the elimination of stress change-influenced components of pore-water pressure and not a state of zero water pressure or zero water content. Note also that, in this example, equilibration is to a completely new water-pressure distribution, and is not merely a return to pre-existing conditions. Where equilibration is through a decrease in pore-water pressures, the term *consolidation* is used. This covers the whole process including the ground strains which accompany the migration of the soil moisture. *Swelling* is the corresponding term for pore-pressure equilibration in the increasing sense, although mineral/pore-water interactions discussed in Chapter 9 demonstrate the true complexity of these latter processes.

3.4.1 *Undrained pore-water pressures in embankments and cuttings*

Embankments placed on soft subsoils (see Chapter 11) will significantly increase the pore-water pressures in their foundations. The increase in pore pressures may be calculated from the total stress changes using pore-pressure parameters (Skempton, 1954; Henkel, 1960).

The dissipation of these water pressures with the escape of water from the soil leads not only to settlement but also to a gain in strength in the soil. Thus the end of construction marks the most serious stability condition, with steady improvement thereafter as consolidation proceeds. Exceptions to this rule occur when the soil has significant permeability anisotropy (such as laminated silts and clays) and the water migrates laterally from the centre-line of the embankment out to its toe, causing some temporary loss of strength. On the whole, however, the foundation may be expected to increase in strength with consolidation. This leads directly to the concept of raising an embankment in layers or lifts, with time for consolidation allowed after each. The pore-water pressures set up under the early lifts can wholly or partly escape before the additional loading is applied. In this way it is possible to build higher and steeper embankments than would otherwise be the case. It is also possible to 'build out' some of the settlements that occur.

The behaviour of the fills is more complex. Pore-pressure response in these is heavily influenced by placement water content and by the type and amount of compaction. Indeed, some soils may exhibit substantial pore-water suctions after placement and compaction dry of optimum, although these may be lost and the pore-pressure response become positive as the fill is more deeply buried. Another source of depressed pore-water pressures arises when, in an attempt to balance cut and fill, spoil from deep cuttings is used in small embankments. The individual blocks of soil then respond to a net stress decrease. Where this is the case, the moisture content changes required for equilibrium lead to a loss in strength, and hence to the possibility of delayed failures.

Exactly the same principles apply to cut slopes, but in reverse. Here the moisture content increases as equilibration takes place and both strength and stability decrease with time. The installation of drains speeds this process and might therefore be thought, paradoxically, to be undesirable. However, the equilibration, although more rapid, is to a steady seepage regime with lower pore-water pressures than if the drains were not installed. Bishop and Bjerrum (1960) discuss the principles of this process, with further important contributions on the timescale of equilibration in cut slopes being given by Vaughan and Walbancke (1973) and by Eigenbrod (1975).

Pore-water pressure changes can, in many cases, be calculated from changes in the vertical total stress, $\Delta\sigma_v$, and these in turn are approximately derived from the thickness of soil placed or removed. This will provide a reasonable estimate provided that the ratio between the stress components remains constant and that the vertical stress is close to being the major principal stress. Hence

$$\Delta u = \bar{B}\Delta\sigma_v \qquad (3.12)$$

and $\bar{B}$ can be expressed in terms of the other pore pressure parameters, although it is more usual to measure it in a triaxial test with controlled stress ratios. In saturated soils it can be expected to be approximately 1.0.

Overconsolidated soils are those which have previously experienced consolidation under stresses higher than those existing at present. This may be due, for instance, to the removal of overlying soil and rock strata. This tends to lock in high lateral stresses. When excavated into, the lateral stress relief in these soils can be the major component. Not only the pore-water pressure effect, but also the ground strains, can occur distant from the crest of the slope as well as underneath the exposed slope face. However, where the slope is created by slow erosion (river or coastal erosion being prime sources for this) the stress relief occurs so gradually that the equilibration is synchronous. In this case the lateral stress relief is *drained* and only the vertical stress relief (under the exposed face itself) is *undrained*. Such a receding cliff will carry back within itself a localized depressed pore-water pressure zone. This may not be detectable using piezometers back from the crest of the slope, but it (and its subsequent gradual equilibration) is of vital importance to the long-term performance of any stabilization works that might be constructed.

3.4.2 *Steady seepage pore-water pressures*

With the passage of time, the pore-water pressures in a slope come into equilibrium with the mean external hydraulic boundary conditions. The term 'mean' is used advisedly, since the air face of the slope may be subject to seasonal moisture variations or the more permeable soils in the slope may be connected hydraulically to a source of fluctuating pressure, for example a river or the sea. This equilibration may be speedy or very slow depending on the scale of the problem (drainage path lengths) and the permeability of the materials involved. In broad terms, equilibration times for a 10 m high slope in gravel or coarse rockfill could be measured in few minutes; in sands, an hour to a day; in silts, weeks to months. For clays, the equilibration times can not only be measured in years and decades, but may even stretch into centuries for larger slopes.

Space precludes a detailed treatment here of the procedures for seepage analysis. Instead, this section will be restricted to noting some of the more important general points. (In natural slopes the pore-water pressures can be measured, rather than predicted. This changes the nature of the problem.)

First, it is crucial to identify and properly represent the hydraulic boundary conditions. This is not always as easy as it sounds; capillary action on the exposed face of a clay slope (or evapotranspiration) may be difficult to predict quantitatively or to measure, but may totally dominate the pore-pressure regime.

Second, it is vital to understand the permeability zonation of the slope section. The exact values of permeability are of no significance to the pore-pressure distribution, although they do have an immediate bearing on the through-flow, because it is the relationship between the permeabilities that controls the seepage pattern. Anisotropy in the permeability coefficients caused by bedding in natural soils, or by rolling in fills, can easily be catered for with modern analytical and computational techniques, and can in many cases be shown to influence the results. Obtaining the basic data is the prime, and possibly insoluble, obstacle to the use of this refinement.

Where the permeability varies in the *direction* in which flow takes place, particularly when this change is a decrease, the effect on the seepage pattern is more pronounced than when the permeability changes in the other direction. A cause for permeability to decrease in the direction of flow is the dependence of permeability on effective stress. Where this effect exists, normal methods of pore-water pressure prediction lead to *unsafe* results. Accounts of work using this concept are given by Morgenstern and Guther (1972), De Mello (1976) and Bromhead and Vaughan (1979).

In natural slopes it is usual to measure the pore-water pressures by means of piezometers. The installation plan must be sufficiently generous to measure the boundary conditions (shallow and deep) as well as the pore pressures in the body of the slope. Provision must be made for extra piezometers so that special cases such as shear zones or seepages into the borehole may be instrumented.

Shallow piezometers are influenced by seasonal effects and require reading

throughout at least one complete calendar year or longer if the late winter levels are
to be obtained. Longer records can show effects that would otherwise be missed.
Hutchinson and Gostelow (1976), for instance, obtained their highest piezometric
levels seven years after installation. Deep piezometers, especially those into stiff clays,
may take a very long time to equilibrate and need to be monitored until this is complete.
This may reveal long-term trends—in the coastal landslides at Folkestone Warren,
pore-pressure changes of only 0.15 m head per annum were detected in a series of
readings (Hutchinson *et al.*, 1980). Seasonal effects in clay slopes are unlikely to be
significant below about 5 m depth, so this may be taken as roughly the boundary
between shallow and deep piezometers. Where the pore pressures are tidally-
controlled, the response and time lag (if any) should be observed throughout a tidal
cycle.

3.4.3 *Representation of pore-water pressure information*

The end result of a field investigation or theoretical study of seepage is a set of pore-
water pressures. These values represent the real distribution of pressures throughout
the seepage domain. Ideally, the stability analysis program should accept this
distribution directly as input data.

Two other concepts are used to describe pore-pressure information: these are the
piezometric line and the *pore-pressure ratio* (r_u). Take a soil slope with a through flow of
water. There will in all probability be a boundary between the region in which flow
takes place and the soil above it. This is the phreatic surface (or line in a cross-section).
Unfortunately, it cannot be used to define the pore-water pressures except in the
limiting case of no flow when this line is horizontal and the water-pressure head is given
by the depth beneath this line. The water-pressure head in a piezometer will rise to some
different level—below the phreatic surface if there is downflow, or above it where there
is upflow. This level is called the piezometric level.

If a slip surface is sketched on to a cross-section and the pore-water pressure heads
are found at a series of points along it (derived from a flow net, or measured in the field)
then the piezometric levels for each of these points can be drawn. The line joining these
is the piezometric line, and it will readily be appreciated that the position of this is
related to the slip surface drawn originally: take a different slip surface and a different
piezometric line will be found. The piezometric line is thus a useful means of
interpolating between a few, limited, piezometric observations for a restricted range of
slip surfaces, but it cannot be a realistic method for describing a pore-pressure regime of
any complexity.

Suction in the pore water can be represented by a piezometric line *below* the slip
surface, although the natural tendency is to treat the piezometric line as a phreatic line
in cases such as this, that is, to take zero pore-water pressure. It is a wise precaution to
check what implicit assumption is made about this in any computer program being
used for the first time.

The pore-pressure ratio, symbolized by r_u, which is the main alternative to the
piezometric line in commercial programs, is a much abused concept. At any point in the
soil it is the ratio of pore-water pressure to vertical total stress, that is

$$r_u = \frac{u}{\gamma_z} \tag{3.13}$$

An average value for this ratio can be found along a slip surface, through a particular zone of soil, or even for the whole of a slope cross-section. Programmers like this tidy approach: one r_u value for each soil along with a cohesion, an angle of shearing resistance and a unit weight keeps the input data simple and compact. The programming is easy too, as the steps which yield the weight of each slice and select the shear parameters for use on its segment of slip surface also produce the pore pressure. However, for the user this creates problems. The subdivision of the slope into zones of approximately equal r_u may differ radically from the subdivision on the basis of soil type. A great deal of averaging is implicit in the process, and on back-analysis it will often be found that systematic but unintentional changes in the pore-pressure distribution have been made: high pore pressures are smoothed out, low ones are rounded up. In turn this shifts the emphasis in the results from the more serious deep-seated modes of sliding to shallow surface sliding modes of lesser overall importance. This shift is of greater importance to the design process than the 'errors' in the numerical results.

It is easy to confuse r_u with $\bar{B}$ (equation 3.12). The latter is an incremental response of pore pressure to a change in vertical stress; the former is the cumulative effect of all the elements of pore pressure response. Take as an example a point 10 m deep with an 8 m piezometric head. The unit weight of soil is about twice that of water, so r_u is approximately $8/(2 \times 10)$ or 0.4. Now add 2 m of fill, which will increase the pore pressure in this saturated soil with a $\bar{B}$ of 1.0 by some 4 m of head, so that r_u becomes $(4 + 8)/(2 \times (10 \times 2))$ or 0.5.

3.4.4 Seepage erosion

Seepage erosion involves the removal of solid particles from the soil by an outflow of water, often leading to the collapse of overlying strata (Hutchinson et al., 1981). Hutchinson (1982) reviews this as a slope failure mechanism and concludes that soils most prone to seepage erosion are cohesionless fine sands with median diameters between 0.09 mm and 0.25 mm with a uniformity coefficient of between 1.8 and 3.1. This should not be taken as implying immunity from this mechanism for other soils.

The stream power of the outflowing water has a direct bearing on the rate at which seepage erosion acts, and so control of the hydraulic gradients and/or retention of the solid particles with suitably graded and weighted filters are appropriate remedial measures.

3.4.5 Conclusion

It is of paramount importance to use realistic pore-water pressures in a stability assessment, and to understand how these pore-water pressures are likely to change with time. Errors in the pore pressures can make a more significant difference to the calculated strength and stability than can any other single factor, and need to be treated with respect. In addition to the influence of consolidation or swelling on pore-water pressures in the longer term, both submergence of the ground surface or its exposure by drawing down a reservoir for instance, can have adverse effects on stability. Whereas an external water load ordinarily helps to support a slope (see section 2.2.4), the opposite effect may result from changes to the local groundwater regime: a 'draw-down' case is critical if significant pore-water pressures remain in the slope.

3.5 'Brittleness' and its implications

Some soils and rocks experience a loss of strength if deformed past their maximum load-carrying capacity, and this loss of strength is called 'brittleness'. Some brittleness is to be expected in those materials which dilate during shear to a critical voids ratio at which deformation can continue, but the major cause of brittleness is the forced alignment of plate-shaped particles (usually clay minerals) in the direction of shear. Very large strains may be required to complete this process, which is irreversible and leaves a surface or surfaces in the soil with a permanently impaired residual shear strength (Skempton, 1964). The U.S. term *ultimate strength* is a more straightforward alternative description which avoids conflict with other uses of the term 'residual'.

The peak strength of a soil, or its maximum shear load-carrying capacity, is largely dependent on the particle size and shape distribution, on the state of packing and on any interparticle bonds. These latter factors reflect the stress and environmental 'experience' of the soil. Shearing to a critical voids ratio eliminates the effect of initial packing, so wiping out much of the stress-history effect. It does still, however, leave a strength that owes more to the particle shapes than to the friction between them. Only in the case of the residual strength of argillaceous soils and mudrocks is the interparticle friction a significant factor. These types usually have a significant brittleness, which can be quantified in terms of a Brittleness Index or percentage loss in shear strength from peak to residual (Fig. 3.5). Such a low residual strength is only mobilized in soils which have a sufficiently high fraction of plate-shaped particles to form a continuous shear surface (Lupini *et al.*, 1981; Skempton, 1964).

In broad terms, therefore, the brittleness is commonly a function of the mineralogy of the clay fraction, reaching its largest magnitude with soils that have a low residual strength. Montmorillonite species have residual strengths in the range 4° to 9°, illites

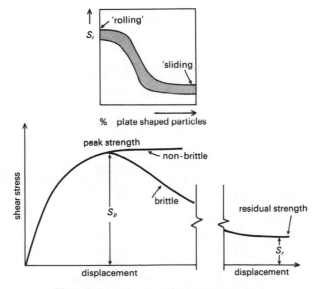

Figure 3.5 Brittle stress displacement in shear.

from 9° to 15°, kaolinites 14° to 22° and, for comparison, silts and sands (which show much less brittleness) have residual strengths in the range 28° to 35°.

3.5.1 *Progressive failure*

One of the most important aspects of brittleness (the other being the formation of slip surfaces) is the role it plays in progressive failure (Bishop, 1966). If a slope is brought into a state of local overstress—by building too high or steep, for example—then the soil elements in this condition must shed load as they deform because the load-carrying capacity is diminished as the strains increase. This load is passed to adjacent soil elements until equilibrium is re-established. It may be that equilibrium cannot be achieved, in which case the shear zone grows, possibly by the coalescence of originally-separate growing shear zones, until it is continuous through the whole slope. Further deformation occurs until the sliding mass moves into equilibrium, with the residual strength acting all around the sliding surface.

A change in pore-water pressure conditions, such as the swelling in a cutting side slope (Vaughan and Walbancke, 1973), can be the trigger for such a failure mechanism. In addition, pore-water pressure responses to shear strains during failure can influence the rate at which movements develop and continue. Such a failure is then truly progressive, both in time and space (Bishop, 1966). A failure of this type, in a slope behind a retaining wall in a cutting in Uxbridge (Watson, 1956), was arrested before a continuous shear surface had formed. This was a clear indication of the progressive growth of shear surfaces during failure.

The magnitude of the movements which take place is related to the brittleness index, and the rate at which they occur to the post-peak gradient of the load-deformation curve. In non-brittle soils, embankment construction may be controlled by monitoring movements and slowing down fill placement if the deformation rate increases unacceptably. A short period of consolidation then provides the gain in strength needed to allow filling to be resumed. This method of site control cannot be used where the soil has appreciable brittleness, since by the time the rate of displacement has been observed to change significantly it is likely that progressive failure is well advanced.

One of the more important direct consequences of progressive failure is that, at the start of major displacements, some of the soil is carrying load with its peak strength, but some of the soil is losing its strength as it has been deformed past its peak. Using the empirical results from a series of case histories of slides in small cut-slopes in London and Lias clays, Skempton (1970) and Skempton and Chandler (1975) suggest that this mean strength, intermediate between peak and residual, is approximated by the peak strength of normally consolidated specimens of the same soil. This has an angle of shearing resistance equal to that for the peak strength, and a very low cohesion (in the 0 to 1.5 kN/m² range). Until it becomes possible to analyse progressive failure in full, this empirical criterion has to be used to select design parameters for cut slopes in all brittle clays.

3.5.2 *Flow slides*

Other forms of brittleness are a manifestation of pore-fluid pressure effects, including liquefaction and fluidization. Many granular soils exist with their particles in a metastable arrangement. Disruption of this, for instance by vibration or shear strain, causes collapse into a denser state with an increase in the pore-fluid pressure arising through compression of the voids. There is thus a loss of strength in the soil. Bishop (1973) reviews many examples of this in the context of the stability of tips and spoil heaps where careless placement (loose tipping) can give rise to the necessary metastable structure. He points out that dissipation of the excess pore-fluid pressure, which is made easier after sliding when the soil is spread out, allows many of the materials susceptible to this form of failure to regain their strength. Characteristic features of these *flow-slides* include their long run-out and high speed, factors which contribute to the hazard posed by these forms of mass movement. They include slides in colliery discard in South Wales with run-outs of more than 1000 m which took place in a matter of minutes, and similar movements in fly ash, china-clay waste and even coarse discard from a limestone quarry. Some of Bishop's examples are drawn from natural slopes, which are not immune to this phenomenon.

Flow movements of extreme violence have been experienced in many mountainous areas (Voight, 1978). The German term *Sturzstrom* has been utilized to describe these.

Submarine landslides often take the form of flows, particularly in the unconsolidated sediments at river mouths or on the continental margin.

Bjerrum (1955) describes quick clay landslides in the geologically-recent (Pleistocene) sediments of Canada and Scandinavia. These may be related phenomena. When deposited in a marine environment, the clay particles flocculate or stick together in clumps, a direct result of the chemistry of the pore water. After uplift, and leaching of the clay by fresh water, the interparticle bonding is destroyed and a metastable structure remains. In areas subject to the devastating quick clay slides that can be triggered by even small slope movements in river banks, there is a high level of public awareness of the hazards. This is rare in comparison to the average popular perception of landslide hazard elsewhere in the community.

3.6 Sources and rates of movement

Strains in a slope take place before, during and after the process we term 'failure'. These strains reveal themselves (usually at the surface) as deformations or ground movement. Some movement, particularly close to the surface, occurs even in slopes considered stable, as a result of seasonal moisture content variation; creep rates of a centimetre or so per annum are easily reached. Because of its shallow nature it is only likely to affect poorly-founded structures.

Pre-failure deformations are almost synchronous with the formation of the slope unless influenced by progressive failure or time effects (for example consolidation) in the pore water. Typically, in cut slopes, crest deflections of up to 2–3% of the slope height may be experienced, but this depends greatly on the stiffness of the materials involved and may cause the opening of tension cracks. Where long-term groundwater level changes are under way, such deformations may be spread over many years.

Much larger movements can occur during brittle failure. In argillaceous soils and mudrocks the main displacements take place over a time period varying from about an hour to several days. Deformations of between 10 and 50 per cent of the slope height have been observed. This depends partly on the brittleness and partly on the shape of the failure surface. Where, for instance, base failure is involved, material collects at the slope toe and will lead to lesser deformations than would occur in an otherwise similar slide involving a slope type of failure (Fig. 3.1).

Flow slides (and all related mass movements) may be thought of as failures with extremely high brittleness. Accordingly, they have very large displacements, and take place at high speed. Many flow slides in the world's mountain belts are initiated by rockfalls from mountain peaks and have run-outs of up to 5 km and speeds of the order of 200–300 km/h. Avoidance of the hazard would seem to be the only practical solution here. Even for the smaller movements (for example, flow-slides in granular soils of spoil heaps) no satisfactory analysis is possible. It is, however, possible to perform controlled load and cyclic loading tests (Seed, 1976; Taylor *et al.*, 1978), to differentiate those 'aggregates' which are susceptible to liquefaction from those which are not.

These movements cannot reasonably be accommodated by structural design and need to be prevented from occurring. Indeed, the analytical tools needed to predict such movements with any accuracy do not exist, and it is still impossible to predict the nature, let alone the magnitude, of the displacement pattern in other than the broadest terms.

3.6.1 *Reactivation of existing slides*

Existing landslides may be reactivated by a variety of means. Hutchinson *et al.* (1980) discuss a large number of these with reference to the coastal landslides of Folkestone Warren, paying special attention to the up to 30 m displacement that occurred in a single slide event in 1915. Large displacements are incompatible with the nature of the stress-deformation curve for an existing slip surface which should behave in a non-brittle way. Although major causes for reactivation that have been considered included unloading (by marine erosion) at the toe of the slide, pore-water pressure variation, and the reimposition of brittleness through chemical effects, it was concluded that the major effect was the loading at the head of the slide by a number of large rockfalls from the rear scarp.

It is loading, either by the ill-considered placing of fills or by falls and slides from a head scarp, that forms the commonest reactivating agent for existing slides. The problem is worsened if undrained pore-water pressures are set up under the loads—in such cases even small loads can have a disproportionate effect on stability. In contrast, toe erosion (unless on a massive scale) has a lesser impact. Base failure modes (Figs. 3.1 and 3.3) require improbable deep scour at the toe to remove passive restraint, whereas slope failure modes, which can lose significant amounts of material in a short duration by secondary slides or falls, are relatively insensitive to the loss of their toes.

Often it will be found that porous and permeable landslide debris can be subject to large increases in unit weight when the groundwater level is raised. This, in conjunction with the water-pressure increase, can be a major destabilizing influence. It may well be brought about where normal drainage is disrupted or dammed, as it can be by the freezing of natural or artificial outfalls.

On the whole, pore-water pressure changes in soil strata of low permeability in response to hydraulic stimuli tend to be gradual, and therefore not a source of rapid movement. They tend, rather, to be responsible for slow movements which, while equally damaging to structures, pose rather less hazard to life.

Sometimes depressed pore-water pressure conditions in slopes caused by the stress changes following excavation even survive the onset of sliding. The continued equilibration of these can be a source of further movement. It is conceivable that an alteration to the chemistry of the pore water in the vicinity of a slip surface could reintroduce brittleness.

The formation of a slip surface eliminates the brittleness. Subsequent movements take place immediately in response to external stimuli. If these are applied rapidly, then the slope will move rapidly: the response to gradual changes is gradual. Major destabilizing events give rise to displacements similar in magnitude to brittle failures, and indeed, may be mistaken for them. Ordinarily, however, rates of less than 100 mm per annum are normally experienced in the largest slopes, even those subject to marine attack.

3.7 Remedial and preventive measures

In the design of remedial measures for unstable natural slopes the engineer is more constrained by the materials present than in the design of embankments where he can choose if not the nature of the materials to be used then at least how they are placed. Often, too, the scale of natural slope instability is massive, and the remedial measures that are constructed reflect the finance available rather than a design to suit some overall factor of safety requirement.

A partial safety factor approach is sometimes used. This demands an assessment of the various design parameters to assign to each its own safety factor. For example, soil strengths may be very variable or poorly measured: hence apply a safety factor to them. By the time this has been done for each parameter and these partial factors all summed or multiplied together, than the resulting overall safety factor is likely to be quite large. Paradoxically, it will be smaller for a major dam, say, than for a landscaped hump in an ornamental park! Since this is an unacceptable result, and the possibility of all the worst cases happening together is extremely small, then there is a temptation to select an arbitrary value, probably in the range 1.4 to 1.8.

An alternative strategy is to construct primary remedial measures of such a scale as to enable the works to survive the worst anticipated destabilizing event with sufficient reserve to allow reconstruction or the prevention of further change: secondary remedial measures are used to control the severity of such events. For example, a toe load and sea defence for a coastal landslide must, following a bad storm or series of storms during which part of the fill is scoured away, still prevent movement of the landslide for sufficient time to allow the fill to be replaced. Surface drains will prevent infiltration, and maintain the reserve stability until the works are restored.

3.7.1 Cut and fill solutions

When design calculations show that it is not feasible to construct a slope as originally conceived, the first course is normally to include drains to eliminate construction pore-water pressures. If this cannot bring about the required improvement in stability, then attention must be given to lowering the overall slope angle or height. Battering the slope faces to a flatter angle (Fig. 3.6a) is usually most effective for shallow potential modes of failure, typical of slopes in non-cohesive soil, but does little for deeper-seated modes. Figure 3.6b shows how a berm can bring loads to bear in the optimum position

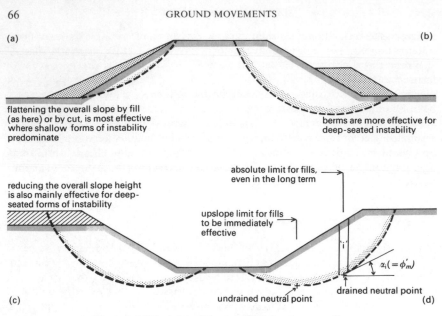

Figure 3.6 Principles of slope stabilization by re-grading.

to stabilize such deeper slip surfaces, which are often found where a cohesive soil forms the slope or its foundation. Reduction in overall slope height (Fig. 3.6c), although an alternative solution for this latter case, is rarely a viable option in practice.

It may be possible to adopt a compromise by, say, designing a bermed profile for stability against deep-seated slips during the critical period after the end of construction. Extra fill may be placed later, to bring an embankment to its design cross-section when its burden of extra pore-water pressures have a lesser effect on stability.

The positioning of fills, and to a lesser extent cuts, to stabilize actual slides is aided by the 'neutral line' concept (Hutchinson, 1977). Consider a simple rotational slide, in Fig. 3.6d, with a small load ΔW placed on slice i. The factor of safety changes from F_0 where $F_0 = M_R/M_0$ to F_1, where

$$F_1 = \frac{M_R + (\Delta W \cos \alpha_i - \Delta P_{b_i}) \tan \phi_i'}{M_0 + \Delta W \sin \alpha_i} \tag{3.14}$$

in which ΔP_{b_i} is the pore pressure resultant arising from the increase of stress under the load ΔW. If the pore pressure response is proportional to the increase in vertical total stress, that is equation 3.12 applies, then

$$\Delta P_{b_i} = \bar{B} \Delta W \sec \alpha_i \tag{3.15}$$

It is possible to place the load such that no change in factor of safety occurs ($F_1 = F_0$), in which case

$$\tan \alpha_i = (1.0 - \bar{B} \sec^2 \alpha_i) \frac{\tan \phi'}{F} \tag{3.16}$$

Placing a load to either side of this position (the 'neutral point') will have an adverse or helpful effect on stability. The neutral point lies on the vertical section at which the slip surface has an inclination α_i which satisfies this formula, and this depends on the pore-water pressure response in the soil.

Where several sections have been investigated, the trace of the neutral points in plan forms the neutral line. Knowledge of the position of this is useful in preliminary planning, for instance in highway alignment where the corridors in which the road must be in fill or cut can form major constraints on the geometric design.

Since a pore-pressure response of say 70 per cent has the same effect as a 30 per cent dissipated, 100 per cent response, it is possible (through manipulation of the $\bar{B}$ term) to explore what happens to the neutral point during dissipation. It is found that it migrates from the position where $\alpha = 0$ to $\alpha = \phi'_m$ as $\bar{B}$ alters from 1.0 to 0. Hence these limiting positions are termed the *undrained* and *drained* neutral points since $\bar{B} = 1.0$ is the likely maximum pore-water pressure response for undrained loading of saturated soil, and $\bar{B} = 0$ is equivalent to the effect of full drainage. There is therefore a zone between these two points where fill would have a stabilizing effect in the long term, but is deleterious in the short term. Careful programming of fill placement is therefore needed to obtain maximum effectiveness from the remedial measures.

Complex slides with a number of components may have a series of neutral points for each element or series of elements. This imposes extra constraints. Great care with construction in landslide zones is nevertheless called for in every case, even those with relatively simple shape.

In addition to their effect on stability, cuts and fills can alter the hydrological pattern with potentially adverse effects. This not only applies to surface water courses, but also to natural subsurface drainage patterns which are removed, blocked (sometimes by compaction) or which may be unable to cope with diverted flows.

3.7.2 Drainage

Drainage can greatly influence the stability of slopes by reducing the pore-water pressures and hence increasing the strength of the soil. Drainage measures may be concentrated in the control of runoff, shallow drainage or deep drainage; and may be intended to provide a permanent control of seepage or to merely allow excess pore-water pressures set up during construction, for instance, to escape. A selection of these drainage measures is shown in Fig. 3.7.

Surface drains to control runoff are usually lined or unlined ditches dimensioned to safely carry the flow from fairly extreme climatic events. Regular maintenance of these is essential, but is often neglected. Seasonal moisture content changes can disrupt flimsy linings, and vegetation growth can trap sediment or otherwise decrease the ditch capacity. It is difficult to design self-cleansing falls as the slopes of the earthworks predominate. Surface drains act as run-off concentrators and it is essential that the collected water should be discharged safely.

Some improvement in the stability of natural slopes, particularly in respect of shallow modes of instability, can be achieved through control of surface water. This involves re-routing streams, draining ponds and so on.

Shallow drains can also be used to control water movement in surface layers of higher permeability. Typically these drains are rubble- or gravel-filled trenches, and the use of geotextiles as filters to prevent the gravel clogging is recommended.

Deep drainage is installed to control pore-water pressures adjacent to actual or

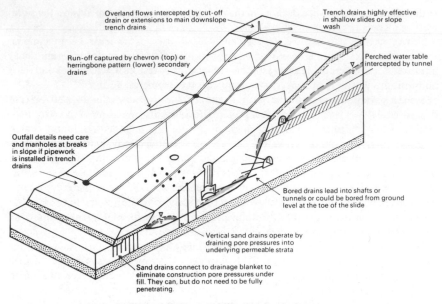

Overland flows intercepted by cut-off drain or extensions to main downslope trench drains

Trench drains highly effective in shallow slides or slope wash

Run-off captured by chevron (top) or herringbone pattern (lower) secondary drains

Perched water table intercepted by tunnel

Outfall details need care and manholes at breaks in slope if pipework is installed in trench drains

Bored drains lead into shafts or tunnels or could be bored from ground level at the toe of the slide

Vertical sand drains operate by draining pore pressures into underlying permeable strata

Sand drains connect to drainage blanket to eliminate construction pore pressures under fill. They can, but do not need to be fully penetrating.

Figure 3.7 Slope stabilization by drainage.

potential slip surfaces. Adits and tunnels may be constructed through the sliding mass as at Folkestone Warren (Hutchinson *et al.*, 1980) where the headings are driven from the seawall back into the slide debris (in one case through into *in-situ* material). Following construction of one of these, the groundwater level in a nearby observation well was observed to fall nearly 20 m. The effectiveness of this form of drainage can be increased with additional bored drains (kept open by a porous pipe lining). In order to set up the drilling equipment for these inside a tunnel it is usually necessary to form an enlarged working chamber.

However, all tunnelling work in active landslides is extremely hazardous and it may be preferable to tunnel outside the slipped material, penetrating the slip surface with bored drains if need be. If the source of water is outside the slide, such a tunnel can act as a cut-off drain, or interceptor. A large tunnel has been used for such a purpose at the Marine Colliery, Merthyr Tydfil, to intercept flows in the Pennant Sandstone before they entered a large pre-existing and unstable tip of colliery discard.

Vertical shafts may also be used, as at Herne Bay (Berkeley-Thorne and Roberts, 1981). Although the discharge from bored drain arrays needs to be pumped from a sump, and this can impose a maintenance burden, the discharges themselves need not be high, especially in clay soils, for the drains to be effective, and the running costs can be small. If vertical drains can penetrate an underlying permeable stratum which contains lower pore-water pressures the drains can be self-emptying, and bored drain systems using this principle have been employed effectively. Several proprietary systems exist for backfilling the drain holes including the use of sand-filled tubular woven filter fabric 'sand wicks'.

Bored vertical sand-filled drain wells can also be used to help speedily eliminate construction-induced pore-water pressures from the foundations of large embankments. Here they do not need to be pumped out nor to penetrate underlying permeable strata, since the water will escape from the top of the drains. However, pumping can be used as a cheaper alternative to fill construction in field trials on the efficacy of such drains at a particular site.

One of the major differences between permanent and construction drains is the filtration requirement in the former case. Traditionally, a layer or layers of soil with intermediate gradings are placed between the earth fill and the drain, the grading chosen to hold back fines from the earthfill, but not to lose its own fines into the drain. It is sometimes impossible to satisfy both of these conditions simultaneously, thus requiring two or more graded filter layers.

Design rules for filters are based on the ratio between the various particle size fractions in both the filter and the protected soil. Such rules may not be entirely satisfactory in all soils or when protecting clays from erosion, and Vaughan and Soares (1982) propose a more involved treatment based on the average flocculated 'particle' size. This is dependent on the pore-water chemistry.

Geofabrics make effective filters but need protection from UV light during placing and may be torn by sharp particles or by strains in fill. Continual improvements in technology, however, make the use of geofabrics as filters (and soil reinforcement) an increasingly attractive proposition.

3.7.3 *Restraining structures*

Attempts have been made to restrain slides with retaining walls; piles, grouted, heat-treated or frozen zones; and with shear keys, ground anchors or similar structures. The restraining forces involved are invariably high, and many of these methods have proved unsuccessful. An extensive review by Hutchinson (1977) includes many unorthodox schemes as well as listing successes and failures in the more traditional techniques. For brevity, only ground anchors will be considered here.

Stressed ground anchors are a method of concentrating the high forces needed to restrain landslides. Rock bolts and rock anchors are commonplace in rock slope work, and soil anchors are used as a substitute for shores in retaining wall support, but in soil slopes their use is less common. This may well be due to some problems associated with their use. First, there is the problem of obtaining an anchorage at depth capable of sustaining very high loads. This is only overcome by anchoring into rock, since underreaming is often impractical and even in the stiffest of soils anchorage lengths tend to be very long indeed. Second, a reaction pad or beam has to provided at ground level and this is often inconvenient to locate: available positions dictate the location and inclination of the anchors to a certain extent, preventing the optimum use of the available forces. The applied load stresses the soil and builds up pore-water pressures and leads to consolidation of the zone between the anchorage at depth and the load pad. Hence there is a loss of prestress in the anchor tendon in addition to that which occurs due to creep in the anchor components. This is better overcome by periodic re-stressing until consolidation is complete than by the provision of extra initial load in the anchor.

An anchor should ideally be installed at an angle to the slip surface equal to the mobilized angle of shearing resistance of the soil. This can readily be shown by resolving the forces on a typical slice and differentiating with respect to the anchor inclination angle. This maximizes the coupled effects of decreasing the net destabilizing effect while increasing the sliding resistance. It is often found, however, that geological constraints, as well as those of land use at the surface, may prevent the use of this optimum, thus reducing the effectiveness of the installation.

Finally, some methods of stability analysis as they are formulated in published work may simply not allow the use of additional force vectors. This is especially the case with the *horizontal* component of an anchor force but is not the case with the method outlined on page 51.

3.7.4 *Rock slope stabilization.*

Rock slopes may be stabilized by removing or anchoring unstable blocks. Where this proves impractical, then catch fences or ditches, or surface meshes, may be used to control any blocks that do fall. In exceptional cases, avalanche sheds or similar structures can protect vulnerable installations.

Drainage measures are confined to the joints of the rock mass, which may need to be raked out if infilled by erodible material. 'Dental' masonry or concrete may be used to support the sides of large cavities in rock face.

The subject is extensively reviewed by Fookes and Sweeney (1976).

3.8   Special subsurface instrumentation and monitoring techniques

Virtually all the orthodox ground investigation techniques are used in the investigation of slope instability, and it is only in the detection and location of slip surfaces that special methods and skills are required. A large number of these, including many not in common use, are listed by Hutchinson (1982). Broadly, the methods fall into two groups: observational methods applicable to all slides, and instrumental methods which can only be used if the slide is moving.

In the former group there is the direct logging of trial pits, shafts or adits. Trial pits are used for shallow slides (less than 5 m deep). For deeper pits the costs of excavation and strutting become prohibitive. Shafts are again expensive, and thus rare, although they were used in the investigations at Bury Hill, Staffordshire (Hutchinson *et al.*, 1973). Other exploratory excavations are still more expensive and less common.

Slip surfaces can be recognized from their polished and striated or fluted (slickensided) appearance. They may be isolated, but can occur thickly in a shear zone leaving lenticular zones of clay between them. Shallow slip surfaces often have a different colour from the ambient rock or soil as groundwater passing through the shear zone carrying weak organic solutions reacts with the minerals present. The migration of pore water towards a very old slip surface on which movement no longer takes place can transform the slickensiding into a thin soft clay gouge or laminar infilling.

As well as logging exposed faces, slip surfaces can be found in split cores from boreholes. Continuous core recovery is most desirable as slip surfaces can easily be lost between samples or core runs. Their original presence can then only be inferred from changes in lithology, colour, texture and so on (see Fig. 3.1) in adjacent samples. The position of slip surfaces can also be confirmed by gaps in the fossil succession, but this demands microfossil work since macrofossils are rarely sufficiently abundant in small cores used in site investigation. Most of these methods will only work if sliding occurs across the bedding for only then is a discontinuity formed. It is therefore vital to understand the geological structure of the site as well as to have logged the undisturbed succession. This calls for detailed mapping outside the landslide area, and possibly additional boreholes to ensure that the undisturbed sequence has been logged.

Surface mapping is, of course, an essential adjunct to subsurface investigation, but lies outside the scope of this section. Guidance on techniques may be found in the first instance from the Working Party Reports on the Preparation of Maps and Plans in terms of Engineering Geology, and on Terrain Evaluation, published by the Engineering Group of the Geological Society (1972, 1982).

Sounding equipment has been used in the softer soils in the search for slip surfaces. Electrical penetrometers equipped for resistivity measurement can detect, for instance, high water-content zones at a slip surface, or discontinuities in salinity where sliding has disturbed a salt-water/fresh-water interface.

In moving slides, instruments can be installed close to the surface or in boreholes (Fig. 3.8) to reveal the position of the slip surface. These can vary in sophistication from inclinometers which, when read successively, reveal the complete displacement profile throughout the access tubing provided; to simple slip-indicating systems which rely on slide movements kinking a flexible access tube at a depth which can be found by plumbing from the surface. Where the latter concept is used in the access tube of a standpipe type piezometer, a very cheap and effective combination is obtained. Standpipe type piezometers are cheap, simple, self de-airing and in many ways almost foolproof. Their main drawback is that they stabilize slowly and have poor response times. Other piezometers are based on hydraulic, pneumatic or electrical principles and are more suited to remote or automatic reading. There is a similar diversity in the types of extensometers which are marketed to measure surface and subsurface earth movements. Most of the manufacturers will offer detailed advice on the use of their

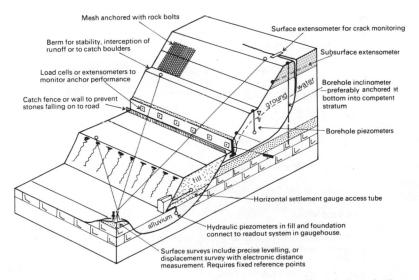

Figure 3.8 Monitoring for rock and soil slopes. Treatment for unstable rock slopes is discussed extensively by Fookes and Sweeney (1975).

products in a particular situation, although in cases where the choice is difficult, guidance is given by Hanna (1973) or by Wilson and Mikkelsen (1978).

Slip surface positions can be inferred from surface movement vectors which tend to parallel the slide displacements at depth. Extensive disruption of the ground surface of a slide mass is usually a reflection of the internal shearing which takes place above a part of the slip surface where it abruptly changes direction.

Notation

a	coefficient (see equation 3.8)
$\bar{B}$	pore-pressure response parameter
b	width of a typical slice
c', c_u	effective cohesion, undrained strength (cohesion)
E, E'	normal, normal effective, components of interslice force
F, F_0, F_1	factor of safety
f	function controlling relative inclinations of interslice forces
H	resultant of all horizontal forces on a slice
h	height, depth
i	index (used as a subscript, relates to slice i)
j	index (used as a subscript, relates to slice j)
k, k_{cr}, k_h, k_v	seismic acceleration (cr = critical, h = horizontal, v = vertical)
l	length
M	resultant of all moments
M_R, M_0	resultant moment causing, resisting, instability
N, N'	normal, normal effective, force on a slip surface
n	number of slices
P	force, usually a water-pressure resultant
P_b, P_w	water force on the base, side, of a slice
Q	externally applied force (e.g. by ground anchor)
q	load intensity
R	moments of the normal effective component of an interslice force about the position of the slip surface
r_u	pore-pressure ratio (see equation 3.13)
S	shear force

u	pore-water pressure
V	resultant of all vertical forces on a slice
$W, \Delta W$	weight, increment of weight
X	interslice shear force component
x, y	coordinates
z	depth
α, β	angle
θ	angle
γ, γ_w	unit weight of soil, or of water
ϕ', ϕ'_r	angle of shearing resistance (peak, residual)
σ, σ_v	normal stress, vertical total stress
σ'	normal effective stress
τ	shear stress
λ	coefficient controlling inclination of interslice forces
$\sum$	summation
Δ	increment

References

Berkeley Thorne, R. and Roberts, A.G. (1981) *Sea Defence and Coast Protection Works—a Guide to Design* (3rd edn.) Telford, London.

Bishop, A.W. (1955) The use of the slip circle in the stability analysis of earth slopes. *Géotechnique* **5**, 7–17.

Bishop, A.W. (1966) Progressive failure, with special reference to the mechanism causing it. *Proc. Geotechnical Conference, Oslo*, **22**, pp. 142–150.

Bishop, A.W. and Bjerrum, L. (1960) The relevance of the triaxial test to the solution of stability problems. *Proc. Res. Conf. Shear Strength of Cohesive Soils, Boulder, Colorado*, (Special Pub. ASCE) pp. 439–501.

Bishop, A.W. (1973) The stability of tips and spoil heaps. *Q. J. Engineering Geology* **6**, 335–376.

Bjerrum, L. (1955) Stability of natural slopes in quick clay. *Géotechnique* **5**, 101–119.

Bromhead, E.N. (1978) Large landslides in London Clay at Herne Bay, Kent. *Q. J. Engineering Geology* **11**, 291–304.

Bromhead, E.N. and Vaughan, P.R. (1979) Solutions for seepage in soils with effective stress dependent permeability. *Proc. 1st International Conf. on Numerical Methods for Non-linear Problems, Swansea*, Pineridge Press, pp. 567–578.

De Mello, V.F.B. (1977) Reflections on design decisions of practical significance to embankment dams. *Géotechnique* **27**, 279–355.

Eigenbrod, K.D. (1975) Analysis of pore pressure change following the excavation of a slope. *Canadian Geotechnical J.* **12**, 429–440.

Engineering Group of The Geological Society: Working Party (1982) Land surface evaluation for engineering practice. *Q. J. Engineering Geology* **15**, 265–316.

Engineering Group of the Geological Society: Working Party (1972) The preparation of maps and plans in terms of engineering geology. *Q. J. Engineering Geology* **5**, 293–381.

Fookes, P.G. and Sweeney, M. (1976) Stabilisation and control of local rockfalls and degrading rock slopes. *Q. J. Engineering Geology* **9**, 37–56.

Hanna, T.H. (1973) *Foundation Instrumentation*. Trans Tech Publications, Cleveland.

Henkel, D.J. (1967) Local geology and the stability of natural slopes. *Proc. ASCE*, (SM4) 437–446.

Henkel, D.J. (1960) The shear strength of saturated remoulded clays. *Proc. Res. Conf. on Shear Strength of Cohesive Soils, Boulder, Colorado* (Special Pub. ASCE) pp. 533–554.

Hoek, E. and Bray, J.W. (1977) *Rock Slope Engineering* (2nd edn.) Institute of Mining & Metallurgy, London.

Hutchinson, J.N. (1968) 'Mass movement', in *Encyclopaedia of Geomorphology*, (ed. R.W. Fairbridge,) Publ. Reinhold, New York, pp. 688–695.

Hutchinson, J.N. (1977) Assessment of the effectiveness of corrective measures in relation to geological conditions and types of slope movements. *Bull. Int. Assn. of Engineering Geology* **16**, 131–155.

Hutchinson, J.N., Bromhead, E.N. and Lupini, J.F. (1980) Additional observations on the Folkestone Warren Landslides. *Q. J. Engineering Geology* **13**, 1–32.

Hutchinson, J.N., Chandler, M.P and Bromhead, E.N. (1981) Cliff recession on the S.W. coast of the Isle of Wight. *Proc. 10th Int. Conf. on Soil Mechanics and Foundation Engineering, Stockholm*, pp. 429–434.

Hutchinson, J.N., Somerville, S. and Petley, D.J. (1973) A landslide in periglacially disturbed Etruria Marl at Bury Hill, Staffordshire. *Q. J. Engineering Geology* **6**, 377–404.

Hutchinson, J.N. and Gostelow, P. (1976) The abandoned cliff at Hadleigh, Essex. *Phil. Trans. Royal Society London* **A283**, 557–604.

Hutchinson, J.N. (1982) Methods of locating slip surfaces in landslides. *Proc. Symp. Landslides: Theory & Practice, Bled, Yugoslavia*, (Reprinted as BGRG Monograph No. 30, Geo Books, Norwich.)

Hutchinson, J.N. (1982) Damage to slopes produced by seepage erosion in sands. *Landslides and Mudflows: Reports of Int. Seminar on Water related Exogenous Processes and Prevention of their Negative Impact on the Environment, Alma Ata, USSR*, (Unesco and UNEP), ed. A. Sheko. Moscow Centre of International Projects GKNT, pp. 250–268.

Janbu, N. (1973) Slope stability computations, *Embankment Dam Engineering: Casagrande Memorial Volume*, eds. Hirschfield & Poulos, John Wiley & Sons, New York, pp. 47–86.

Little, A.L. and Price, V.E. (1958) The use of an electronic computer for slope stability analysis. *Géotechnique* **8**, 113–120.

Lupini, J.F., Skinner, A.E. and Vaughan, P.R. (1981) The drained residual strength of cohesive soils. *Géotechnique* **31**, 181–214.

Maksumovic, M. (1970) A new method of slope stability analysis. Private communication.

Morgenstern, N.R. and Price, V.E. (1965) The analysis of the stability of general slip surfaces. *Géotechnique* **15**, 79–93.

Morgenstern, N.R and Price, V.E. (1967) A numerical method for solving the equations of stability of general slip surfaces. *Computer Journal* **9**, 388–393.

Morgenstern, N.R. and Guther, H. (1972) Seepage into excavations in a medium possessing stress dependent permeability. *Proc. Symp. on Percolation through Fissured Rock* Int. Soc. Rock Mechs and Int. Assoc. Engineering Geol., Stuttgart, T2-C, pp. 1–15.

Pilot, G. and Moreau, M. (1973) *La Stabilité des Remblais sur Sols Mous*. Editions Eyrolles, Paris.

Sarma, S.K. (1973) Stability analysis of embankments and slopes. *Géotechnique* **23**, 423–433.

Seed, H.B. (1976) Evaluation of soil liquefaction effects on level ground during earthquakes. in *Liquefaction problems in geotechnical engineering*, ASCE National Convention, Philadelphia, Pa., pp. 1–104.

Sharpe, C.F.S. (1938) *Landslides and Related Phenomena*. Columbia Univ. Press, New York.

Skempton, A.W. and DeLory, F.A. (1957) Stability of natural slopes in London Clay. *Proc. 4th Int. Conf. Soil Mechanics and Foundation Engineering, London*, **2**, pp. 378–381.

Skempton, A.W. (1954) The pore pressure coefficients A and B in saturated soils. *Géotechnique* **4**, 143–147.

Skempton, A.W. (1964) Long term stability of clay slopes. *Géotechnique* **14**, 77–101.

Skempton, A.W. (1970) First time slides in overconsolidated clays. *Géotechnique* **20**, 320–324.

Skempton, A.W. and Hutchinson, J.N. (1969) 'Stability of natural slopes & embankment foundations,' in *Proc. 7th Int. Conf. Soil Mechanics and Foundation Engineering, Mexico, State-of-the-art volume*, pp. 291–340.

Skempton, A.W. and Chandler, R.J. (1975) The design of permanent cutting slopes in stiff fissured clays. *Géotechnique* **25**, 425–427.

Spencer, E.E. (1967) A method of the analysis of the stability of embankments assuming parallel inter-slice forces. *Géotechnique* **17**, 11–26.

Taylor, D.W. (1937) Stability of earth slopes. *J. Boston Soc. Civil Engineers* **24**, 197–246.

Taylor, R.K., Kennedy, G.W. and Macmillan, G.L. (1978) Susceptibility of coarse grained coal-mine discard to liquefaction. *Proc. 3rd International Cong. Int. Assoc. Eng. Geol., Madrid, Special Sessions*, pp. 91–100.

Varnes, D.J. (1958) Landslide types and processes, *Landslides and Engineering Practice*, ed. E.B. Eckel, U.S. National Academy of Sciences, Highway research Board, Special Report 29, pp. 20–47.

Vaughan, P.R. and Soares, H.F. (1982) Design of filters for clay cores of dams. *Proc. ASCE* (GT1) 17–31.

Vaughan, P.R. and Walbancke, H.J. (1973) Pore pressure changes and delayed failure of cutting slopes in overconsolidated clay. *Géotechnique* **23**, 531–539.

Voight, B. (ed.) (1978) *Rockslides and Avalanches* (2 vols.). Elsevier, Amsterdam.

Watson, J.D. (1956) Earth movements affecting LTE Railway in deep cutting east of Uxbridge. *Proc. Inst. Civil Engineers*, **11**, (5) 320–323.

Whitman, R.V. and Bailey, W.A. (1967) Use of computers for slope stability analysis. *J. Soil Mechs. Div., ASCE*, **93**, 475–498.

Wilson, S.D. and Mikkelsen, P.E. (1978) Field instrumentation. *Landslides: Analysis and Control*, eds. R.L. Schuster and R.J. Krizek, U.S. National Acad. Sciences, Transportation Research Board, Spec. Report 196. pp. 112–138.

Zaruba, Q. and Mencl, V. (1968) *Landslides and their Control*. Elsevier, Amsterdam, (translated from Czech edition, Academia, Prague).

4 Deep trenches and excavations in soil

M.R. HURRELL and P.B. ATTEWELL

4.1 Introduction

Excavations may conveniently be classified into three types:
shallow trenches, ranging from about 1.2 m to 1.8 m deep,
deep trenches, ranging from 1.8 m to about 6 m deep,
deep excavations, generally exceeding 6 m in depth.
The relative magnitudes of the above categories are shown diagrammatically in Fig. 4.1. There is a further intrinsic criterion of classification—that of excavation time and project duration which may range from a few hours in the case of shallow trenches to several months for deep excavations.

Shallow trenches are used for the installation of and maintenance access for water, sewage, gas, electrical and telecommunication utilities. These trenches are considered in Chapter 5. In this present chapter, deep trenches and deep excavations are discussed in terms of design, excavation, and their effects on adjacent ground and structures. Deep trenches, for gravity sewers and storm drains, are considered separately from deep excavations used for the construction of rapid transit systems, water storage facilities, and deep basements.

Inward yielding of the ground support system, accompanied by settlement of adjacent ground, is an inevitable result of all excavation operations, no matter how carefully the excavation-support process is executed. It is therefore necessary for the engineer to provide a project design which enables ground movement to be controlled and minimized. The engineer must also recognize that his design responsibilities may extend far beyond the site boundary, and understand that under the U.K. Institution

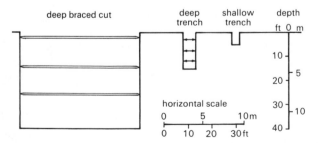

Figure 4.1 Diagram showing the relative sizes of the principal types of excavation (generally after Harris and O'Rourke, 1983).

of Civil Engineers (5th Edition) form of contract, the contractor is not required to insure for damage which is the unavoidable result of the construction of the works in accordance with the contract.

Excavation-induced ground movements cannot be predicted to any degree of accuracy because the designer is not able to specify the character and mechanical response of the ground, nor to control with any degree of precision the variable construction procedures which inevitably occur at a site. It should be possible, however, to make reasonable estimates of upper and lower limits of consequential movements and the effects of those movements, particularly when experience, based on field measurements, has been acquired on similar work in a similar locality.

4.2 Deep trenches

4.2.1 Introduction

While studies into deep excavations provide an insight into the general nature of ground movements related to trench construction, they cannot be directly applied to the estimation of ground movements induced by trench construction in specific cases.

Important differences between the two types of excavation relate to construction practices. Deep excavations, generally deeper and wider than trenching operations, are subject to phenomena such as base heave, which rarely arises as a pressing problem during trenching operations. Compared with deep excavations, the methods of support and backfilling specific to trenching often lead to relatively large horizontal ground movements and concentrations of ground movement close to the edge of the trench, these concentrations generally being restricted to a distance equal to 50% of the trench depth around the excavation. Deep trenches require the rapid installation of wall support, in the form of braced sheet piling or timber, not only to permit safe man-entry and to prevent wall collapse in weak ground, but also to ensure that the ground movements caused by the excavation are restricted to the acceptably low level required for minimizing disruption to buried services and structures in the vicinity. In the short-term, deformations consist of movements associated both with ground stress relaxation and with the subsequent mobilization of the full restraint of the propping system.

In clay soil, ground movements attributable to the removal of lateral (horizontal) ground constraint may be increased by the presence of water in tension cracks created by the inward wall movement, and often by the placing of excavated material too close to the trench sides.

In overconsolidated clays, quasi-elastic strains, the opening of fissures and discontinuous shearing processes may result in progressive bulk softening of the super-adjacent ground. When temporary support is withdrawn on completion of excavation and any in-trench construction, movements of this ground may continue in the longer-term as the compacted backfill consolidates (densifies) and settles. Backfilling procedures, aimed at minimizing these effects, are outlined in the U.K. British Standard CP2003 (1977).

4.2.2 Trench excavation, support and backfilling

Excavation of deep trenches in soil will usually be carried out by means of a mechanical excavator traversing the line of the trench. For long trenches, excavation, support,

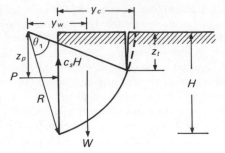

Figure 4.2 Calculation of thrust necessary for deep trench support.
Total thrust, P, assessed from a simple total stress stability analysis for a case history where a dry
tension crack, 2m deep, is assumed to cut through an overlying crust of firm to stiff clay.
From considerations of moment equilibrium

$$Pz_p = Wy_w - c_uR^2\theta_1 - c_sH(R - y_c)$$

where
W is the weight of the potential sliding soil and acts at the centre of its mass,
H is the trench depth (5.3 m in this instance),
c_u is the undrained shear strength of the soil,
c_s is the adhesion between soil and sheeting,
R is the radius of the slip circle, and
y_c, y_w, z_p and θ_1 are defined in the diagram.

service installation, and backfilling would be carried out in stages, or bays, 4 m to 6 m
long, in order to reduce the need for excessive quantities of shoring and bracing
materials on site. Proprietary waling and support systems, often used with more
common timber walings and adjustable steel props, are quite expensive to hire.

In spite of restrictions on open excavation length, and with support that is deemed
adequate for the trench proper, construction procedures are all-too-often insufficiently
rigorous to minimize movements in adjacent ground.

At the design stage, the thrust necessary at the trench supports to maintain
excavation stability and minimize ground movements may be assessed by means of
simple total stress stability analysis, as shown in Fig. 4.2. Terzaghi and Peck (1968)
suggest that the point of application of resultant thrust be taken as 60% of the trench
depth from ground surface. Calculated thrusts are very sensitive to the adopted value of
soil undrained shear strength, c_u, and so the strut loads evaluated on this basis should
be considered as order of magnitude approximations.

After excavation to formation level, preparation of the foundation, and placement of
the service utility, trench backfilling should be completed in stages, with mechanical
compaction of successive layers of backfill soil. The *in-situ* density of the backfill may be
checked against the design specification by means of the sand replacement method
(British Standard CP 2003: 1977).

4.2.3 *Ground movements associated with deep trenching*

General magnitudes of ground movement caused by excavation may be summarized
on the basis of soil type (generally after Harris and O'Rourke, 1983) as:

(i) *Medium to dense sands and gravel above the water table.* Vertical and horizontal movements vary from a maximum of 0.5% of trench depth at the edge of the trench, to a negligible value at a distance of 1.5-times the depth of excavation.

(ii) *Soft clay.* Where a major fraction of the soil profile comprises soft clay, large ground movements may be expected. Horizontal and vertical movements at the trench walls may be as high as 3% of the trench depth and affect a zone having a width more than twice the excavation depth. Large displacements may occur subsequently as the result of traffic loads and consolidation of poorly-compacted backfill, generally developing within a zone equal to approximately 50% of the trench depth from the trench walls.

4.2.4 *Prediction of ground movements associated with deep trenches*

A corpus of case history data currently provides the basis for the prediction of these movements. Figure 4.3 summarizes Peck's (1969) study of the distribution of vertical settlement around deep trench excavations and includes the results of more recent parametric finite element studies by Crofts *et al.* (1980). These curves should be perused in association with those of Rumsey and Cooper given in Chapter 5.

The ground displacement field close to and parallel to a trench wall may be investigated using the predictive method proposed by Harris and O'Rourke (1983). Displacement fields normal (at right angles), v, to an excavated sidewall, and vertical, w, are evaluated empirically from case history data on trenching in similar material. Such predictions are usually only applicable to short-term movements, *before* removal of ground support and subsequent backfilling.

Assuming, as per Fig. 4.4, that the inward movement normal to the trench walls as the trench advances can be described by means of a sine curve, and applying this same function to the vertical movement, then

$$\frac{d}{d_{max}} = \frac{1}{2}\left\{\sin\left(\frac{cx}{H}\right) + 1\right\}$$
(4.1)

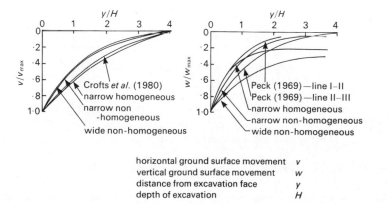

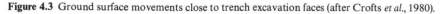

Figure 4.3 Ground surface movements close to trench excavation faces (after Crofts *et al.*, 1980).

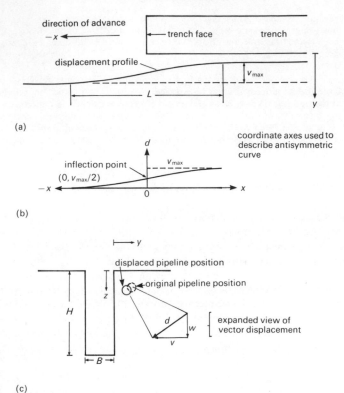

Figure 4.4 Resolution of movement towards a trench (generally after Harris and O'Rourke, 1983). (a) Plan view showing idealized horizontal displacement. (b) Displacement profile on reference axes. (c) Cross-section showing pipeline movement.

or,

$$\frac{cx}{H} = \sin^{-1}\left\{\frac{2d}{d_{max}} - 1\right\} \tag{4.2}$$

where c is an empirical coefficient evaluated from relevant case history data (in units of radians),

         x is the *longitudinal* distance of the excavation face from the point in question,

         H is the trench depth,

         d is the resultant ($\sqrt{v^2 + w^2}$) ground movement at any point along the transverse displacement development profile,

and d_{max} is the maximum resultant displacement on the fully-developed profile.

Constant c in equation 4.2 is evaluated from case-history data plots of $\sin^{-1}\{(2d/d_{max}) - 1\}$ against x/H. The gradient of the best-fit curve through observed data plotted in the above manner is the value of c to be used in equation 4.2.

An example of the technique is shown in Fig. 4.5. The case-history plots are based on

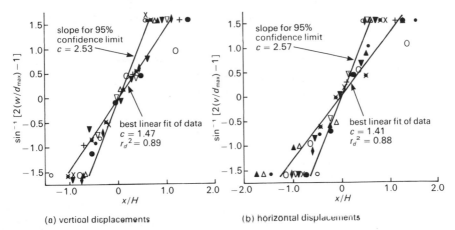

(a) vertical displacements   (b) horizontal displacements

Figure 4.5 Measured soil displacements plotted according to a transformed sine function (after Harris and O'Rourke, 1983).
r_d^2 is specified as 'the regression coefficient of determination'.

o • + O o   Bays 1, 2, 3, 4, and 5, respectively (Chard and Symons, 1982)
x △ ▲ ♦ ▽   Bays 1, 2, 3, 4, and 5, respectively (Symons, Chard, and Carder, 1981)
     ▼   S-line (Gumbel and Wilson, 1980)
     ✖   A-line (Gumbel and Wilson, 1980).

trenches in very stiff London Clay, in medium to stiff clay overlying very soft to soft clay, and in medium to stiff London Clay.

By defining a compounded (transverse, v, and vertical, w) displacement field in this way it is easier to discuss, in the next section, its effects on adjacent structures.

4.2.5 *Effects of trenching on nearby structures*

Ground movements have been specified for the vertical plane transverse (normal) to an excavation sidewall. These movements subject a structure to shear and bending strains. Ignoring the third component of movement parallel to the trench sidewall, the maximum resultant movement into the trench after the face has advanced for at least a distance L (see Fig. 4.4) beyond the y, z plane in question comprises:

(a) Movement of the unsupported vertical wall of the trench.
(b) After installation of trench sidewall support, movement of the trench wall to bring it into contact with the supports.
(c) Deflection of the supports as a result of the mobilization of full propping system support to retain the integrity of the trench sides.
(d) Movement occurring during consolidation of the backfill used to reinstate the trench.

It is convenient to class effects (a) to (c) above as short-term, and component (d) as being long-term, time-dependent.

4.2.5.1 *Effects on pipelines.* Excavation and backfilling of a trench close to a pipeline which lies parallel to the line of excavation is one of the most common causes of pipeline failure. Some pipe fractures occur at the time of trenching, but pipes often fracture several months after completion of the backfilling operation.

The magnitudes of pipeline bending and joint-rotational distortions are closely dependent on the closeness of the pipe to the trench and the positions of the pipeline joints along the x-axis of the inward displacement development profile. Obviously, for an advancing trench, the ground displacement profile progressively affects previously-undisturbed pipe sections, so subjecting the entire pipeline to conditions of both maximum strain and joint rotation. Since damage to a pipeline can be caused by both excessive strains *and* joint rotation, both conditions require appraisal.

Very local measurements of pipe strain made to determine the effect of adjacent-trenching should be interpreted with care. Such measurements will tend to under-estimate permanent strains unless they are taken close to the start or end of the length of trench. Reference should be made to Chapter 6 for information on pipe material properties.

A further potentially critical interaction between deep trench and pipe occurs when an excavation crosses and undermines a pipeline, and particularly if the pipe is allowed to remain unsupported over the trench width until the trench is reinstated.

Any such sensitive zones in cast-iron pipelines (gas or water mains) can be strengthened by inserting sections of steel pipe into the line before trenching begins. Good site practice is needed in order to minimize excavation-induced ground movements and the settlements associated with backfill consolidation. Otherwise, under severe conditions even a strengthened section of pipe could be at risk from local buckling and pull-out at joints. Figure 4.6 shows diagrammatically how a historical sequence of vertical relaxation caused by excavation, followed by consolidation of poorly-compacted backfill, can lead to permanent loss of pipe support in the trench area and pipe flexure within the adjacent ground.

Based upon the work of Tarzi *et al.* (1979) into the horizontal movement of continuous and jointed pipelines buried parallel to a trench, and assuming that the pipe moves conformably with the ground in which it is bedded, the maximum *short-term* pipe displacement may be considered as that resulting from the components (a) to (c) listed above and as per Fig. 4.7.

Having estimated in this way the maximum pipeline displacement towards the trench, the maximum induced bending strain may be evaluated from equation 4.3 below:

$$\varepsilon_{\text{short-term}} = D\lambda^2(v_1 + v_2 + v_3)F \cdot G \cdot A_{\text{short-term}} \tag{4.3}$$

where the factors D, λ, F, G and A are parameters resulting from an *elastic* model analysis, and where D is the depth of pipe embedment (m), λ is equal to $\sqrt{Dk_u/2EI}$ and is in units of m^{-1} (see also the use of this parameter in Chapter 6 for pipe

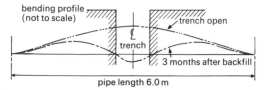

Figure 4.6 Changes in flexure of a pipe crossing a trench (after Howe *et al.*, 1980).

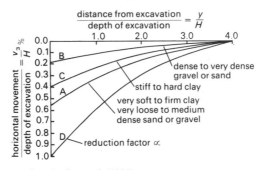

Figure 4.7 Movements after Crofts *et al.* (1980).
Short-term displacement $v_s = v_1 + v_2 + v_3$
where

$v_1 = \alpha$ (see graph) $\times u_b$ (for a long trench: see Table)
$v_2 = \alpha$ (see graph) $\times$ (10 mm + [between 2 mm and 18 mm])

$$v_3 = \left[\frac{\text{Value from graph}}{100} \times H \times \frac{L}{2H} \right] \text{ for trenches where } L < 2H$$

Immediate bulging u_b of excavated face of trench:

Type of ground*	Depth of excavation	Bulging of the excavation face, u_b
very soft clay	0–1.5 m	0–15 mm
very soft sandy clay	Depths greater than 1.5 m should not	
very soft gravelly clay	be excavated in one stage	
soft clay	0–2 m	0–8 mm
soft sandy clay	2–4 m	8–15 mm
soft gravelly clay	Depths greater than 4 m should not be	
soft gravelly clay	excavated in one stage	
firm clay	0–2.5 m	0–5 mm
firm sandy clay	2.5–5 m	5–10 mm
firm gravelly clay	5–7.5 m	10–15 mm
stiff clay	0–3 m	0–5 mm
stiff sandy clay	3–6 m	5–9 mm
stiff gravelly clay	6–10 m	9–15 mm

* For very stiff clays, dense sand or gravel or clean dry sand which will run, bulging does not apply.

displacements), k_u is the coefficient of subgrade reaction for the pipeline foundation (kN/m³—see section 4.2.7), EI is the flexural stiffness of the pipe (kN·m²), v_1, v_2, v_3 are the assumed pipe movements, as per Fig. 4.7, in metres and are entered as such in equation 4.3, A is a maximum bending moment (dimensionless) factor for continuous and jointed pipelines (see Fig. 4.8) and is equal to $M_{max}/2EI\lambda^2 v_s$, M_{max} being the maximum bending moment and v_s being the maximum short-term displacement $(v_1 + v_2 + v_3)$, G is a bending moment (dimensionless) factor, defined by Fig. 4.9, and F is a bending moment (dimensionless) factor, defined by Fig. 4.10.

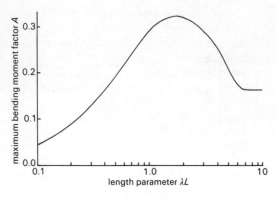

Figure 4.8 Envelope to maxima of maximum bending moment parameter A plotted against length of trench parameter λL (after Crofts *et al.*, 1980).

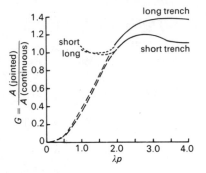

Figure 4.9 Relationship between maximum moment factors for continuous and jointed pipelines for varying λp, where p is the length between articulating joints (after Crofts *et al.*, 1980).

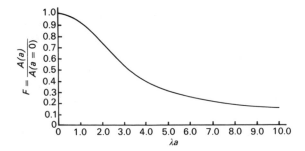

Figure 4.10 Bending moment reduction factor F for length of end restraint parameter λa where a may be taken as equal to $y/2$ (after Crofts *et al.*, 1980).

It is recognized that simple on-site ground movement measurement operations may record a single short-term displacement v_s (and $w_{s\,max}$, assumed to be at the wall), and perhaps a long-term displacement v_l (and $w_{l\,max}$), without resolving the constitutive elements v_1, v_2, v_3 (see the comment by O'Rourke, 1978, on Crofts et al., 1977). Thus, empirical data purporting, for example, to relate to v_3 may also include v_1, and in many cases v_2. Strict adoption of equations 4.3 and 4.9 could then magnify the real ground movement picture by including the same displacement(s) twice, and so promote an over-conservative prediction if used empirically.

The Harris and O'Rourke (1983) method of ground movement prediction (section 4.2.4) can be assumed to apply also to buried pipelines adjacent and parallel to the trench. However, its application to pipes at distances exceeding one-half the trench depth from the trench side would be expected to produce rather conservative results.

On the basis of equation 4.1, theoretical solutions may be obtained for maximum bending strain and joint rotation, that is, for the effects of excavation on perfectly flexible pipes having joints of zero rotational stiffness. In theory, maximum bending strain is developed at the point of maximum pipe curvature, K, which, from equation 4.2, may be derived as

$$K = \frac{-d_{max}}{2H^2} c^2 \sin\left(\frac{cx}{H}\right). \tag{4.4}$$

Since, from elastic beam theory, and assuming that the position of the neutral axis remains constant,

$$K = \frac{\varepsilon_{max}}{R} \tag{4.5}$$

where ε_{max} is the extreme fibre bending strain and R is the radius of curvature, then

$$\varepsilon_{max} = \frac{R_0 d_{max} c^2}{2H^2} \tag{4.6}$$

where R_0 is the extreme fibre radius of the pipe.

Similarly, the theoretical maximum joint rotation occurs at the point shown in

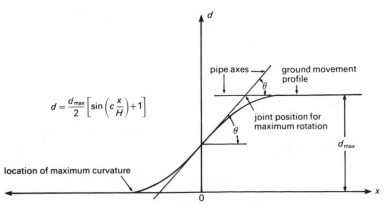

Figure 4.11 Sine curve showing maximum curvature and rotation (after Harris and O'Rourke, 1983).

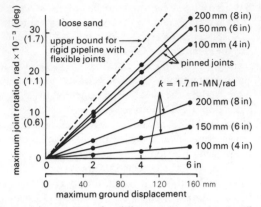

Figure 4.12 Maximum joint rotations for different joint rotational stiffnesses k and pipeline diameters in loose sand (after Harris and O'Rourke, 1983).
Note that 1 rad $= 360/2\pi$ degrees.

Fig. 4.11 and is determined by taking the second derivative of equation 4.2, setting $x = 0$ to give

$$\theta_{max} = \tan^{-1}\left\{\frac{d_{max}c}{2H}\right\}. \tag{4.7}$$

Figure 4.12 shows the results of finite element investigations into the effects of pipe diameter, pipe stiffness and soil state on the above theoretical expressions.

It is possible to combine the effects of critical bending and joint rotation into a working allowable ground movement envelope once critical strain and rotation criteria have been formulated for a particular type of pipe. Figure 4.13 gives an example of one such application relevant to cast-iron gas pipelines with an assumed unserviceability state arising at a bending strain of 1000 $\mu\varepsilon$ and a joint rotation of 5×10^{-3} rads.

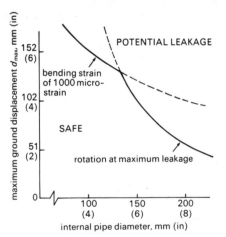

Figure 4.13 Combined effects of strain and rotation on pipeline performance (after Harris and O'Rourke, 1983).

Table 4.1 Definition of category as a proportion of breaking strain (after Crofts *et al.*, 1980).

Risk category	Assumed proportion of either breaking strain or maximum strain for unserviceability developed by proposed excavation and other causes
Category (1): Acceptable risk because breakage or damage unlikely	0.0 up to 0.25
Category (2): Border zone because breakage or damage can probably be avoided if special care is taken with all aspects of the work and its supervision	Above 0.25 and up to 0.60
Category (3): Unacceptable risk of breakage or unserviceability	Above 0.60

It is envisaged that standard design curves could be formulated for the more common combinations of pipe material and joint characteristics and so form the basis of a unified method for predicting and, hopefully, mitigating the effects of such excavation relaxation phenomena on adjacent buried pipelines.

The proximity of a pipe to a trench clearly determines its risk of failure or loss of serviceability. Crofts *et al.* (1979) propose that risk be defined in terms of the proportion of pipe breaking strain induced by excavation and predicted by available methods (see section 4.2.4). Table 4.1 lists three categories of risk and Figure 4.14 relates risk directly to pipeline location.

Whilst attempts may be made to predict the distribution of ground movements around trenches in the short-term, quantitative considerations of *long-term* consolidation effects have not been established to the same degree. Although the problem is complex, and very much dependent upon the quality and effectiveness of the backfilling operation, long-term ground movements may be considered to arise substantially from the re-establishment of at-rest earth pressures from the active state. Crofts *et al.* (1977) propose an estimate of consolidation movements based on the assumption that the soil forming the walls of the excavation exerts active earth pressure against the backfill,

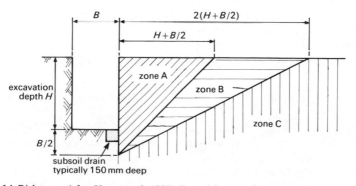

Figure 4.14 Risk zones (after Howe *et al.*, 1980). Zone A is zone of long-term high risk, Zone B is zone of intermediate risk, Zone C is zone of no risk.

Table 4.2 Soil parameters suggested by Crofts *et al.* (1980).

Soil description		Coefficient of horizontal subgrade reaction k_u (kN/m³)	Coefficient of active pressure K_a	Coefficient of earth pressure at rest K_0		Backfill compressibility $C_c/1 + e_0$	
Clay	Sand			Clay	Sand	Clay	Sand
Very soft to soft	Very loose to loose	4 000	0.41	0.5	0.5	0.30	0.15
Firm	Medium dense	8 000	0.33	0.8	0.45	0.15	0.10
Stiff	Dense	16 000	0.27	1.0	0.4	0.10	0.05
Very stiff or hard	Very dense	32 000	0.22	1.5	0.4	0.05	0.03

which consequently densifies and stiffens, so offering progressively greater resistance until such time as the earth pressure rises to its at-rest value.

Extending the notation of section 4.2.5.1, the long-term component of transverse horizontal pipe movement, v_4 (metres), may be estimated as

$$v_4 = \alpha \frac{B}{2} \frac{C_c}{1 + e_0} \log_{10}\left(\frac{K_0}{K_a}\right) \qquad (4.8)$$

where B is the trench width,

$\dfrac{C_c}{1 + e_0}$ is the compressibility of the backfill,

α is a factor given by the 'D-line' of Fig. 4.7,

and K_a, K_0 are, respectively, the coefficients of active and at-rest earth pressure for the soil in which the excavation is made.

It is reasonable to question whether parameters C_c, e_0, K_a and K_0 can ever be known with sufficient confidence to render an estimate of v_4 reliable. However, to allow very approximate calculation, typical values for a range of soil types are given in Table 4.2. In a similar manner to equation 4.3, the *total* long-term bending strain may then be estimated as

$$\varepsilon_{\text{long-term}} = D\lambda^2(v_1 + v_2 + v_3 + v_4)F \cdot G \cdot A_{\text{long-term}} \qquad (4.9)$$

4.2.5.2 *Effects on buildings.* Buildings adjacent to the excavation would be expected to deform in the hogging mode (ground curvature with convexity upwards), undergoing differential settlements and differential lateral distortions. Structural damage criteria are discussed briefly in Chapter 6, but because the subject is considered more completely in Chapter 8 (to which reference should be made), it is not developed here (see, however, some further comments in section 4.3.6.1).

Table 4.3 U.K. Water Research Centre recommended ranges of values for E_s and K_{0T} when predicting ground movements (after Fry, 1982). K_{0T} is the total stress 'at rest' coefficient of earth pressure.

Soil type	$E_s(MN/m^2)$	K_{0T}
Soft clays and clayey silts	2–4	0.6–1.2
Firm clays	4–10	0.8–1.5
Stiff clays	10–20	1–3
Very stiff hard clays	20–40 +	2–3.5
Loose sand and gravel	4–8	0.4–0.6
Medium dense sand and gravel	8–25	0.4–0.7
Dense sand and gravel	25–50 +	0.4–1.0

4.2.6 Measurement of soil parameters for the analytical prediction

For the successful application of numerical methods of analysis, suitably representative values for describing the soil behaviour are required. Elastic modelling of soil-structure interaction requires an assessment of soil stiffness (elastic modulus E_s), the in-situ total stress or effective stress 'at-rest' coefficient of earth pressure (K_{0T} and K_0, respectively), and the coefficient of sub-grade reaction k_u.

Field values are best derived via the following tests:

(i) Dutch cone tests for E_s (see Mitchell and Gardner, 1975). A minimum of two probe runs is required to obtain reasonable accuracy.
(ii) Push-in pressure cell techniques to derive K_{0T}. At least two measurements should be made at critical locations and at least 24 hours allowed for each test. Assuming the groundwater table to be at ground surface, then

$$K_{0T} = K_0 + (1 - K_0)\frac{\gamma_w}{\gamma_b} \qquad (4.10)$$

where γ_b is the bulk unit weight of the soil and γ_w is the unit weight of water (9.81 kN/m³).

Methods available for determining vertical k_u are imprecise and poorly understood. As an approximation, the horizontal coefficient of subgrade reaction may be adopted. Digioia et al. (1975) suggest that the coefficient be determined as

$$k_u = \frac{E_s}{1.35D_p} \qquad (4.11)$$

where E_s is obtained from down-the-hole pressuremeter tests and D_p is the pipe diameter. In ground where in-situ tests are not practicable, then recourse may be made to Table 4.3.

4.3 Deep excavations

4.3.1 Introduction

In urban areas, and because of inherent space restrictions, major foundation excavations usually require vertical cuts and a system of lateral support. In particular, limited surface space has led to the proliferation of underground space projects, notably the provision of underground car parking and basements in inner-city areas. Allied to this is the requirement to take new relief roads with minimal disturbance through densely-populated areas. This often involves deep excavation and the construction of high retaining walls close to existing buildings.

The means of supporting the sides of a deep excavation is governed by the purpose of the excavation. If the cut is of a temporary nature, then, for a building foundation, support is required only while construction takes the building from formation level to above ground level. In such cases, cross-bracing of steel or timber support, or driven sheet piling (with or without a ground anchoring system), may be appropriate. Excavation for retaining walls or building foundations often makes the provision of anchored cast-*in-situ* reinforced concrete or diaphragm walls desirable, these then being integrated into the permanent structure.

4.3.2 Braced excavations

4.3.2.1 *General.* Given suitable location and ground conditions, braced steel piling wall support is a cheap and reliable method. The economic attraction of sheet piling lies in its capability of re-use, typically five times or more in favourable ground. This system, however, has the environmental disadvantage that its installation is accompanied by noise and vibration (see Chapter 13) which may preclude its use in urban areas. In non-cohesive soils these detrimental effects may be minimized by the use of rapid placement vibro-rigs rather than conventional single- or double-acting hammer rigs.

Vertical restraint to buckling may be provided by vertical bracing panels within the excavation. However, additional freedom of movement for plant within the excavation may be gained by employing king piles driven to sufficient depth below the base of the excavation in order to prevent wall uplift. Reinforced concrete struts and walings may be used to best advantage when incorporated in the permanent works, especially ring walings supporting cofferdams. Employed in this manner no cross bracing is necessary, so giving completely unobstructed space within the cofferdam. Construction of the south pier of the Forth Road Bridge (Anderson *et al.*, 1965) provides a notable example of such use.

Thus, in general, a braced excavation should meet the following requirements:

(i) The sheeting members must be capable of withstanding lateral (horizontal) loads resulting from soil and hydrostatic water pressure, and from surcharge loads, without buckling.
(ii) Sheet piles should be capable of being extracted for re-use.
(iii) Structural framing for supporting the sheeting should be strong enough to resist buckling or excessive yielding. The framing should be braced to restrain movement in the horizontal and vertical planes.
(iv) Bracing frames should be placed at sufficiently wide intervals to allow working space for mechanical excavation plant and for easy handling of materials.
(v) To avoid excessive movement of adjacent ground, a means of jacking must be provided between struts and walings.
(vi) Bracing frames must be capable of being adjusted to transfer load safely to the permanent structure, and of being dismantled without damaging the permanent structure or placing the operatives at risk.
(vii) The advantage of using components of the permanent structure for excavation support, for example, the floors of multi-level basements, should be considered.

4.3.2.2 *Support design of braced excavations.* As the depth of excavation of a braced cut increases, the soil outside the cut tends to settle and to drag the piles down by negative skin friction. To minimize such settlement the piles must be driven far enough into a bearing stratum to develop point resistance at least equal to the downdrag forces. Sheet piles are often subjected to downward forces from the self-weight of the bracing system and, in urban areas, the weight of decking and street traffic that must be accommodated during excavation. For these reasons, the depth of embedment of sheet or soldier piles must be increased beyond that otherwise required. Indeed, it may be advantageous and economical to extend the piles to reach a firm stratum even if the depth of embedment would appear to be unnecessarily great when compared with the results of design calculation.

The required horizontal stiffness of the supporting system is imposed by struts spanning the inside of the cut. These struts are normally of steel fabrication. They are jacked into position against support walings which are set out from the vertical members to permit wedges or other blocking to be placed in order to make up for sheeting irregularities.

Peck (1969) produced design pressure diagrams for strutted excavations (see Fig. 4.15). These diagrams do not represent the real distribution of earth pressures at any vertical section in the excavation, but are envelopes idealizing maximum strut loads which may be approached at any particular level.

Pre-loading of struts is essential to promote a stiff bracing system and thereby restrict wall movement. In most cases struts are pre-stressed to between 40% and 70% (say 50%) of the design load, being sufficiently rigid to restrict further movement of the ground at the particular level of support, and sufficiently low to avoid being overstressed as additional excavation occurs. Practices which call for very high pre-loads should be avoided, since they are likely to cause concentrations of earth pressures that may exceed design pressures. Placing limits on the depth of excavation below the lowest brace is important for preventing ground loss during open cutting. For close

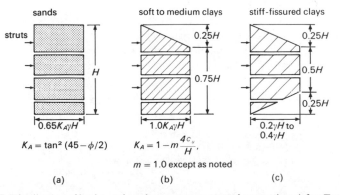

Figure 4.15 Distribution of horizontal earth pressure on strutted excavations (after Terzaghi and Peck, 1967).
Note:
(b) $m = 0.4$ for excavations wholly in soft clays
   $m = 1.0$ for excavations in soft clay where stiff clays or other materials exist at or close to the base of the excavation.
(c) is applicable where N (stability number $\gamma H/c_u$) ≤ 4.

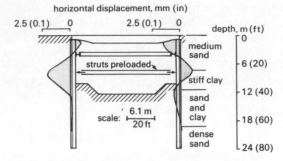

Figure 4.16 Wall displacements from preloading struts (after O'Rourke, 1981).

control over ground movements, excavation depths should be limited to 5.5 m below the lowest braces for deep cuts in medium to dense sands and in sands interbedded with stiff clay. Figure 4.16 shows a typical system response to pre-loading, the example being taken from the Washington D.C. metro construction.

The effective stiffness, K_{Es}, of a strut at a given level of support is given by

$$K_{Es} = \frac{P_s}{\Delta s} \tag{4.12}$$

where P_s is average pre-load force in the strut,
and Δs is the average apparent strut deformation at each strut level.
On the other hand, the ideal elastic stiffness, K_s of an individual strut is given by

$$K_s = \frac{E_s A_s}{l} \tag{4.13}$$

where E_s is the modulus of elasticity of the strut material,
       A_s is the cross-sectional area of the strut,
and l is the length of the strut.
The higher the ratio of effective to ideal strut stiffness, the greater is the effectiveness of the pre-loading technique in developing rigid support.

Great care should be exercised when adopting the common procedure of installing raking struts between the wall and a portion of the completed foundation. Even though

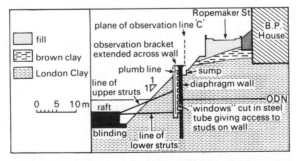

Figure 4.17 Section through north diaphragm wall after bulk excavation and before excavation for struts (Britannic House, Ropemaker Street, City of London, after Cole and Burland, 1972).

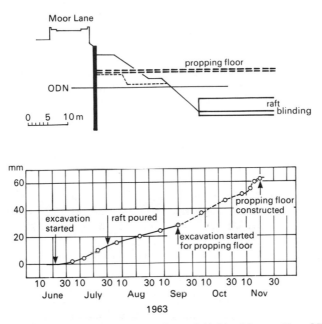

Figure 4.18 Movement of west diaphragm wall on Grid Line N at +10 m ODN towards excavation, against time (Britannic House, Ropemaker Street, London, after Cole and Burland, 1972).

pre-loaded, this type of bracing system may lack the restraint to prevent significant inward ground movement resulting from cantilevering of the wall. Such an example is shown in Fig. 4.17 where a diaphragm wall was supported by pre-loaded, raked struts while excavation, followed by construction of the foundation raft and a subsequent basement floor (to act as a permanent wall support strut), was carried out. The wall showed steady movement under temporary support (Fig. 4.18) until such time as horizontal bracing in the form of the propping floor was installed. Ground movements allowed by the yielding support were recorded up to 40 m away from the excavation.

An excavation procedure which may be used to limit deformations in soft to medium clays and in cut and cover works involves the inclusion of temporary berms in the cut (Clough and Denby, 1977). Slipping and creeping leads to the gradual distortion of berms, and to control the ground movements the time lag between excavation at the centre of the cut and the installation of stiff braces should be minimal.

4.3.3 *Tied wall excavations*

4.3.3.1 *General.* The sheeting of wide excavations, if supported by walings and struts, requires long, heavily-braced compression members supported by king piles or hangers. The use of tied-back supports avoids the need for long compression members and provides a completely unobstructed space within the excavation. This method may be somewhat restricted in its application to urban areas because of the difficulty in finding an anchor alignment unobstructed by buried cables, gas and water mains, and sewers. Unless these services have been mapped carefully and can be positively located, a safe tie alignment cannot be specified with confidence.

Several general points may be noted:

 (i) A tied-back wall does not inherently perform better than a braced wall.

 (ii) Movement of a tied-back wall, and soil settlements behind the wall, can be substantially reduced by a judicious choice of tie-back configuration and pre-load.

 (iii) Over-excavation often occurs during the construction of braced walls, the cutting of an irregular profile resulting in voids between the excavated sides and the subsequently installed supporting walls. A consequential doubling of the ground movement magnitudes can be avoided by the adoption of pre-stressed tied-back walls.

Ties may be employed both for temporary support, in conjunction with sheet piles, and as permanent features with sheet piles or cast-*in-situ*, diaphragm or contiguous pile walls. Diaphragm walls are usually designed as permanent features to criteria applicable to retaining walls. The working stresses in the concrete and steel reinforcement are those pertaining to permanent works.

4.3.3.2 *Design of tied walls.* In terms of stability, similar considerations apply to both tied-back sheet walls and tied-back diaphragm or contiguous pile walls. For cutting walls and bulkhead walls the only difference is wall stiffness. However, in building construction the latter will be supplemented for long-term stability by the presence of propping floors.

Lateral pressures on wall members supported by anchorages are affected by yielding anchors and walings in a similar way to the pressures on braced excavations. In a properly designed tie-back system, the inward deflection of the wall will be a maximum at the ground surface and a minimum at the bottom of the excavation. Peck (1969) suggests that a triangular pressure distribution is appropriate for such a response. If some yielding is permitted, the wall and waling pressures will not be under-estimated if the trapezoidal diagrams of Fig. 4.15 are adopted, but assuming uniform pressures below the one-quarter depth level (0.25 H). If no yielding, or only very small inward deflections, can be permitted, then the soil behind the wall requires tie precompression with consequent increase in ground pressures. Pressure distributions for the stability design of anchored flexible walls are shown in Figure 4.19.

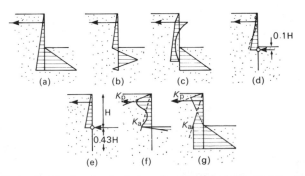

Figure 4.19 Pressure distributions for the design of anchored flexible walls (after Milligan, 1983): (*a*) free earth support; (*b*) fixed earth support; (*c*) Danish regulations; (*d*) after Terzaghi (1943); (*e*) after Tschebotarioff (1949); (*f*) after Ohde (1938); (*g*) after Verdeyen and Roisin (1953).

In the case of inclined anchorages, where vertical loads are transferred to the wall, sheet piles or soldier piles must be taken to sufficient depth below the base of the excavation to carry the vertical loads in skin friction and end-bearing.

The design of ground anchors is essentially a bearing-capacity soil-mechanics problem. Simple design procedures have been noted in Attewell and Farmer (1976) and are outlined in Fig. 4.20. Excavation-induced ground movements may be effectively controlled by staged excavation with progressive installation and stressing of ground anchors.

As part of a relief scheme for London's North Circular road, a 1 km underpass in a cutting up to 10 metres deep was planned as the Neasden By-Pass (Sills *et al.*, 1978). Various types of retaining structures were considered, both conventional retaining walls and diaphragm walls being adopted in the final design. The choice of diaphragm walls, with multiple level anchor supports, tackled the problem of constructing high retaining walls close to existing buildings.

Based upon the soil profile (200 mm of top soil overlying stiff, brown, fissured London Clay to an average depth of 8 m, itself underlain by grey-blue fissured London

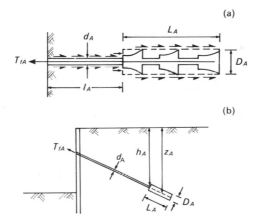

Figure 4.20 Simple procedure for anchor design.
(*a*) Clays:

$$T_{fA} = \pi D_A L_A c_u f + \frac{\pi}{4}(D_A^2 - d_A^2)N_c c_u + \pi d_A l_A c_a$$

where

T_{fA} (pull-out capacity) = side resistance + end bearing,
c_u is the undrained shear strength, and
c_a is the apparent cohesion = $c_u f$,

where $0.3 < f < 0.6$, and is usually taken as 0.45.

(*b*) Gravels and sands:

$$T_{fA} = A_A \bar{\gamma}' z_A \pi D_A L_A \tan \phi' + B_A \bar{\gamma}' h_A \frac{\pi}{4}(D_A^2 - d_A^2)$$

where

$$A_A = \frac{\text{soil-to-anchor contact pressures}}{\text{Effective overburden pressure}} \text{ and has a value of 1 to 2 (often 1.3),}$$

$B_A \simeq N_q$ (function of ϕ),
$\bar{\gamma}'$ is the average submerged (or effective) unit weight of the soil, and
$\bar{\gamma}' z_A$ is the average effective overburden pressure at the centre of the plug.

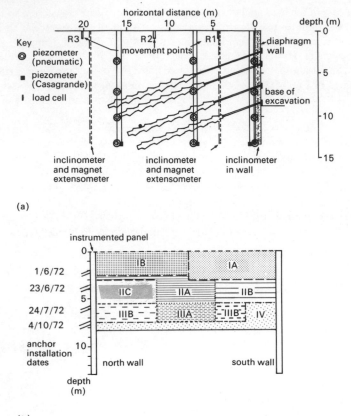

(a)

(b)

Figure 4.21 Instrumentation and excavation on the North Circular relief road, Neasden by-pass, London (after Sills *et al.*, 1977). (*a*) Section through instrumentation. (*b*) Progress of excavation. Stage I completed by end of February 1972, Stage II completed by mid June 1972, Stage III completed by early July 1972, Stage IV completed by early August 1972.

Clay) the diaphragm wall design was of the form shown in Fig. 4.21(*a*). Anchor lengths of 16 metres to 18 metres, raked at an angle of between 20° and 40° (to avoid encroachment below nearby houses), were adopted in four rows with eight anchors per 0.6 m × 4.57 m wall panel. Before excavation, underreamed anchors, constructed to design specification, were inserted vertically in the vicinity of the workings. This facilitated *in situ* testing of the support mechanism and verification of the design procedure.

The ground was excavated in stages, as shown in Fig. 4.21(*b*), and anchors installed progressively. Each anchor was stressed to 70% of the ultimate tendon strength and monitored, followed by monitoring at 115% of the design working load, before being relaxed to its final working load after 24 hours. One panel section was instrumented with borehole extensometers, inclinometers and vibrating wire load cells to maintain a close check on system performance.

The data acquisition operation proved to be successful and similar procedures are recommended for all major deep excavation works, not only for monitoring performance, but also to provide guidance for similar work in the future.

4.3.4 *Prediction of ground movements associated with deep excavations*

As part of the design for a deep excavation the ground movements likely to occur must be assessed with respect to their wider effects. This is particularly important in the proximity of other buildings and buried services. There are two options available to the designer. First, he may perform an approximate empirical prediction based upon the increasing body of case-history data. This will tend to yield order of magnitude displacements and influence fields. Such an approach may be acceptable in some cases where the workmanship and ground control are known to be of high standard, where the construction procedure is well-tried and proven, and where the location is reasonably free from sensitive adjacent structures and services. The second option involves the use of numerical modelling techniques, such as the finite element method. An application of this is discussed subsequently.

Since deep excavation generates horizontal and vertical soil displacements behind the supporting walls, and heave at the base, it is useful to consider the empirical evidence that is available for predicting these movements.

4.3.4.1 *Empirical prediction.* Observational data were classified by Peck (1969) in terms of four broad soil types: cohesionless sand, cohesive granular soil, soft to medium clay, and stiff clay. Figure 4.22 relates some case history evidence on ground settlement and supporting sheet movement to these categories of ground, and Table 4.4 lists some

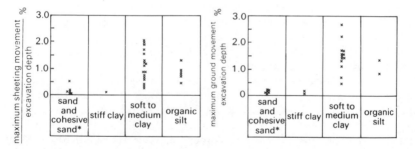

Figure 4.22 Frequency distributions of maximum soil movements caused by excavation employing soldier piles and lagging or steel sheeting; data from Peck (1969) and others (after D'Appolonia, 1971). *Does not include cases where large movements have been caused by subsurface erosion and piping.

Table 4.4 Factors controlling lateral (horizontal) wall movements that occur above excavation level (after D'Appolonia, 1971).

Horizontal and vertical brace spacing
Depth of excavation below brace level before brace is installed
Length of excavation parallel to wall made at any one level prior to installing braces at that level
Elapsed time between excavation and brace installation
Details of prestressing and wedging braces
Details of excavating and placing lagging between soldier beams

---

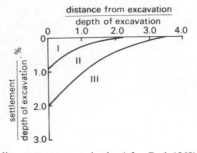

Figure 4.23 Settlements adjacent to open cuts in clay (after Peck 1969). *Zone I*: Soft to hard clay, average workmanship. *Zone II*: (*a*) Very soft to soft clay. (1) Limited depth of clay below bottom of excavation (2) Significant depth of clay before bottom of excavation but $N_b < N_{cb}$; (b) Settlements affected by construction difficulties. *Zone III*: Very soft to soft clay to a significant depth below bottom of excavation and with $N_b \geq N_{cb}$

of the main factors controlling inward movements. Based on field data and finite element studies Mana *et al.* (1981) produced a method of estimating movements within braced cuts in clay soil. It involved the determination of several factors including the wall stiffness, strut stiffness, excavation depth, and the factor of safety against basal heave. Such a method of prediction—involving coefficients and multipliers based on the above factors—can be particularly valuable for design engineers needing to make rapid assessments of the influence of design changes.

The following empirically-based summary offers general guidance.

(i) For *sand and cohesive granular soils*, ground movements are small if the groundwater level is below the excavation base, or otherwise brought under control, and if the braces are adequately pre-stressed. Maximum settlement in loose sands and gravel may be of the order of 0.5% of the excavation depth. Movements are less laterally extensive than in the case of clays. Allow for a disturbance zone equal to the excavation depth from the side of the excavation.

(ii) For *soft to medium clays*, maximum inward movement of sheeting and maximum surface settlements are commonly both of the order of 1% to 2% of the excavation depth (see, for example, Fig. 4.23). On the other hand, settlements in soft clays may extend to a distance of three to four times the excavation depth. Analytical work based on velocity field methods shows that the maximum ground movement should be approximately equal to the maximum lateral (horizontal) displacement of the wall, indicating that the soil deforms at constant or near-constant volume.

Maximum settlement and the lateral extent of movement are related to the stability number, N_b, of the excavation base ($N_b = \gamma H / c_u$, where γ is the unit weight of the soil, H is the depth of cut, and c_u is the undrained shear strength of the soil). If N_{cb} is the critical stability number for base failure (see Fig. 4.24), then ground movements noticeably increase as the ratio N_{cb}/N_b tends to unity. For a normally consolidated clay, yield occurs when $N_{cb}/N_b \simeq 1.5$ to 2.0. The presence of rock-head at shallow depth reduces the magnitude of the otherwise expected ground movements by preventing deep movement below the cut.

Figure 4.25 is a case history of horizontal wall movements in an excavation which adopted sheet piles and cast-in-place concrete for support at adjacent wall sections of a subway cut in medium clay at Boston, U.S.A. Soil profiles, wall depth, and brace

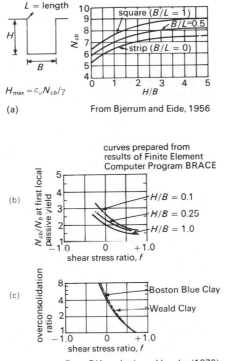

$H_{max} = c_u N_{cb}/\gamma$

(a)   From Bjerrum and Eide, 1956

curves prepared from
results of Finite Element
Computer Program BRACE

(b)

(c)

From D'Appolonia and Lambe (1970)

Figure 4.24 Bottom stability of an excavation in clay (after D'Appolonia, 1971). (a) Bearing capacity numbers. (b) Values of N_{cb}/N_b at which first passive yield occurs. (c) OCR $v. f$ (note $f = $ (original in situ shear stress)$/c_u = (1 - K_0)/(2c_u/\bar\sigma_v))$.

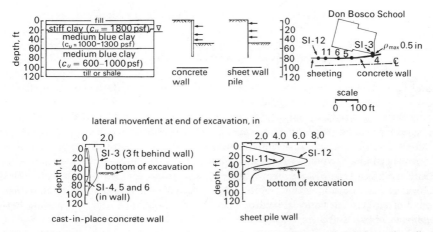

Figure 4.25 Comparison of wall deflections for cast-in-place concrete wall and sheet pile wall at nearby sections of excavation in medium clay: subway construction in Boston, Mass., U.S.A. (after D'Appolonia, 1971).

spacings were sensibly identical for both support systems, the only difference being the wall stiffnesses. A maximum horizontal displacement of 12 mm was recorded for the concrete wall, but movements of between 112 mm and 175 mm were noted for the sheet piles. This example demonstrates the point that empirical prediction of ground movement must always take account of the support system characteristics.

(iii) *Stiff clay*: ground movements induced by excavation in clays generally reduce to a fraction of 1% of excavation depth with increasing soil stiffness. Greater and more widespread movements *can* occur in firmly-supported excavations within deep deposits of stiff clay.

Several indicators emerging from measurements in overconsolidated London Clay may be applicable to similar soils elsewhere. Summarized, these are (after Burland *et al.*, 1978):

(a) Movements around an unsupported or partially-supported excavation are time-dependent and the provision of early resistive support is essential.

(b) The value of K_0 is far from unique with depth for laterally-extensive clay beds. Any assessment of initial *in situ* values of horizontal earth pressures must take account of stress history.

(c) Significant horizontal and vertical ground movements may extend to a distance of three or more times the excavation depth from the walls.

(d) Settlements beyond the boundaries of an excavation in London Clay seldom exceed 0.15% of the depth of excavation, given conditions of average quality of workmanship. Further, the ratio of horizontal to vertical displacements depends upon the support methods. This ratio may exceed 3:1 in open excavation. On the other hand, the ratio will normally be less than unity, given early effective support.

(e) Lateral displacement fields may extend to more than five-times excavation depth in tied-back wall-supported cuts.

(f) Basal heave and swelling continue for many years as steady-state stress and pore-pressure regimes re-establish.

(g) Young's deformation modulus for London Clay increases noticeably with depth, and its *in situ* values are three times greater than those obtained from laboratory tests on high-quality samples.

(h) Movements attributable to stress relief can generate a progressive slip surface having a shear strength equal to or close to the residual value in unsupported cuts.

4.3.5 *Soil mechanics aspects of excavation-induced deformation*

Soil shear strength is mobilized by relaxation deformation of the ground. Regimes of dilatancy and compression would be expected to develop quickly in permeable non-cohesive soils. Readjustment is time-dependent in clay soils, and several years may be needed for ground strains to equilibrate and pore pressures to achieve a steady state.

Figure 4.26 illustrates, in a semi-qualitative manner, the stress changes and strains experienced in a normally consolidated plastic clay. The critical factor in determining both the magnitude of settlement outside the excavation *and* the scale of base-heave effects is the inward movement of the sides of the cut below excavation level. If local yield of the base elements is avoided, that is, the unloading stress path for the base elements of clay soil does not approach the soil failure envelope, heave effects will be small and so too will be the lateral movement of the soil below excavation level. Movements will be large if the effective stresses in the basal elements approach the failure envelope, resulting in passive yield conditions. A more detailed discussion on

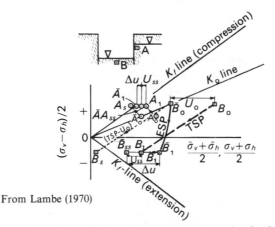

From Lambe (1970)

Figure 4.26 Stress paths for soil elements near to an excavation in clay (after D'Appolonia, 1971). TSP, total stress path; ESP, effective straight path.
Note: Initially there is no flow and ultimatly there is steady-state seepage.

	Element A	Element B
Initial (static) pore pressures, U_0	$A_0\bar{A}_0$	$B_0\bar{B}_0$
Pore pressure at steady-state flow, u_{ss}	A_1A_{ss}	$B_1\bar{B}_{ss}$
Pore pressure upon unloading	Decreases	Decreases
Pore pressure during consolidation	Decreases	Increases
Strain upon unloading	Vertical compression	Vertical extension
Strain during consolidation	Vertical compression	Vertical extension
Undrained shear strength during consolidation	Increases	Decreases

these matters may be found in Morgenstern *et al.* (1977).

In the analysis of stress relief movements in clay, the initial 'at rest' coefficient of earth pressure, K_0, is of great importance. The 'at rest' stress conditions style the strain field to be expected when soil confinement is relieved by excavation. Overconsolidated clays may be subject to excessive heave as a result of large stress-relief-induced lateral ground movements below the excavation. In addition, passive shear conditions below the base, aided and accompanied by high negative pore-water pressures, facilitate migration of pore-water from the wall elements (in active shear). This causes a reduction in soil strength beneath the base of the excavation (in addition to that caused by strain softening effects) and an increase in basal deformations.

A reliable estimation of K_0 is an essential pre-requisite for finite element analytical treatments. However, K_0 is notoriously difficult to determine with any accuracy. In a review of the subject, Kyrou (1982) suggests that *in-situ* techniques, especially self-boring pressuremeters, provide the best available engineering practice.

It is well known that the strength and deformation modulus properties of soft to medium clays can vary by as much as a factor of 4 depending upon the loading direction. Work by Clough (1981), which incorporates the effects of soil strength anisotropy into the standard Terzaghi isotropic analysis, shows that an assumption of inherent isotropy may lead to several errors, in particular:

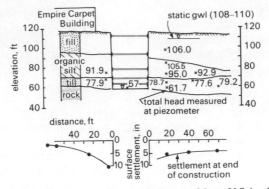

Figure 4.27 Cross-section at a subway excavation in Boston, Mass., U.S.A., showing changes in total head and surface settlements caused by excavation and de-watering (after Lambe, 1970).

   (i) The calculated basal-heave factor of safety may be as much as 50% above the actual one in clays which are strongly anisotropic and may overestimate it by 10% to 30% in moderately anisotropic clays.

  (ii) Lateral wall movements and ground surface settlements may be larger than expected.

 (iii) Potentially detrimental effects, arising from a neglect of structural (and strength) anisotropy, are most likely to occur when the calculated basal heave stability factor is below about 1.4.

In practice, basal stability of a deep excavation may be affected by several factors:

   (i) Upward heave, resulting from stress relaxation in soft cohesive soils.

  (ii) Upward heave caused by hydrostatic pressures in permeable layers below the base.

 (iii) Piping of the base as a result of upward water seepage forces.

The former effect (i) must be mitigated by design based on bearing capacity theory. Reduction of hydrostatic pressure (ii) by groundwater lowering may result in ground settlements over a quite extensive area. Lambe (1970) presents a case history (see Fig. 4.27) of a 16.7 metre deep cut supported by steel sheeting and cross-bracing. Settlements as large as 250 mm were measured at the edge of the excavation, a major portion of the settlement being ascribed to consolidation promoted by groundwater lowering.

Instability arising from boiling, or quick conditions, at the excavation base can be controlled either by inserting sheet piling or concrete walling into an impermeable layer to form a cut-off, or by extending these members to a depth calculated to increase the length of seepage path necessary to reduce the upward flow velocity to a tolerable level. Standard procedures of flow net construction and analysis are covered comprehensively in Cedergren (1977) to which reference should be made.

The presence of a saturated layer of London Clay with sand partings some 20 m below ground level had a major influence on the basic design conditions for the 17 m-deep underground car park at the House of Commons, London (Burland and Hancock, 1977). Doubts as to the likely long-term effectiveness of relief wells led to the decision to take the diaphragm walls down to an intact clay layer some 30 m below ground level, so cutting off all horizontal seepage along the sand partings. Under this design configuration, stability against hydraulic uplift was analysed and found to be satisfactory.

4.3.6 *Damage to structures*

The ground movements caused by deep excavation, and comprising both horizontal and vertical differential displacements in soil, can affect both above-ground (buildings) and in-ground (pipes) structures near to the excavation.

4.3.6.1 *Buildings.* Angular distortions of a hogging character and lateral tensile distortions may create structural problems and affect the serviceability of the structure. For a conservative assessment of potential damage before excavation it is probably wise to assume that the predicted ground distortions will be transferred to the structure, that is, to assume that the structural foundation will deform compliantly with the ground. Some of this conservatism will then be removed by applying the self-weight damage criteria discussed by Burland and Wroth (1974) and Burland *et al.* (1977). Reference may also be made to the summary criteria given in Chapter 6, and to Chapter 8 of the present book. Such assessments before excavation will allow any protective measures (typically strengthening against tensile or shear failure at points of structural weakness, provision of underpinning and the use of progressive jacking for a sensitive building) to be initiated. Furthermore, it is sensible to instrument and continuously monitor the behaviour of all vulnerable buildings, one or both parties (the excavation project promoter and the building owner) having also charged building surveyors with the task of preparing pre-excavation building schedules. All this information forms an objective basis from which to begin any assessment of damage claims.

4.3.6.2 *Pipes.* Consideration of the risk of damage to buried pipelines induced by excavation is considered in section 4.2.5.1 of the present chapter and also with respect to tunnels in soil in Chapter 6.

4.3.7 *Use of numerical methods in prediction*

Deep excavations usually provide ideal modelling scenarios for application of the finite element method. Stratified deposits having differing geotechnical properties are easily handled as is the process of staged excavation, but there remains the general problem of

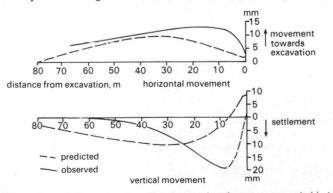

Figure 4.28 Comparison of observed and predicted ground surface movements behind the south wall of the House of Commons car park, Westminster, London (after Burland and Hancock, 1977).

defining ground deformation moduli. Choice of values (E, v) should be based on engineering judgement stemming from a distillation of evidence derived from laboratory tests on large soil samples, perhaps down-the-hole tests (for example, screw plate), perhaps field tests (for example, plate bearing) and on case history knowledge of the performance of buildings adjacent to excavations in similar ground.

The House of Commons Car Park project (Burland and Hancock, 1977), excavated in London Clay and noted earlier, provides an example of how intelligent application of the finite element method can prove fruitful. Based on *ad hoc* test results and modified case history data, the analysis of the excavation process produced some predictions in good agreement with the observed behaviour (see Fig. 4.28) Nevertheless, some misleading predictions did emerge due (inevitably) to the over-simplifications in the model. The degree to which an excavation is realistically modelled must be determined by reference to many factors, including the reliability of the soils data, the degree of accuracy required and the cost of analysis. A comprehensive numerical investigation of the retaining walls of the Bell Common Tunnel (Potts and Burland, 1982) amplifies the necessity for relatively complex, realistic analysis, and the danger of over-simplification, by comparing elastic and elastoplastic treatments of the problem.

The finite element method is most generally applicable to stiff clays. Recent investigations suggest that these soils may reasonably be approximated as porous elastic materials, particularly when modelled two-dimensionally in plane strain.

Notation

A	maximum pipe bending moment factor (after Crofts *et al*, 1980)	(dimensionless)
A_A	soil/ground anchor contact pressure factor	(dimensionless)
a	length of pipe end restraint	(m)
B	deep-trench width	(m)
B_A	Bearing capacity factor	(dimensionless)
C_c	coefficient of soil compressibility	(dimensionless)
c	empirical ground movement coefficient	(radians)
c_a	apparent cohesion	(kN/m^2)
c_s	adhesion between soil and sheeting	(kN/m^2)
c_u	undrained cohesive soil strength	(kN/m^2)
D	pipe diameter	(m)
D_A	ground-anchor plug diameter	(m)
d	resultant ground movement	(metres)
d_A	ground-anchor diameter	(m)
E	Young's modulus of elasticity	(kN/m^2)

E_s	constrained soil modulus	(kN/m^2)
e_0	initial soil void ratio	(dimensionless)
F	pipe bending moment reduction factor (after Crofts et $al.$, 1980)	(dimensionless)
f_A	cohesion reduction factor	(dimensionless)
f	shear stress ratio	(dimensionless)
G	bending moment factor for continuous and jointed pipelines (after Crofts et $al.$, 1980)	(dimensionless)
H	trench and deep excavation depth	(m)
h_A	depth of ground-anchor grout-plug burial	(m)
I	second moment of area	(m^4)
K	pipe curvature	(m^{-1})
K_a	coefficient of active earth pressure	(dimensionless)
K_0	coefficient of at-rest earth pressure	(dimensionless)
K_{0T}	coefficient of at-rest earth pressure in terms of total stress.	(dimensionless)
K_p	coefficient of passive earth pressure	(dimensionless)
k_u	coefficient of sub-grade reaction	(kN/m^3)
L	length of trench excavation	(m)
L_A	length of ground-anchor plug	(m)
l_A	length of ungrouted tendon	(m)
M_{max}	maximum bending moment	(kN.m)
m	coefficient of active earth pressure reduction factor	(dimensionless)
N_b	deep excavation bottom stability number	(dimensionless)
N_{cb}	deep excavation $critical$ stability number	(dimensionless)
$\left.\begin{array}{c} N_c \\ N_q \end{array}\right\}$	Terzaghi's bearing capacity factors	(dimensionless)
P	resultant trench support thrust	(kN)
p	pipe section length	(m)
R	slip circle radius	(m)
R_0	external pipe radius	(m)
T_{fA}	ground-anchor pull-out capacity	(kN)

u	ground/pipe displacement in x-direction	(metres)
u_b	immediate 'bulging' displacement of trench-face proper	(metres)
u_0	initial pore-water pressure	(kN/m^2)
Δu	change in pore-water pressure	(kN/m^2)
W	weight of soil wedge	(kN)
v	ground/pipe displacement in y-direction	(metres)
$\left.\begin{array}{c} v_1 \\ v_2 \\ v_3 \\ v_4 \end{array}\right\}$	ground/pipe displacement components in y-direction	(metres)
v_s	sum of short-term ground/pipe displacement components in y-direction	(metres)
v_l	sum of all ground/pipe displacement components in y-direction	(metres)
w	ground/pipe displacement in z (vertical) direction	(metres)
x	cartesian coordinate parallel to direction of excavation ($-$ve in direction of excavation)	
y	cartesian coordinate transverse (normal) to excavation line ($y = 0$ at sidewall of trench)	
y_c	horizontal distance from excavation edge to slip circle	(m)
y_w	horizontal distance to weight vector	(m)
z	cartesian coordinate ($+$ve vertically down from ground level)	
z_A	depth of anchor-plug centre burial	(m)
z_p	depth to horizontal thrust vector	(m)
z_t	depth of tension crack	(m)
α	pipe movement reduction factor (after Crofts *et al.*, 1980)	(dimensionless)
γ	bulk unit weight of soil	(kN/m^3)
$\bar{\gamma}'$	average submerged (or effective) unit weight of soil	(kN/m^3)
ε	extreme pipe fibre bending strain	(dimensionless)
ε_{max}	maximum extreme fibre bending strain	(dimensionless)
θ	pipe joint rotation	(radians)
θ_1	angle in slip circle analysis	(radians)

λ	pipe stiffness (damping) factor	(m^{-1})
σ_H	horizontal total ground stress	(kN/m^2)
σ_V	vertical total ground stress	(kN/m^2)
ϕ	friction angle	(degrees)
ϕ'	friction angle in terms of effective stress	(degrees)

References

Attewell, P.B. and Farmer, I.W. (1976) *Principles of Engineering Geology.* Chapman and Hall, London.

Bjerrum, L. and Eide, O. (1956) Stability of strutted excavations in clay. *Géotechnique* **6** (1) 32–47.

Burland, J.B. (1975) Some examples of the influence of field measurements on foundation design and construction. In *Proc. 6th Conference for Africa on Soil Mechanics and Foundation Engineering, Durban, S. Africa*, Vol. 2, pp. 51–71.

Burland, J.B., Broms, B.B. and De Mello, V.E.B. (1977) The behaviour of foundations and structures. In *9th International Conference on Soil Mechanics and Foundation Engineering, Tokyo, Japan*, Vol. 1, pp. 495–546.

Burland, J.B. and Hancock, R.J.R. (1977) Underground car park at the House of Commons, London. Geotechnical aspects. *The Structural Engineer* **55** (2) 87–100.

Burland, J.B., Longworth, T.I. and Moore, J.F.A. (1977) A study of ground movements and progressive failure caused by a deep excavation in Oxford clay. *Géotechnique* **27** (4) 557–591.

Burland, J.B. and Moore, J.F.A. (1973) The measurement of ground displacement around deep excavations. In *Proc. Symp. on Field Instrumentation*, British Geotechnical Society, London, pp. 170–184.

Burland, J.B. and Simpson, B. (1979) Movements around excavations in London Clay. In *Proc. Conf. on Design Parameters in Geotechnical Engineering*, British Geotechnical Society, London, Vol. 1, pp. 13–29.

Burland, J.B., Simpson, B. and St John, H.D. (1979) Movements around excavations in London Clay. In *Proc. 7th European Conf. on Soil Mechanics and Foundation Engineering, Brighton*, Vol. 1, pp. 18–19.

Burland, J.B. and Wroth, C.P. (1974) Settlements of buildings and associated damage. *General Report on Session V, Conference on the Settlement of Structures*, British Geotechnical Society, London, Pentech Press, London, pp. 611–654.

Carder, D.R. Taylor, M.E. and Pocock, R.J. (1982) *Response of a pipeline to ground movements caused by trenching in compressible alluvium.* TRRL Report No. 1047, Transport and Road Research Laboratory, Crowthorne, Berks., England.

Cedergren, H.R. (1977) *Seepage, Drainage and Flownets.* John Wiley and Sons Inc., N.Y.

Chapman, K.R., Cording, E.J. and Schnabel, H. (Jnr) (1972) Performance of braced excavation in granular and cohesive soils. In *Proc Speciality Conf. on the Performance of Earth and Earth Supported Structures, Purdue University, Indiana*, Vol. 3 ASCE, N.Y., pp. 271–293.

Chard, B.M. and Symons, I.F. (1982) *Trial trench construction in London clay: A ground movement study at Bracknell.* TRRL Report No. 1051, Transport and Road Research Laboratory, Crowthorne, Berks., England.

Chard, B.M., Symons, I.F., Toombs, A.F. and Nagarkatti, A.S. (1983) *Trial trench construction in alluvium: A ground movement study at Avonmouth.* TRRL Report No. 1077, Transport and Road Research Laboratory, Crowthorne, Berks., England.

Clough, G.W., and Denby, G.M. (1977) Stablising berm design for temporary walls. *J. Geotech. Eng. Div., ASCE*, **1** (GT9) 75–90.

Clough, G.W. and Hansen, L.A. (1981) Clay anisotropy and wall behaviour. *J. Geotech. Eng. Div., ASCE*, **103** (GT7) 813–913.

Clough, G.W., Hansen, L.A. and Mana, A.I. (1979) Prediction of supported excavation movements under marginal stability conditions in clay. In *Proc. 3rd Int. Conf. on Numerical Methods in Geomechanics*, A.A. Balkema, Rotterdam, Netherland, pp. 1485–1502.

Clough, G.W. and Tsui, Y. (1974) Performance of tied-back walls in clay. *J. Geotech. Eng. Div.,* *ASCE,* **100** (GT12) 1259–1274.

Cole, K.W. and Burland, J.B. (1972) Observations of retaining wall movements associated with a large excavation. In *Proc. 5th European Conference on Soil Mechanics and Foundation Engineering, Madrid,* pp. 445–453.

Crofts, J.E., Menzies, B.K. and Tarzi, A.I. (1977) Lateral displacement of shallow buried pipelines due to adjacent deep trench excavation. *Géotechnique* **27** (2), 161–179.

Crofts, J.E., Menzies, B.K. and Tarzi, A.I. (1980) Bending of pipelines in laterally deforming soil adjacent to trench excavations. In *Proc. Conf. on Ground Movements and Structures, UWIST, Cardiff,* ed. J.D. Geddes, Pentech Press, London, pp. 857–876.

D'Appolonia, D.J. (1971) Effects of foundations on nearby structures. In *Proc. 4th Panamerican Conf. on Soil Mechanics and Foundation Engineering, San Juan, Puerto Rico,* ASCE, N.Y., pp. 189–236.

D'Appolonia, D.J. and Lambe, T.W. (1971) Performance of four foundations on end bearing piles. *J. Soil Mech. and Found. Div., ASCE,* **97** (SM1) 77–94.

Darwent, T.J. (ed.) (1975) Diaphragm walls and anchorages. In *Proc. Conf. on Diaphragm Walls and Anchorages, London,* Inst. of Civil Engineers, London.

Fry, R.H. (1982) *Measurement of soil parameters for prediction of ground movement and associated strain in buried pipes.* External Report 77E, WRC Engineering Centre, Swindon, Wilts., England.

Gumbel, J.E. and Wilson, J. (1982) Observations of ground movements around a trench excavation in London Clay. In *Proc. 2nd Conf. on Ground Movements and Structures, Cardiff,* ed. J.D. Geddes, Pentech Press, London, pp. 841–856.

Hanna, T. (1978) Diaphragm walls. In *Developments in Soil Mechanics,* Applied Science, London, pp. 213–249.

Harris, C.W. and O'Rourke, T.D. (1983) Response of jointed cast iron pipe lines to parallel trench construction. Geotechnical Engineering Report 83–5, School of Civil and Environmental Engineering. Cornell University. Ithaca, N.Y.

Henkel, D.J. (1970) Geotechnical considerations of lateral stresses. State of the Art paper. *Speciality Conference on Lateral Stresses in the Ground and Design of Earth-Retaining Structures, Cornell Univ. N.Y.,* ASCE, pp. 1–49.

Howe, M., Hunter, P. and Owen, C. (1980) Ground movements caused by deep excavations and tunnels and their effect on adjacent mains. In *Proc. 2nd Conf. on Ground Movements and Structures, UWIST, Cardiff.* ed. J.D. Geddes, Pentech Press, London, pp. 812–840.

Huder, J. (1969) Deep braced excavation with high water table. In *Proc. 7th Int. Conf. on Soil Mechanics and Foundation Engineering, Mexico City,* Vol. 2, pp. 443–448.

Kyrou, K. (1981) *Report on the at-rest earth pressure coefficient K_0: Measurement and evaluation.* External report 76E, WRC Engineering Centre, Swindon, Wiltshire, England.

Lambe, T.W. (1970) Braced excavations. In *Proc. ASCE Speciality Conf. on Lateral Stresses and Earth Retaining Structures, Cornell University, Ithaca, N.Y.,* pp. 52–64.

Lambe, T.W. (1973) Predictions in soil engineering. 13th Rankine lecture. *Géotechnique* **23** (1) 143–202.

Lambe, T.W., Wolfskill, L.A. and Wong I.H. (1970) Measured performance of braced excavation. *J. Soil Mech. and Found. Div., ASCE,* **96** (SM3) 279–293.

Laurence, G.J.L. (1977) *Trench wall jack : An apparatus to measure the equivalent elastic modulus of soil.* TRRL Supplementary Report No. 347, Transport and Road Research Laboratory, Crowthorne, Berkshire, England.

Leigh, W.J.P., Scoble, M.J. and Young, G.J. (1980) Ground movements adjacent to a deep excavation. In *Proc. 2nd Conf. on Ground Movements and Structures, UWIST, Cardiff,* ed. J.D. Geddes, Pentech Press, London, pp. 759–777.

Littlejohn, G.S. and McFarland, I.M. (1974) A case history of multi-tied diaphragm walls. In *Proc. Conf. on Diaphragm Walls and Anchorages,* Inst. of Civil Engineers, London, pp. 113–121.

Mana, A.I. and Clough, C.W. (1981) Prediction of movements for braced cuts in clay. *J. Geotech. Eng. Div., ASCE,* **107** (GT6) 759–777.

Meyerhoff, G.G. (1951) The ultimate bearing capacity of foundations. *Géotechnique* **2** (2) 301–332.

Milligan, G.W.E. (1983) Soil deformations near anchored sheetpile walls. *Géotechnique* **31** (1) 41–55.

Mitchell, J.K. and Gardner, W.S. (1975) *In situ* measurement of volume change characteristics. in *Proc. Speciality Conf. on In situ Measurement of Soil Properties, North Carolina State University*, Vol. 2, pp. 279–346.

Morgenstern, N., Blight, G.E., Jambu, N. and Rezendiz, D. (1977) Slopes and excavations. In *Proc. 9th Int. Conf. on Soil Mechanics and Foundation Engineering, Tokyo, Japan*, Vol. 2, pp. 547–581.

Morton, K., Leonard, M.S.M. and Carter, R.W. (1980) Building settlements and ground movements associated with construction of two stations of the modified initial system of the mass transit railway, Hong Kong. in *Proc. 2nd Conf. on Ground Movements and Structures, UWIST, Cardiff*, ed. J.D. Geddes, Pentech Press, London, pp. 788–802.

Ohde, J. (1938) Theory of earth pressure with special reference to earth pressure distributions. *Bautechnik* **16**, Nos 10, 11, 13, 19, 25, 37, 42, 53, 54. Translated for the Earth Retaining Structures Code Committee of the Institution of Structural Engineers, London, 1951.

O'Rourke, T.D. (1978) Lateral displacement of shallow buried pipelines due to adjacent deep trench excavations, Discussion: *Géotechnique* **28** (2) 214–216.

O'Rourke, T.D. (1981) Ground movements caused by braced excavations. *J. Geotech. Eng. Div. ASCE*, **107** (GT9) 1159–1178.

Peck, R.B. (1969) Deep excavations and tunnelling in soft ground. In *Proc. 7th Int. Conf. on Soil Mechanics and Foundation Engineering, Mexico City*, State of the Art Volume, pp. 225–290.

Peck, R.B., Hanson, W.E. and Thorburn, T.H. (1974) *Foundation Engineering*. 2nd ed., John Wiley and Sons Inc., N.Y.

Rowe, P.W. (1952) *Anchored sheet pile walls.* Proc. Inst. Civil Engineers, London, Paper No. 6107.

Sills, G.C., Burland, J.B. and Czechowiski, M.K. (1977) Behaviour of an anchored diaphragm wall in stiff clay. In *Proc. 9th Int. Conf. on Soil Mechanics and Foundation Engineering, Tokyo*, 1977, **2**, 147–154.

Symons, I.F. (1980) Ground movements and their influence on shallow buried pipes. *The Public Health Engineer (J. Inst. of Public Health Engineers)* **8** (4) 149–153.

Symons, I.F., Chard, B. and Carder, D.R. (1981) Ground movements caused by deep trench construction. In *Proc. of the Conf., Sewerage '81*, Inst. of Civil Engineers, London, pp. 49–61.

Tarzi, E.I., Menzies, B.K. and Crofts, J.E. (1979) Bending of jointed pipelines in laterally deforming soil. *Geotechnique* **29** (2) 203–206.

Terzaghi, K. (1943) *Theoretical Soil Mechanics.* J. Wiley and Sons Inc., N.Y.

Terzaghi, K. and Peck, R.B. (1967) *Soil Mechanics in Engineering Practice.* 2nd ed., John Wiley and Sons Inc., N.Y.

Tomlinson, M.J. (1970) Lateral support of deep exacavations. In *Proc. Conf. on Ground Engineering*, Inst. of Civil Engineers, London, pp. 55–64.

Tomlinson, M.J. (1975) *Foundation Design and Construction.* 3rd ed., Pitman, London.

Toombs, A.F., McCaul, C. and Symons, I.F. (1982) *Ground movements caused by deep trench construction in an urban area.* TRRL Report 1040, Transport and Road Research Laboratory, Crowthorne, Berkshire, England.

Tschebotarioff, G.P. (1949) *Large scale earth pressure tests on model flexible bulkheads.* Final Report to U.S. Navy Bureau of Yards and Docks, New Jersey, USA, Princeton University.

Tschebotarioff, G.P. (1973) *Foundation, Retaining and Earth Structures.* 2nd ed., McGraw-Hill, N.Y.

Vaughan, P.R. and Walbancke, H.J. (1974) Pore pressure changes and the delayed failure of cutting slopes in overconsolidated clay. *Géotechnique* **23** (4) 531–540.

Verdeyen, J. and Roisin, V. (1953) *Soutènement des élastiques.* In *Proc. 3rd Int. Conf. on Soil Mechanics, Zürich*, Vol. 2, pp. 188–192.

5 Trenches in soil

P.B. RUMSEY and I. COOPER

5.1 Introduction

The three most important aspects of soil behaviour in trenching are:

1. Strength (that is the likelihood of trench wall collapse).
2. Deformation (that is the amount of movement which the trench walls will undergo if collapse does not occur).
3. Load imposed on a support system placed in the trench.

Records of field measurements of ground movements and methods of predicting movement adjacent to trenches are described later in this chapter.

5.2 Shearing resistance and deformation

The rate at which water can enter or leave the soil is governed by the soil permeability. In highly permeable soils such as sands. drainage takes place rapidly (say in hours) whereas in low-permeability soils such as clays drainage is very slow (days or months). Whether or not 'drained' conditions can be said to apply will depend very much on the permeability of the soil and the time since load application.

In the context of soil strength and trench excavation, the major concern is whether a trench wall is going to collapse and, if so, how long the process might take. In order to examine this subject further a potential surface within the soil can be imagined, along which the soil might eventually fail. Failure of the wedge is resisted by the shear strength of the soil (τ) along the potential failure surface. If the force (p) trying to move the wedge downwards (i.e. the resolved component of its self-weight) becomes greater than the force (τ) resisting movement, failure then occurs.

It is evident that the soil shear strength depends on the permeability of the soil and on time, or on whether the soil is effectively 'drained' or 'undrained'. Examples of deformation usually assume that the soil is loaded. However, when a hole or trench is excavated in the ground, the soil in the trench wall is unloaded. Before excavation, *in-situ* stresses in the ground act, in effect, on the front of the excavation face. The process of excavation removes these stresses and therefore unloads the soil in the wall.

The effect of unloading on the shear strength of the soil is analogous to the effect of loading, except that the process is reversed. If the soil is loaded, and the excess pore-water pressure generated by the loading is allowed to dissipate, the shear strength of the soil then increases. If the soil is unloaded, the opposite will happen; pore-water

110

pressure will decrease, the soil will react by sucking in water from the surrounding ground to restore equilibrium, and the shear strength will decrease.

Again, the time required for completion of this process will depend on the permeability of the soil. However, one important conclusion can be drawn; the shear strength (and therefore the degree of stability) of a trench wall is at a maximum immediately after excavation. With time, the stability decreases. If the trench is sufficiently shallow, and the soil sufficiently strong that collapse of the trench does not ensue, the stability will continue to decrease until a minimum is achieved at some time after excavation, this time being governed by the rate of restoration of equilibrium pressure in the water within the soil voids. In the case of sands and gravels, the time required for long-term equilibrium may be very short, often less than one day. In the case of many clays it may take weeks or months for the water pressures to reach equilibrium.

The risk of instability, or collapse, of a trench wall has been shown to increase with the amount of drainage which takes place from the soil. The permeability of many soils is significantly affected by structural features such as fissures and sand layers. Such features in clays may increase the mass permeability considerably, allowing relatively rapid drainage and increasing the rate of decrease of stability with time. It is therefore dangerous to assume that all clays will effectively be 'undrained' during the construction period—an assumption which could, in some circumstances, create a risk of unexpected collapse.

The strength of the soil is of direct relevance if it is proposed to excavate a trench without support, as might be the case for shallow vertical trenches or for unsupported deeper trenches with battered sides. However, for the majority of deep trenches, support is provided from within the trench and the soil forming the trench wall may never approach failure. In the latter condition the soil will tend to move inwards and downwards, exerting a load on the support within the trench.

When load is removed from the ground (as in the case of trench excavation) the soil deforms. If the load is removed sufficiently quickly that the permeability of the soil does not allow a significant flow of pore-water to take place, the resulting deformation is 'undrained'. The amount of movement will depend on the original state of stress in the ground and on the load deformation properties of the soil under undrained conditions.

An idealized load deformation relationship for the case of trench excavation is

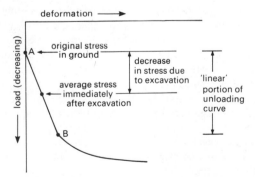

Figure 5.1 Idealized load-deformation relationship.

shown in Fig. 5.1. Upon excavation (under undrained conditions) the soil deforms and moves downwards and towards the excavation as shearing takes place. If the strength of the soil is low. as in the case of a soft clay, the deformations may be very large and failure may ensue unless adequate support is provided. In a strong soil, such as a very stiff clay, the deformation may be very small, as represented by a point somewhere along the line A − B (Fig. 5.1), the 'linear' portion of the load deformation curve.

In the case of a strong soil, which does not approach failure immediately after excavation, the reduction in stress in the ground brought about by the excavation will cause an immediate reduction in pore-water pressure. The soil will react by sucking in water from the surrounding ground. The inflow of water into the ground immediately adjacent to the trench will cause the soil structure to expand or swell, resulting in further movement downwards and towards the excavation. The time taken for water to flow into the soil mass, and thus the rate at which swelling occurs will depend on the permeability of the soil (compare with physico-chemical swelling, Chapter 9).

There are, then, two kinds of movement which take place when a trench is excavated, one due to shear deformation of soil in the trench wall, and the other caused by changes in water-pressure distribution giving rise to swelling. In both cases, the resulting movement of the soil towards the trench will be associated with settlement of the adjacent ground. Shear deformations will take place in all soils and frequently cannot be avoided since they occur as soon as the excavation is made. The movement caused by swelling is time-dependent and will occur if the trench is open for a sufficiently long time to permit the water pressures in the soil to equalize, or if the original *in-situ* stresses are not re-established by the backfilling process.

The time required for swelling to take place will vary enormously from one soil type to another. However, in many cases, the magnitude of this latter form of movement can be considerable; hence the importance of properly reinstating a trench as soon as possible to minimize the likelihood of damage to adjacent pipelines and other structures.

5.3   Construction and control of ground movement

5.3.1   *Effects of contruction procedure*

Construction procedure can have an important effect on movement of trench walls and the possibility of collapse. Several ways in which the likelihood (and magnitude) of ground movement caused by trench excavation can be minimized by careful workmanship during construction are summarized in Table 5.1, and discussed in the following paragraphs. In critical situations, and where warranted by the possible cost consequences of excessive movement, ground treatment techniques such as grouting or ground freezing may be considered. In all cases there will need to be a balance, as in all ground engineering operations, between the cost of improved and specialized techniques and the consequences of allowing the movement to occur. In 'green field' situations for example, where construction will have no influence on adjacent plant or buildings, economies can be effected by adopting support procedures which allow safe working within the trench but which offer little restraint to the trench walls. Support systems in this category will be described later in this chapter.

Table 5.1 Construction procedure and ground movement

Construction stage	Procedures to minimize movement	Type of ground movement
Excavation and installation of support	1. Install support as excavation proceeds (or before excavation—predriven sheets) 2. Minimize space behind support 3. Ensure all slack taken up in struts 4. If (1) is not possible, excavate as quickly as possible, ensure smooth vertical walls, install support immediately.	1. Short term movement due to shear deformation (provided excavation and support installation done quickly—otherwise some swelling) 2. Possible trench wall failure if unstable
Pipelaying	1. As (2) and (3) above.	1. Swelling due to water pressures causing movement towards the trench and settlement 2. Possible trench wall failure if support inadequate.
Withdrawal of support and backfilling	1. Minimize the unsupported areas during backfilling (e.g. the exposed face between bottom of sheets and top of backfill should be kept to a minimum) 2. Use good quality backfill 3. Compact to high density	1. Swelling due to equalization of water pressures (amount dependent on degree to which trench is unsupported during this stage, and on the nature and quality of backfill). 2. Possible trench wall failure (either 'undrained' or 'drained' depending on soil type and geometry)
General	1. Ensure good supervision. 2. Ensure good trenching practice. 3. In critical situations employ ground treatment (e.g. dewatering, grouting) where practicable, and/or review the need for a predriven support left in place, or other suitable (proprietary?) system.	

Consider the construction of a new sewer in trench. The construction may be performed in three stages:

(i) Excavation and support installation.
(ii) Pipelaying.
(iii) Support withdrawal and backfilling.

Earlier in the chapter attention was drawn to two main sources of soil movement: shear deformation, and time-dependent movement.

The time taken for the latter form of movement to develop may be significantly reduced by the presence of structural features within the soil which give rise to a higher mass permeability. It is virtually impossible to prevent undrained shear deformation taking place at the excavation stage. By definition, such movement occurs as soon as a hole is made in the ground and the *in-situ* stresses are released. However, in practice, procedures can be adopted to minimize the movement.

good support installation

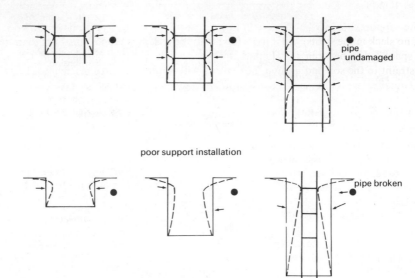

poor support installation

Figure 5.2 Installation of trench support.

The installation of support in stages, as excavation proceeds, will limit movement by reducing the cumulative effect which occurs if the excavation is taken to full depth before support is installed. The principle is illustrated in Fig. 5.2. The benefits of adopting this procedure from the standpoint of safety should be obvious. An alternative method commonly employed in weak soil, where large ground movements are expected, is the use of pre-driven sheet pile trench support. Excavation then proceeds, struts and walings being placed at suitable intervals. (In critical situations, the sheets are left in place and then cut off at a suitable level below ground.)

During this stage of construction, lack of attention to detail and poor workmanship can readily create a situation which leads to greater movement at later stages. Within the limits of cost and practicality it is important to ensure that excavation proceeds as quickly as possible, that the walls are vertical and as smooth as possible, that the support is installed as quickly as possible, that any holes or spaces between the support and the trench wall are filled, and that slack is taken up in all struts. All these requirements assume particular importance during later stages of construction.

While pipelaying is in progress, the trench remains open. On many construction contracts any one section of trench will be open for only a day or less. However, it is not unusual for much longer periods of time to pass before reinstatement. As long as the trench is open, equalization of pore pressures in the trench walls will continue to take place, gradually bringing the ground into equilibrium with the reduced stresses generated by the original excavation. As this process continues, the ground gradually moves inwards towards the trench.

If the support has been installed properly it has two beneficial effects (apart from increasing the safety of men working in the trench):

(1)   It restrains ground movement by taking up load in the struts.
(2)   It induces a load into the ground, tending to maintain its original stress condition.

The support can only be effective in this way if it is rigid relative to the ground (that is, no slack in the struts) and if it fits flush into the hole it is supporting. Furthermore, if space is left between the support and the trench wall the support will offer no restraint to the soil and ground movements will continue to take place until the soil eventually meets the support, or until the support is withdrawn and the trench reinstated.

Even so, high earth pressures commonly associated with weak soil can cause significant deflection of the support under load. The final stages of construction often give rise to the greatest movement. If the support has been effective, it will have prevented a significant amount of potential movement and, in the process, may have restored some of the stress originally removed from the trench wall during excavation. Consequently, as the support is withdrawn, further ground movement can be expected. In critical situations where the expected scale of movement will be unacceptable it may be advisable to adopt a pre-driven and left-in-place sheet pile support as described earlier. Generally, however, the likelihood of large movements can be greatly reduced by ensuring that any gap between the bottom of the trench support and the top of the backfill is kept as small as possible.

The quality of backfilling will have an important effect on long-term movements. The potential for such movement is not eliminated by the act of trench reinstatement. If the quality of backfill material is poor, and compaction in the trench inadequate, movements will continue to occur after reinstatement until such time as the stresses in the backfill and those in the surrounding ground have equalized.

It is noted that trenches backfilled with permeable material can act as sub-surface drains and give rise to long-term settlements resulting from local drawdown of the water table. Suitable precautions involving, for example, clay seals, inserted at suitable intervals along the trench, may be necessary to preclude this possibility.

5.3.2 Dewatering

Any discussion on the control of ground movements would be incomplete without some mention of dewatering. The common use of dewatering techniques emphasizes the important role played by groundwater in trench stability and soil behaviour.

A wide variety of techniques is available for controlling groundwater, and specialist advice in this field may often be required. However, in practice the two most commonly-used methods are sump pumping and well-pointing. The effectiveness of each technique (illustrated in Fig. 5.3) will depend upon the nature of the soil, trench geometry and the degree and rate of groundwater lowering required.

Pumping from a sump is probably the simplest and cheapest method available. It does, however, suffer from two main disadvantages. First, it draws water inwards towards the bottom of the trench and consequently may cause instability due to upward seepage. The second disadvantage lies in the tendency for pumping to remove fine particles of soil from suspension in the groundwater. This process of internal erosion can cause sudden undermining of the trench base or sides, or longer-term movement. It is therefore essential to surround the hose inlet with some form of graded filter to prevent the migration of suspended fines, and to check at frequent

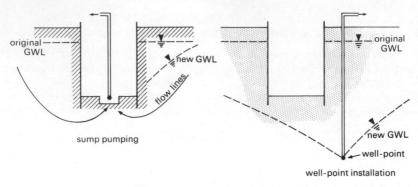

original GWL

new GWL

flow lines

sump pumping

original GWL

new GWL

well-point

well-point installation

Figure 5.3 Methods of dewatering.

intervals (by examining the discharged water at the outlet) that soil is not being eroded in this way.

Well-pointing draws water away from the trench and therefore has a positive stabilizing effect on the trench walls. Its greatest use is in sands. The permeability of clays is usually too low for efficient use, but the method can sometimes be effective in silts. In gravels or other very permeable deposits, the heavy flow of water may require the well-points to be so close together as to be impractical.

All dewatering systems rely on mechanical pumps which are susceptible to breakdown. Whenever dewatering is used it is essential to allow for this possibility in assessing the loads imposed on the trench support, unless effective standby pumping can be guaranteed. A further general point relevant to all dewatering is the settlement caused by groundwater lowering. The extent will depend on the soil type, the rate of pumping, the location of the well-point or sump and the trench geometry. Where uncertainty exists about the effects or extent of settlement it is advisable to obtain specialist advice.

5.4 Trench stability

Figure 5.4 indicates the maximum height at which a vertical unsupported wall of soil can stand without collapsing in the short-term, 'undrained' state. The critical

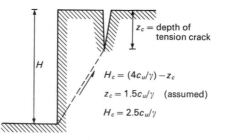

H

z_c = depth of tension crack

$H_c = (4c_u/\gamma) - z_c$

$z_c = 1.5c_u/\gamma$ (assumed)

$H_c = 2.5c_u/\gamma$

Figure 5.4 Critical depth H_c of trench wall in clay (short-term, undrained, immediately after excavation).

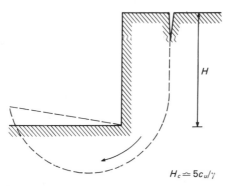

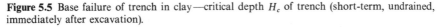

$$H_c \simeq 5c_u/\gamma$$

Figure 5.5 Base failure of trench in clay—critical depth H_c of trench (short-term, undrained, immediately after excavation).

height (H_c) depends upon the soil undrained shear strength (c_u), its bulk unit weight (γ), and the depth (z_c) of any tension crack that develops at the crest. Hence, the critical height of a trench wall can be taken as (2.5 to 4) c_u/γ. It follows that a trench wall higher than H_c would be expected to collapse as soon as it is excavated. It is possible for wide trenches to be adequately supported to prevent collapse of the walls, but to fail by heaving of the trench bottom (Fig. 5.5).

For a strutted excavation, the critical height above which base failure can occur is dependent on the depth: width ratio but can be estimated as $5c_u/\gamma$ where the relevant value of c_u is measured at or below formation level.

If the wall of a trench is battered back, the stability will increase. The most common way of examining short-term stability in this case under undrained conditions is by use of stability curves which can be found in soil mechanics textbooks (see, for example, Taylor, 1937; Terzaghi and Peck, 1967).

It must be emphasized that many trenches may not fulfil the necessary conditions for the assumption of 'undrained' behaviour. When a trench is first excavated, 'undrained' conditions apply. The longer the trench is left open, the greater is the drainage that takes place, causing a reduction in stability and warranting different approaches to analysis and design.

A decision to construct an unsupported, battered trench implies greater supervision and maintenance of the exposed face to ensure that all loose material is removed before it becomes a danger to workmen in the trench. Furthermore, an open excavation requires consideration of the many ways in which failure, or collapse, might occur. Several of the more common failure modes are indicated in Fig. 5.6.

5.5 Design of trench support

This subject is covered more comprehensively in a report by the U.K. Construction Industry Research and Information Association (CIRIA) (1983).

The major uncertainty in design is the load generated on the support by the retained soil and groundwater. If load can be assessed, design of the support to withstand the load is reduced to a straightforward structural problem. Furthermore, the advent of proprietary trenching systems warrants, in some cases, a modified approach to design

1. Sliding (generally clays)

2. Falls (all soils, particularly clays)

3. Bedding planes (mainly rock, also fissured clays)

4. Blocks (mainly rock)

5. Drying/saturation/erosion (mainly sands)

6. Internal erosion (sand over clay)

(clay over sand)

7. Wash-out

(1) Failure by sliding—either rotational or wedge-shaped; will occur if the slopes are too high or too steep.

(2) Falls—pressure relief due to excavation causes cracks to develop in clay soils which can fall into the trench without warning. Many trench collapses are associated with this condition, probably because it arises in ground which is thought to be safe, notably in stiff fissured clays.

(3) Bedding plane failure—sliding on bedding planes; applies particularly to rock, but also to soils containing extensive planes of weakness and can be very sudden.

(4) Block failure—beds of weaker rock can lead to undermining of stronger rock above.

(5) Drying/saturation/erosion—when first excavated, sands appear stable; collapse occurs on drying out or saturation, or through erosion by rain or surface water.

(6) Internal erosion—layered soils are particularly susceptible; water flowing through layered soil deposits can erode, soften, weaken and undermine.

(7) Washout—occurs in sands if there is a constant supply of water (e.g. artesian pressure or from broken mains or sewers) and can be catastrophic.

Figure 5.6 Some modes of trench wall deformation.

based on estimating the loads generated by the soil and checking that the manufacturer's system is capable of safely withstanding the estimated loads.

The type of earth pressure distribution adopted for retaining wall design assumes that the lateral pressure exerted against the back of the wall increases in simple proportion to the depth below surface. However, there are fundamental differences between a trench support system and normal retaining structures which give rise to trapezoidal or rectangular stress distributions behind a strutted trench support.

First, support is usually installed as excavation proceeds. If the support is installed correctly, the upper part of the trench is effectively braced before any appreciable yielding has taken place. As excavation proceeds, the support at upper levels of the trench resists further movement. A second difference lies in the way in which the structure fails. A retaining wall is designed as a structural unit, and generally moves or fails as a unit. However, any strut in a trench support system can fail as an individual unit. Failure of one strut will lead to an increase in load on its neighbours, and could initiate a progressive failure of the whole system. The loads which develop in a support system are therefore highly dependent on the system itself and on the way in which it is installed and reacts, and cannot be estimated to a high degree of accuracy.

In practice, design procedures are based on measurements of strut loads in deep excavations which have been used to define design envelopes for selected soil types. These envelopes vary from one source to another, but all are derived largely from those published by Peck (1969). These are reproduced in slightly different form in Fig. 5.7. The envelopes are acknowledged to be conservative but nevertheless provide the main source of quantitative information. In the case of sands, it is

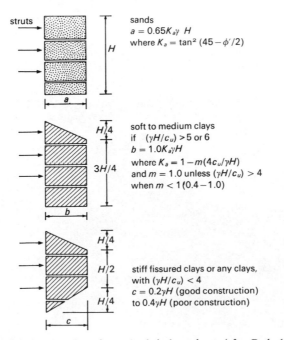

struts

sands
$a = 0.65 K_a \gamma \, H$
where $K_a = \tan^2 (45 - \phi'/2)$

soft to medium clays
if $(\gamma H/c_u) > 5$ or 6
$b = 1.0 K_a \gamma H$
where $K_a = 1 - m(4c_u/\gamma H)$
and $m = 1.0$ unless $(\gamma H/c_u) > 4$
when $m < 1 (0.4 - 1.0)$

stiff fissured clays or any clays,
with $(\gamma H/c_u) < 4$
$c = 0.2\gamma H$ (good construction)
to $0.4\gamma H$ (poor construction)

Figure 5.7 Computation of strut loads in braced cuts (after Peck, 1969).

emphasized that the empirical rules assume a water table below the base of the excavation. If this condition is not fulfilled, additional load from hydrostatic pressure must be taken into account. Caution is also required since changes in the groundwater regime may occur with time.

5.6 Support systems in trench excavations

5.6.1 *General*

Trench support is usually a temporary, but major, works element of pipeline construction or repair work. Depending on the technical expertise and experience of the contractor it may be desirable for the design engineer to maintain a major part in specifying and designing the support system. In any event, both the engineer and contractor should be fully aware of the limitations to be placed on construction by safety, and by the presence of adjacent pipes, pavements and buildings. Careful selection of the correct support system, and of the construction procedure, are required if danger to operators and the risk of damage due to ground movement are to be minimized.

5.6.2 *Proprietary support systems*

5.6.2.1 *General.* Support systems can be considered in two groups; those based closely on traditional trenching methods and materials, and those systems of a proprietary nature.

Traditionally, timber sheets, walings and struts have been used to support

Figure 5.8

excavations for sewers. Nowadays it is more common to see steel trench sheets being used in conjunction with timber walings and mechanical or hydraulic metal struts (Fig. 5.8). This traditional approach to shoring retains significant advantages over some of the proprietary systems, notably in respect of flexibility in dealing with crossing services and other obstructions, in capital cost, and in the space needed for storage of support materials. However, it is always necessary for operatives to enter the trench for the purpose of installing the support, thus increasing the risk to life in the event of trench collapse. Furthermore, there is concern over the amount of 'design' input (be it office or site) on many of the trenches supported by traditional methods, and over the sometimes inadequate size and spacing of walings and struts which are often installed on a 'trial and error' basis. Notwithstanding this, a properly designed and installed traditional support can be as effective as any other method in restricting ground movements adjacent to a trench, and in providing safe working conditions for men in the trench. It can be particularly effective in preventing ground movement in poor ground when used in conjunction with pre-driven sheets which are subsequently cut off at a suitable level and left in place.

The main advantage offered by proprietary systems is in eliminating the need for men to be exposed to the danger of an unsupported trench. It is important to note that some of the methods are used to provide protection for men in the trench rather than to provide support for preventing collapse or excessive ground movement. Nevertheless, once operators are skilled in the use of these systems they are generally quicker to instal than traditional methods of support.

A report by CIRIA on proprietary trench support systems (Mackay, 1982), from which some of the information in this section derives, divides the systems into the following groups:

Figure 5.9

 (i) Hydraulic struts, walings and shores.
(ii) Shields or drag boxes.
(iii) Boxes or lining systems.
(iv) Special systems.

5.6.2.2 *Hydraulic shoring systems.* Hydraulic shoring systems are constructed of lightweight high-strength aluminium alloy with hydraulic cylinders which are used to brace the shoring rails apart, either directly on to the ground or on to timber or steel trench sheets (Fig. 5.9). The installed system is thus similar in some respects to traditional support and retains the flexibility of the latter in dealing with crossing services. However, both walings and vertical shores can be installed and pressurized from outside the excavation, thus greatly improving the safety of operators.

Vertical shores can only be used on their own in strong, cohesive ground. If used without sheeting they need to be placed at relatively close centres, restricting access for pipe lowering. The problem can be overcome in good ground conditions by removing the shores temporarily, provided that there are no men in the unsupported section of the trench and that the temporary removal of support will not put adjacent pipes and buildings at risk from inward ground movements.

Figure 5.10

5.6.2.3 *Trench shields.* Shields (or 'drag boxes') are protective metal boxes which provide safety for personnel in the trench (Fig. 5.10). They are not generally used to support the sides of the trench and are therefore most suited to use in 'green field' situations where ground movement due to excavation will not put adjacent plant at risk. Construction proceeds by excavating slightly wider than the shield and then lowering it into the trench. As each section of pipe is completed, the shield is dragged forward with an excavator.

Shields are easy to make, use and maintain, and can achieve very high production rates in open ground. However, they are heavy, require the use of a powerful excavator, and are unable to accommodate crossing services. Difficulties are also experienced where compacted backfilling is required.

5.6.2.4 *Trench boxes.* Trench boxes consist of two metal support walls of fixed dimensions separated by adjustable struts (Fig. 5.11). The boxes are assembled vertically as excavation proceeds to form a module of the required depth. In contrast to shields, trench boxes are not intended to be dragged horizontally through the ground, but are used to provide a solid wall of sheeting against the side of the trench. In general, three or four modules are used in an excavation/pipelaying operation,

Figure 5.11

the rearmost box being extracted and moved to the front of the line as work proceeds. The correct operation procedure is to excavate a shallow trench slightly wider than the unit, place the box in position, and alternately excavate within the box and push down the corner posts until the required depth is reached. Once the pipes have been installed, the box is gradually raised as backfilling and compaction take place.

Regrettably it is not unusual for the correct installation procedure to be ignored, and for excavation wider than the box to proceed to considerable depth (often full depth) before the box is lowered into position. In this situation large voids are frequently left behind the support and this, coupled with the lack of support during excavation, can allow significant ground movements to take place. Again, this practice may be acceptable in open ground but should be resisted where damage to adjacent plant may arise. The application of boxes to open ground situations rather than in congested urban environments is also suggested by their limited ability to deal with crossing services and the working space required for handling and storage. In the right environment (for example urban roadways), if properly installed, boxes provide a quick and safe method of shoring.

5.6.2.5 *Plate lining systems.* Plate lining (or 'sliding rail') systems employ solid panels sliding into support rails driven vertically into the trench line. The system has the same advantages as the box system, with the additional benefits of lesser weight and easier handling and storage. However, as with trench boxes, the effectiveness of plate lining systems in restricting ground movements is dependent upon the procedure adopted for installation and withdrawal of support. The installed system provides continuous lining to the trench wall without the small gaps which can develop between individual boxes. It has particular advantages where deep trenches are specified or where a stepped trench is required to accommodate two adjacent pipes at different levels. The main disadvantages are that crossing services can interfere with the lining route and necessitate local sheeting or similar support. Care is required in setting up the initial line of support rails.

5.6.2.6 *Special systems.* Apart from specially-designed support systems for particular excavations, there are two special proprietary systems. Both drive interlocking sheet piles hydraulically (one vertically and the other horizontally) to provide trench support to depths down to about 7 m, and are particularly suited to situations where deep excavations are required but where noise and vibration must be minimized. One system is completely self-contained and self-propelled and has a facility for compacting backfill by horizontal thrust; the other is used in conjunction with other trench support but is able to accommodate crossing services. Both are large and very expensive machines requiring a lengthy set-up and dismantling time, but offer high rates of advance in situations to which they are specially suited.

5.7   Prediction of ground movements

5.7.1 *Introduction*

Much work has recently been carried out in the U.K. to monitor ground movements caused by trench excavation and, in particular, their effect on buried pipelines. Records of ground movements associated with trench excavations published to date include those obtained by the British Transport and Road Research Laboratory (Symons, 1980; Symons *et al.*, 1981; Toombs *et al.*, 1982), British Gas (Howe *et al.*, 1980; Gumbel and Wilson, 1980) and Water Research Centre (WRC) Engineering (Rumsey *et al.*, 1981; Rumsey and Cooper, 1982). There is a considerable body of evidence on ground movements arising from deep, wide excavations (see, for example, Peck, 1969: Burland *et al.*, 1979), and from trench excavations for structures such as diaphragm walls (see, for example, Hanna, 1978). These latter excavations are, however, considerably deeper than those dealt with in this chapter. Several of the references listed above also report measurement of changes in strain in buried pipes adjacent to the excavation, for example Howe *et al.* (1980), Carder *et al.* (1982), Rumsey *et al.* (1981) and Rumsey and Cooper (1982).

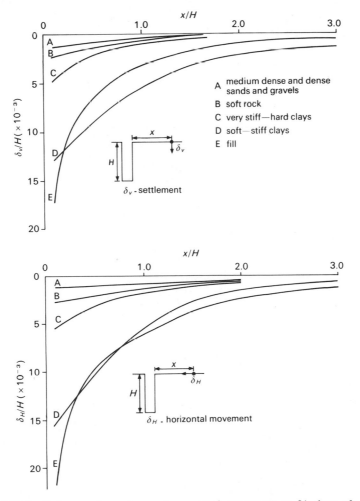

Figure 5.12 Predictive envelopes for surface ground movement at 21 days after trench reinstatement.

5.7.2 *Analytical procedures*

An empirical method of predicting horizontal ground movement caused by trench excavation was proposed by Crofts *et al.* (1977). The method was based partly on case-history data of Peck (1969) for deep wide excavations (Chapter 4). A design procedure utilizing finite element techniques was developed by Kyrou (1980) for predicting the effect of trench excavation—induced ground movements on adjacent

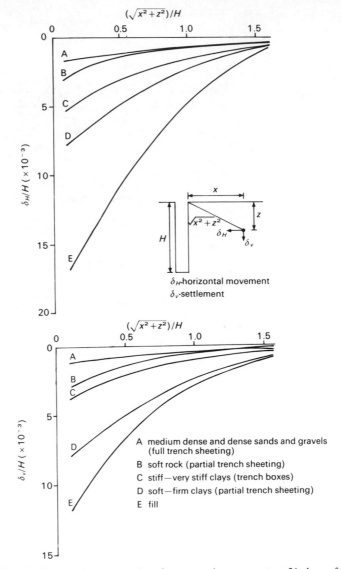

Figure 5.13 Predictive envelopes for sub-surface ground movement at 21 days after trench reinstatement.

buried pipelines. Kyrou's method was extended by WRC Engineering (Fry and Rumsey, 1982).

5.7.3 Predictive procedures

A main source of current information which provides guidance on the prediction of ground movement is compiled by Fry and Rumsey (1983b). It combines the results of field measurements of ground movements at a large number of U.K. Regional Water Authority sites with those published by the U.K. Transport and Road Research Laboratory, and uses them in deriving an empirical method for the prediction of ground movement caused by trench excavation.

The report finds the analytical procedures developed by Kyrou (1980) and described by Fry and Rumsey (1983a) to be more reliable than that proposed by Crofts et al., (1977). The field results are used to calibrate the models described by Fry and Rumsey

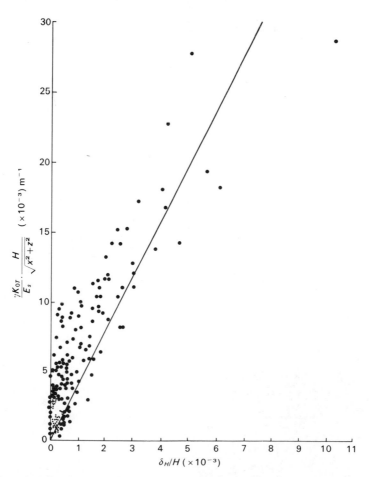

Figure 5.14 Development of empirical prediction for horizontal ground movement.

(1982) in such a way as to develop empirical methods of predicting ground movements. The predictive methods relate horizontal or vertical ground movements at any location in the trench wall to soil parameters and trench geometry. The soil parameters involved include the constrained soil modulus (E_s) and the at-rest earth pressure (K_{0T}). (K_{0T} is defined as the at-rest earth pressure coefficient in terms of total stresses; it is equal to the ratio of pre-excavation total horizontal and vertical stresses at any point in the soil mass. These stresses were measured indirectly at a large number of the experimental sites.)

Predictive envelopes proposed by WRC Engineering for a variety of general soil types are reproduced in Figs. 5.12 and 5.13. The data obtained from field experiments are also plotted on Figs. 5.14 and 5.15. The straight lines indicated on the figures have the same gradient equal to 2.5×10^4. Hence, if these lines are employed as 'reasonable upper bounds', the horizontal and vertical ground movements adjacent to any trench excavation can be estimated using the equation

$$\delta_V = \delta_H = \frac{0.25\gamma K_{0T}}{E_s} \cdot \frac{H^2}{\sqrt{x^2 + z^2}} \tag{5.1}$$

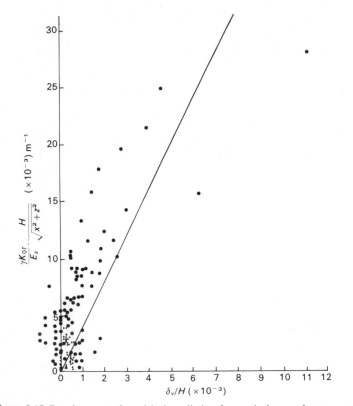

Figure 5.15 Development of empirical prediction for vertical ground movement.

Table 5.2 Soil parameters (E_s, K_{0T}, γ)

Soil type	$E_s(\times 10^3 \, \text{kN/m}^2)$	K_{0T}	$\gamma(\text{kN/m}^3)$
Loose sand and gravel	4–8	0.4–0.6	20
Medium dense sand and gravel	8–24	0.4–0.7	21
Dense sand and gravel	24–50 +	0.4–1.0	22
Soft clay	1–3	0.6–1.2	19
Firm clay	3–10	0.8–1.2	20
Stiff clay	10–20	1–3	20
Very stiff clay	20–40 +	2–3.5	21

(A list of symbols and units is included at the end of this chapter. Typical ranges of soil parameters are given in Table 5.2.)

The foregoing equation and the data reproduced in Figs. 5.14 and 5.15 are based on ground movements measured approximately 1 day after trench reinstatement. To enable estimates of the development of movement in the longer term the following relationship is suggested (based on data reported by WRC, 1983):

$$\frac{\delta_{\text{time}(t)}}{\delta_{\text{time}(t) = 1}} = 1 + 0.28 \log_{10} t \tag{5.2}$$

where $(\delta_{\text{time}(t)} = 1)$ equals either δ_H or δ_V, as calculated by equation (5.1), and time t is in days.

In view of the considerable variation in the properties of natural soils, and the even greater uncertainties associated with differences in workmanship and construction procedure, no single method of prediction can be expected to anticipate all possible conditions. However, the approach outlined above can be used to obtain 'first-order' predictions of practical value to engineers. The typical range of possible error is probably about $\pm 25\%$ provided that measured values of soil parameters are employed in the equations, and that construction procedure is such as to prevent local failure of the trench wall.

Notation

Symbol	Definition	Unit
D	depth factor	unit
E_s	constrained soil modulus of deformation (obtained from Dutch cone test)	kN/m^2
H	trench depth	m
H_c	critical depth	m
K_a	active earth pressure coefficient	unit
K_{0T}	at-rest earth pressure coefficient (total stress basis) obtained from spade cell test	unit
P	disturbing force	kN

c_u	undrained shear strength of soil	kN/m^2
t	time from trench reinstatement	days
x	horizontal distance from trench wall	m
z	depth below ground surface	m
z_c	critical depth of tension crack	m
β	slope angle	degrees
γ	bulk unit weight of soil	kN/m^3
δ_H	horizontal ground movement	m
δ_V	vertical ground movement	m
$\delta_{time\ (t)}$	ground movement at time t	m
ϕ	friction angle	degrees
τ	shear strength	kN/m^2

References

Burland, J.B., Simpson, B., and St John, H.D. (1979) Movements around excavations in London clay. *Proc. 7th European Conf. on Soil Mechanics and Foundation Engineering, Brighton*, British Geotechnical Society, London, vol. 1, pp. 13–29.

Carder, D.R., Taylor, M.E. and Pocock, R.G. (1982) *Response of a Pipeline to Ground Movements caused by Trenching in Compressible Alluvium*. Transport and Road Research Laboratory, Digest LR 1047, Crowthorne, Berks., England.

Construction Industry Research and Information Association (1983) *Trenching Practice*. Report 97, CIRIA, London.

Crofts, J.E., Menzies, B.K., and Tarzi, A.I. (1977) Lateral displacement of shallow buried pipelines due to adjacent deep trench excavations. *Géotechnique* 27 (2) 161–179.

Fry, R.H., and Rumsey, P.B. (1982) Prediction and control of ground movement associated with trench excavation. *Water Pollution Control*, 82 (pt. II) 151–63.

Fry, R.H. and Rumsey, P.B. (1983b) Ground Movements caused by Trench Excavation and the Effect of Adjacent Buried Pipelines. WRC External Report No. 100E, March 1983.

Gumbel, J.E. and Wilson, J. (1980) Observations of ground movements around a trench excavation in London clay. In *Ground Movements and Structures*, Proc. 2nd Int. Conf., Cardiff, April 1980, ed. J.D. Geddes, Pentech Press, London pp. 841–856.

Hanna, T.H. (1978) Diaphragm walls, In *Developments in Soil Mechanics* 1, Applied Science, London, pp. 213–249.

Howe, M., Hunter P., and Owen, R.C. (1980) Ground movements caused by deep excavations and tunnels and their effect on adjacent mains. In *Ground Movements and Structures*, Proc. 2nd Int. Conf., Cardiff, April 1980, ed. J.D. Geddes, Pentech Press, London, pp. 812–840.

Kyrou, K. (1980) The effect of trench excavation induced ground movements on adjacent buried pipelines. Ph.D. thesis, University of Surrey.

Mackay, E.B. (1982) *Proprietary Trench Support Systems*. CIRIA Technical Note 95, Construction Industry Research and Information Association, London, 2nd edn.

Peck, R.B. (1969) Deep excavations and tunnelling in soft ground. *Proc. 7th Int. Conf. Soil Mechanics and Foundation Engineering*, State of Art volume, Mexico City, pp. 225–290.

Rumsey, P.B. and Cooper, I. (1982) Ground movements associated with trench excavation and their effect on adjacent services. *Ground Engineering* 15 (2) 28–30.

Rumsey, P.B., Cooper, I., and Kyrou, K (1981) Ground movement and pipe strain associated with trench excavation. *Proc. Int. Conf. on Maintenance, Repair, Renovation and Renewal of Sewerage Systems.* Institution of Civil Engineers, London, pp. 91–101.

Symons, I.F. (1980) Ground movements and their influence on shallow buried pipes. *The Public Health Engineer* **8** (4) 140–153.

Symons, I.F., Chard, B.M., and Carder, D.R. (1981) Ground movements caused by deep trench construction. *Proc. Int. Conf. on Maintenance, Repair, Renovation and Renewal of Sewerage Systems.* Institution of Civil Engineers, London, pp. 73–82.

Taylor, D.W. (1937) Stability of earth slopes. *J. Boston Society of Civil Engineers* **24**, 197–246.

Terzaghi, K., and Peck, R.B. (1967) *Soil Mechanics in Engineering Practice.* John Wiley and Sons, New York.

Toombs, A.F., McCaul, C., and Symons, I.F. (1982) *Ground Movements Caused by Deep Trench Construction in an Urban Area.* Department of the Environment, and Department of Transport, TRRL Report LR1040, Crowthorne, Berks., England.

6 Tunnelling in soil

PETER ATTEWELL and JOHN YEATES

6.1 Introduction

Excavation below ground causes relaxation of *in-situ* stress. Dilation into the opening accompanies de-compression, and is only partially restricted by the eventual insertion of ground support. The volume of ground relaxing into the excavation is usually expressed in terms of unit distance advance of the excavation that causes the relaxation (i.e. metres^3 per metre advance), and is known as 'ground loss'. Since most tunnels in soil, for transportation or sewage purposes, are of circular cross-section, the ground losses can be partitioned into those associated with radial movement and those with 'axial' movements at the tunnel face proper. The magnitude of these movements is a function of soil type, rate of tunnel advance, tunnel size, and form of temporary and primary support. A discussion of the individual factors contributing to ground loss is contained in the paper by Attewell (1978) to which reference should be made. Since prediction of the effect of ground loss tends to be made without detailed reference to these parameters, they are not re-discussed in the present chapter.

Ground disturbance at the tunnel triggers off a chain of movements up to ground surface. The volume of ground moving inwards at the tunnel may be transmitted quickly to the surface, but because it becomes spread over a much greater horizontal area during its passage upwards, the magnitude of the constitutive movement in any given direction attenuates. In cohesive soils there is some evidence to suggest that the whole of the soil volume lost at the tunnel appears as a settlement depression at ground surface. With non-cohesive soils, this may not be so. Some of the volume lost in a dense granular soil may not be transmitted to ground surface but may be absorbed as a permanent loosening of the ground close to the perimeter of the tunnel. Conversely, tunnelling disturbance in a looser granular soil could encourage a redistribution of particle contacts leading to rather denser overburden and a surface settlement depression that could be of somewhat greater volume than the ground loss volume at the tunnel.

Ground movements at the tunnel and consequential surface settlements are much more difficult to predict on the basis of precedent for non-cohesive soils than for cohesive soils. Non-cohesive ground is much more difficult to control at the tunnel, and so the quality of workmanship at the tunnel face tends to be a factor affecting surface settlement much more than is the case with cohesive ground.

Although ground losses at the tunnel are quite quickly reflected as settlement at ground surface, there may be longer-term vertical movements caused by consolidation of the ground above the tunnel. The tunnel itself, or, even if well-sealed, the disturbed (dilated) ground around the tunnel, provides a drainage route as a result of pore-water pressure reduction for any super-adjacent soil water. Consolidation settlement will be additive to ground-loss settlement, but since there are no established methods of prediction, it cannot be considered here in the detail that it merits. Some guidance on consolidation prediction is, however, given subsequently in this chapter.

An important requirement for the design of a tunnel in soil is that its construction should cause as little damage as possible to overlying and nearby buildings and services. Attempts should be made to predict the possible effects of tunnelling on building foundations and buried pipelines. If the predictions of possible ground movements and structural strain suggest that problems could arise, the tunnel designer is faced with several options: re-route the tunnel away from 'sensitive' buildings or buried services (not always possible in an urban area, and it may be found that a re-routing into ground unfavourable for construction could incur a higher cost penalty than maintaining an original route and added cost for structural protection); ground stabilisation (grouting or freezing) combined with a carefully-specified construction programme on the original route; protection of the existing buildings (by some form of underpinning), and the acceptance of some inevitable compensation damage, together with the quite costly replacing of old (usually cast iron) pipes with new polyethylene pipes or the internal lining of the existing brittle pipes with plastic material; and the re-location of buried pipelines (an expensive option but one which cannot be dismissed in the case of high-pressure gas mains).

This chapter seeks to help the tunnel designer assess the effects of construction upon building foundations and buried pipelines. It also provides a means whereby property owners and pipeline operators, having acquired information from the tunnel designer, can perform their own checks, ask the right questions, and ultimately ensure that their property and services continue to function satisfactorily.

6.2 Prediction of ground movements

As the reader will be aware, prediction of settlement caused by the withdrawal of subsurface support is a much more imprecise operation than is the prediction of settlement caused by surface loading. The latter has benefited from analytical examination and field observation over a period of many years. There is as yet no analytical solution for the former. Either assumptions as to ground behaviour allied to empiricism have to be made or numerical solutions have to be sought. Whichever method is adopted, there is a requirement for the tunnel designer to carefully reassess his predictions in the light of his own experience and that of others. He should ask whether they seem sensible, whether the assumptions (for example, of transverse settlement symmetry) are really valid in the field circumstances that apply, and whether he is being over-cautious in his specification of input conditions (experience suggests that there is a tendency to err perhaps excessively on the side of caution when specifying the input parameters for prediction). Of course, a major requirement is for good site investigation information on ground conditions, knowledge of adjacent building foundations, and knowledge of the location, construction and condition of buried pipelines that could be affected by the tunnel excavation.

To aid ground movement prediction, a body of settlement case history data has been published in tabulated form by Attewell (1978) and updated by Tsutsumi (1983). It includes information on tunnel size and depth, maximum settlement, settlement trough width, volume and slope, ground description, geotechnical properties and

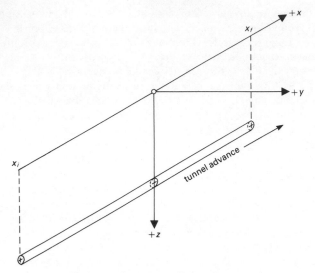

Figure 6.1 Tunnel coordinate system.

method of working. This data table is too lengthy to include in the present chapter, but information can be made available to interested readers on application to the authors.

The following primary notation used in this section of the chapter is listed now for convenience, but a comprehensive symbol list is included at the end of the chapter.

x, y, z are the cartesian coordinates of any point in the ground deformation field. Theoretically, $x = y = 0$ vertically above the tunnel face on the tunnel centre line. Positive $(+)x$ is ahead of the tunnel on the centre line, and horizontal. $\pm y$ is horizontal at right angles to x. Positive $(+)z$ is vertically downwards. This coordinate system is shown in Fig. 6.1.

u, v, w are the ground displacements in the x, y, z directions, respectively. *Increments* $\Delta u, \Delta v,$ are towards the current origin of the cartesian coordinate system. w (settlement) is always positive downwards.

$\varepsilon_x, \varepsilon_y, \varepsilon_z$ are the ground strains in the x, y, z directions, respectively. These strains can change from tensile (positive) to compressive (negative) depending upon position in the deformation field. Tensile ground strains are more likely to have a serious effect upon the brittle foundation of a building or upon a brittle pipe than are compressive ground strains.

γ_{xy} is the ground shear strain in a horizontal plane.

z_0 is the depth of the effective source of ground loss (taken as approximating to the tunnel axis).

R is the excavated radius of a circular cross-section tunnel.

K_R is an empirically-determined parameter.

n is the power of $z_0 - z$ to which i_x, i_y, i are proportional.

V is the volume of the settlement trough per unit distance of tunnel advance, the

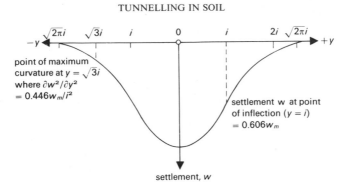

Figure 6.2 Normal probability transverse settlement profile.

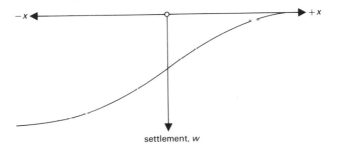

Figure 6.3 Cumulative probability tunnel centre line settlement profile.

settlement being attributable to ground losses and not incorporating any longer-term consolidation movement.

i is a parameter defining the form and span of the settlement trough, on the assumption that the semi-transverse (y axis) settlement profile can be described by a normal probability equation (Schmidt, 1969; Peck, 1969; Attewell, 1978). On the transverse settlement profile (Fig. 6.2), i_y is half the distance between the points of inflexion (greatest slope). On the centre line profile (Fig. 6.3), i_x is the distance from the point of inflexion to the 15.9 per cent (of maximum) settlement point, and twice the distance from the point of inflexion to the 30.9 per cent settlement point. For the numerical examples discussed in the chapter, i_x is taken to be equal to i_y and written simply as i. Case history evidence on i_x and i_y has been appraised by Gordon (1981).

x_i is the initial or tunnel start point ($y = 0$).

x_f is the face or final tunnel position ($y = 0$).

$G(\alpha) =$ $(1/\sqrt{2\pi})\int_{-\alpha}^{\alpha} \exp[-\beta^2/2]d\beta$, and may be determined from, for example, a standard probability table such as Table 6.1. Note particularly that $G(0)$ (e.g. $x - x_f = 0$ directly above the tunnel face) gives a value of $\frac{1}{2}$ and $G(\infty)$ (e.g. $x - x_i \to \infty$ for substantial distances of tunnel advance from the face start position) gives a value of 1. For negative values of α, $G(\alpha) = 1 - G(-\alpha)$.

6.2.1 Analytical/empirical appraisal

This method, using ground deformation and strain equations, starts on the assumption that the transverse settlement profile (yz plane) is of normal probability, or

Table 6.1 Numerical integration of the normal probability curve.

Table of $G[(x - x_f)/i]$

$(x - x_f)/i$	0	1	2	3	4	5	6	7	8	9
0.0	.500	.504	.508	.512	.516	.520	.524	.528	.532	.536
0.1	.540	.544	.548	.552	.556	.560	.564	.567	.571	.575
0.2	.579	.583	.587	.591	.595	.599	.603	.606	.610	.614
0.3	.618	.622	.626	.629	.633	.637	.641	.644	.648	.652
0.4	.655	.659	.663	.666	.670	.674	.677	.681	.684	.688
0.5	.691	.695	.698	.702	.705	.709	.712	.716	.719	.722
0.6	.726	.729	.732	.736	.739	.742	.745	.749	.752	.755
0.7	.758	.761	.764	.767	.770	.773	.776	.779	.782	.785
0.8	.788	.791	.794	.797	.800	.802	.805	.808	.811	.813
0.9	.816	.819	.821	.824	.826	.829	.831	.834	.836	.839
1.0	.841	.844	.846	.848	.851	.853	.855	.858	.860	.862
1.1	.864	.867	.869	.871	.873	.875	.877	.879	.881	.883
1.2	.885	.887	.889	.891	.893	.894	.896	.898	.900	.901
1.3	.903	.905	.907	.908	.910	.911	.913	.915	.916	.918
1.4	.919	.921	.922	.924	.925	.926	.928	.929	.931	.932
1.5	.933	.934	.936	.937	.938	.939	.941	.942	.943	.944
1.6	.945	.946	.947	.948	.949	.951	.952	.953	.954	.954
1.7	.955	.956	.957	.958	.959	.960	.961	.962	.962	.963
1.8	.964	.965	.966	.966	.967	.968	.969	.969	.970	.971
1.9	.971	.972	.973	.973	.974	.974	.975	.976	.976	.977
2.0	.977	.978	.978	.979	.979	.980	.980	.981	.981	.982
2.1	.982	.983	.983	.983	.984	.984	.985	.985	.985	.986
2.2	.986	.986	.987	.987	.987	.988	.988	.988	.989	.989
2.3	.989	.990	.990	.990	.990	.991	.991	.991	.991	.992
2.4	.992	.992	.992	.992	.993	.993	.993	.993	.993	.994
2.5	.994	.994	.994	.994	.994	.995	.995	.995	.995	.995
2.6	.995	.995	.996	.996	.996	.996	.996	.996	.996	.996
2.7	.997	.997	.997	.997	.997	.997	.997	.997	.997	.997
2.8	.997	.998	.998	.998	.998	.998	.998	.998	.998	.998
2.9	.998	.998	.998	.998	.998	.998	.998	.999	.999	.999
3.0	.999	.999	.999	.999	.999	.999	.999	.999	.999	.999

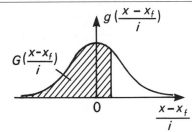

gaussian form. Attempts have been made to match measured profiles with alternative mathematical expressions, but there seems now to be a general consensus of opinion that a normal probability curve match is more appropriate and convenient for predictive purposes. Its application to tunnelling-induced ground settlement is discussed in Schmidt (1969), Attewell (1978), and Attewell and Woodman (1982). Some of the more important elements of this curve (which has been reproduced in several papers by different authors) are shown in Fig. 6.2. These features relate to the surface of a settlement trough but, of course, the real problems, considered later, concern the response of building foundations and buried pipelines within the trough. It is indicated subsequently that the response of in-ground structures to ground movement may not be especially sensitive to the form of ground settlement profile.

The transverse settlement trough represents a terminal state of ground deformation. However, ground movements, settlements and their derivatives are also projected ahead of the tunnel face. These forward movements, although a temporary wave of disturbance as the tunnel face advances, also have an effect on buildings and buried pipes, which are twisted and subjected to torsion in three dimensions.

If the state of ground surface deformation ahead of and to the sides of a tunnel is to be fully predicted, then the form of the settlement profile in an xz plane, parallel to the tunnel centre line must also be specified. A logical extension of the earlier assumption of normal probability form for the transverse (yz) settlement is that the tunnel centre line (xz) profile should be of cumulative probability form (see Fig. 6.3). This assumption has been reasonably validated by examination of several field study reports (*see* Attewell and Woodman, 1982). Preliminary analysis assumes that the origin of the cartesian coordinate system, vertically above the tunnel face, coincides with the 50% of maximum settlement point. Although case history evidence from tunnels in cohesive soil suggest that only 30% to 40% w_{max} will have developed vertically above the tunnel face, this matter is really only academic, since all points at ground surface ahead of and within the boundaries of the transverse settlement trough will at some time realize their ultimate maximum movements for the translating tunnel line source of ground loss.

If the above assumptions as to the form of the ground loss settlement trough apply, then the displacements and strains for any field point, coordinates x, y, z are:

$$w = \frac{V}{\sqrt{2\pi}i} \exp\left[\frac{-y^2}{2i^2}\right] \left\{ G\left(\frac{x - x_i}{i}\right) - G\left(\frac{x - x_f}{i}\right) \right\} \tag{6.1}$$

$$v = \frac{-n}{z_0 - z} yw \tag{6.2}$$

$$u = \frac{nV}{2\pi(z_0 - z)} \exp\left[\frac{-y^2}{2i^2}\right] \left\{ \exp\left[\frac{-(x - x_i)^2}{2i^2}\right] - \exp\left[\frac{-(x - x_f)^2}{2i^2}\right] \right\} \tag{6.3}$$

$$\varepsilon_z = \frac{-nV}{\sqrt{2\pi}i(z_0 - z)} \exp\left[\frac{-y^2}{2i^2}\right] \left\{ \frac{-1}{\sqrt{2\pi}} \left\{ \left(\frac{x - x_i}{i}\right) \right. \right.$$

$$\times \exp\left[\frac{-(x - x_i)^2}{2i^2}\right] - \left(\frac{x - x_f}{i}\right) \exp\left[\frac{-(x - x_f)^2}{2i^2}\right] \right\}$$

$$+ \left(\frac{y^2}{i^2} - 1\right) \left[G\left(\frac{x - x_i}{i}\right) - G\left(\frac{x - x_f}{i}\right) \right] \right\} \tag{6.4}$$

$$\varepsilon_y = \frac{n}{z_0 - z} w \left(\frac{y^2}{i^2} - 1\right) \tag{6.5}$$

$$\varepsilon_x = \frac{-nV}{2\pi i(z_0 - z)} \left\{ \left(\frac{x - x_i}{i}\right) \exp\left[\frac{-(x - x_i)^2}{2i^2}\right] \right.$$

$$- \left(\frac{x - x_f}{i}\right) \exp\left[\frac{-(x - x_f)^2}{2i^2}\right] \right\} \tag{6.6}$$

As noted earlier, reference to Table 6.1 is required for the resolution of the functions $G[(x - x_i)/i]$ and $G[(x - x_f)/i]$. For a tunnel face that has advanced sufficiently to allow the transverse (yz plane) ground loss settlement profile to fully develop, the function $G[(x - x_i)/i]$ can be re-expressed in the above equations as unity. It remains to determine possible input values for the other non-geometric parameters.

Expert advice should be sought on the *volume loss parameter V* for ground movement prediction purposes, but for guidance in different soils the following ranges of values could be appropriate. In firm, stiff and very stiff *cohesive soils*, allow between $\frac{1}{2}\%$ and $2\frac{1}{2}\%$ of the tunnel face excavated area, depending upon the stiffness of the soil and the speed (Attewell, 1978, p. 849) at which the initial support is installed. Prediction may be attempted on the basis of an overload factor (see Fig. 6.4) where the vertical stress at the tunnel axis level is usually taken as the driving pressure for ground loss (the multiple of tunnel depth-to-axis (z_0) and soil unit weight (γ)) and is resisted by the undrained shear strength of the soil (c_u) at the tunnel face. If a building imposes a distributed surface surcharge pressure, then this pressure should be added to the γz_0 term. Internal support (σ_i), usually in the form of compressed air but increasingly with the use of newer technology closed shields in the form of pressurized slurry, obviously resists ground loss at the tunnel. Glossop (1977) has suggested the expression

$$V_s\% = 1.33 \text{ (Simple Overload Factor)} - 1.4,$$

for $1.5 \leq$ Simple Overload Factor (OFS) ≤ 4.

In *non-cohesive soils* (but excluding running or ravelling ground) consider a wide

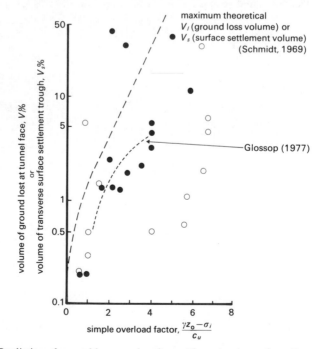

Figure 6.4 Prediction of ground losses and surface settlement volumes from the overload factor for tunnels in cohesive soils. Open circles are Schmidt's (1969) data from shield-driven tunnels. Solid circles are some of present authors' data.

range of between, say, 3% and 10% with a general value of about 5% for V. Decisions as to trial values would be made on the basis of site investigation evidence, and it should be recognized that for very loose soil the 10% value could be exceeded. In structurally-sensitive urban areas the potential ground loss even in loose non-cohesive soil can be reduced by adopting in a segmentally-lined tunnel a build-one-ring/grout-one-ring construction regime.

Man-filled ground must be tunnelled in certain areas and estimation of ground loss in such cases is made difficult by the usually quite variable composition and compaction of the fill. Ground losses of about about 17% have been estimated for a recent household/industrial waste fill in the north of England (Dobson et al., 1979). For guidance purposes, a lower value of about 8% should probably be used for an old fill comprising natural ground material, a value of 10% to 12% for an old, established industrial fill, and a figure of 15% for a recent loose industrial or household waste fill.

More complex expressions for the vertical pressure which drives the ground loss at the tunnel may be found in Széchy (1973). These expressions, attributable to different authors, accommodate c-ϕ soils, surface surcharge pressures, and arching relevant particularly to frictional soils.

Attempts have been made to predict the *settlement trough width parameter i* empirically from case history evidence of settlement through measurements in different types of ground. Whereas the ground loss depends upon the shear strength of the soil at tunnel level, the settlement trough dimensions will be related more closely to the soil strength nearer to ground surface. The equation that has been used, $i = RK_R[(z_0 - z)/2R]^n$, has been discussed by Attewell and Woodman (1982) in the context of a stochastic model and has been shown to be unsatisfactory. Although it would be expected that i should depend to some degree on the excavated radius, O'Reilly and New (1982) have been unable to detect any such dependence in their survey of U.K. tunnel settlements. It is probably advisable (certainly in the case of cohesive clay soils) to assume that both K_R and n tend to unity, which suggests that i approximates to half the tunnel axis depth. Alternatively, O'Reilly and New have suggested, from case history reviews, that the following relations apply for U.K. tunnels having adequate ground settlement records:

$$i = 0.43(z_0 - z) + 1.1 \text{ metres for cohesive soils } (3 \leq z_0 \leq 34),$$

$$i = 0.28(z_0 - z) - 0.1 \text{ metres for granular soils } (6 \leq z_0 \leq 10).$$

These relations, and particularly the latter for granular soils, are not dissimilar to those derived by Atkinson and Potts (1976) from their laboratory model experiments. Reference may also be made to the graphs in Attewell (1978, pp. 884, 885).

Rather than accepting reported values of i (almost invariably i_y) for the development of the above empirical relations, it is preferable to plot the logarithm of the original recorded settlements (log w) against the square of the transverse distance (y^2). A regression line would be drawn through the points. Maximum settlement (w_m) would then be defined by the intercept of the regression line with the axis $y^2 = 0$. The value of i would then be defined by the fact that i^2 is the value of y^2 where $w/w_m = 0.606$.

It is noted that under the normal probability profile assumptions, the width of the fully-developed transverse settlement trough is infinite, but in practice can be taken as $2\sqrt{2\pi}i$ (i.e. 5i). The parallel-to-centre line trough extends a practical distance of 1.7i ahead of and 1.7i behind the tunnel face if the latter is located directly beneath the 50%

maximum surface settlement point. The full form of the settlement distribution is shown graphically later in Fig. 6.7.

Example 1: Settlement calculation. Consider a case history example (Attewell and Farmer, 1973) used in Attewell and Woodman (1982).

> Tunnel axis depth, $z_0 - z = 7.5$ m.
> Settlement trough width parameter, $i = 3.9$ m.
> Maximum surface settlement, w_m (at $y = 0$) = 7.86 mm.
> Ground loss volume, $V = 0.077$ m^3/m (see equation 6.1).
> Let the tunnel face be well advanced from the start point.
> Then $(x - x_i) \to \infty$ and $G\left(\dfrac{x - x_i}{i}\right) = 1$.
> Take a point on ground surface having coordinates $(x - x_f) = 4$ m, $y = 1.5$ m.
> Then $G\left(\dfrac{x - x_f}{i}\right) = G(1.02) = 0.846$ (from Table 6.1).
> It follows from equation 6.1 that the ground settlement at the specified point is 1.13 mm.
> Calculation on equations 6.2 and 6.3 produces values of -0.23 mm for v and -0.90 mm for u. (The negative signs denote displacements horizontally inwards towards the centre of the trough.) A check summation of the three displacements having regard to sign confirms the no-volume-change assumption underlying the analysis.
> Further calculation produces ground strain values of $-110\,\mu\varepsilon$ (compressive), $-130\,\mu\varepsilon$ (compressive) and $+250\,\mu\varepsilon$ (tensile) for $\varepsilon_z, \varepsilon_y$, and ε_x, respectively. Probably because of rounding errors these values do not quite sum to zero in order to satisfy the no-volume-change assumption.

With respect to buildings and in-ground structures, information on ground surface settlement and strain is only part of the requirement. The horizontal u, v ground displacements impose direct strains on a foundation or buried pipe, the upper bound structural strains being equated to ε_x, ε_y by neglecting ground-structure differential stiffness and assuming no ground material shearing. Building foundations, however, settle differentially and incur angular distortions, the magnitudes of which depend on the position of the foundation in the settlement trough (see Attewell, 1978). Strains associated with angular distortions will be additive to the direct lateral distortional strains. It will also be realized that a foundation will be subjected to a wave of torsion before the advance of the tunnel face removes the effects of the forward settlement trough. Similarly, buried pipes within a ground settlement trough will experience direct strain, bending strain (related to ground curvature) and torsion. Reference on these matters may be made to Attewell and Woodman (1982), but they are probably more easily accommodated by use of design curve graphs. There will also be more detailed discussion later in this chapter.

6.2.2 *Design curve graphs*

Use of equations probably implies a degree of precision in prediction that certainly does not exist. The approximate character of the assessment is probably better reflected by expressing the equations graphically in a way that a foundation or buried pipe analysis can be conducted more visibly and perhaps more rapidly.

Figures 6.5 to 6.10 inclusive were constructed originally by desk computer and graph plotter from adaptations of equations 6.1 to 6.6 above. The coordinates for the curves

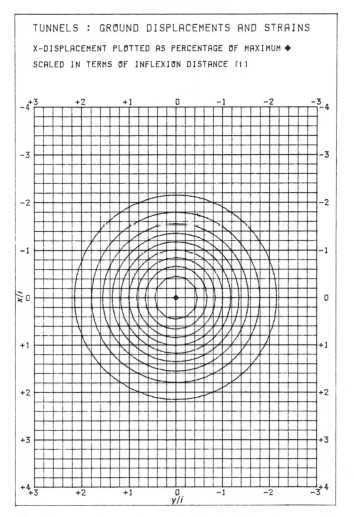

Figure 6.5 10% incremental ground displacement (u) curves, x-axis.

are scaled as x/i (down the graph) and y/i (across the graph), and the positive z-axis is directed downwards into the plane of the paper. On the graphs, the tunnel face is theoretically positioned at the origin. The curves assume that the tunnel is of infinite length (in practice, the distance of advance must be large compared to $(z_0 - z)$) and that the increments of horizontal ground displacement are radial (always towards the current tunnel face origin). Each curve on a particular graph represents 10% of the parameter absolute maximum, each maximum being marked.

The curves show a number of interesting features. For example, the settlement, or z-displacement (w), curves (Fig. 6.7) profile the settlement distribution about the source $(x/i = y/i = 0)$ and show that the transverse ground loss settlement profile becomes fully-developed for all practical purposes when the tunnel face has advanced about $2i$ beyond the profile section. Figure 6.8 shows that the tensile (positive) ε_x regime ahead of

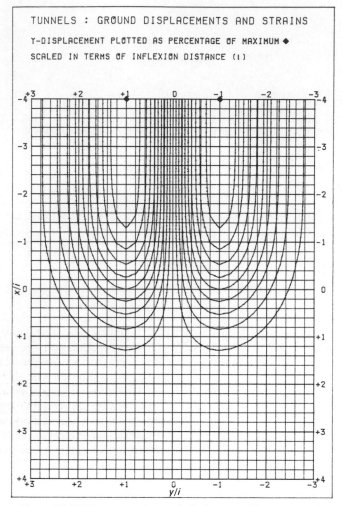

Figure 6.6 10% incremental ground displacement (v) curves, y-axis.

the tunnel face is a mirror image of the compressive (negative) ε_x regime behind the face. Figure 6.9 shows that the transverse ε_y has a compressive maximum above the tunnel centre line which decreases to zero with increasing y to a tensile ε_y maximum that is between 40% and 50% (theoretically 44.6%) of the absolute maximum. The relatively complex pattern of ε_z can be interpreted by remembering that $\varepsilon_z = -(\varepsilon_x + \varepsilon_y)$ as per the underlying assumption of no-volume-change. There is a tensile maximum on the tunnel centre line and behind the face, which decays with radial distance to zero strain and thence to compression. However, vertical strain distribution is of marginal practical importance in the present work, and is not considered further.

Because of restrictions on space, only the six graphs are provided. Although this may be inadequate for some practical purposes there are simple relations and equivalences between parameters which allow the use of these graphs to be extended. For greater

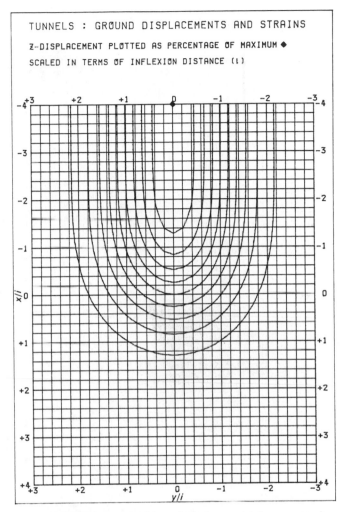

Figure 6.7 10% incremental ground displacement (w) curves, z-axis.

completeness, the i-subscript (x or y) is given, although in most cases an unsubscripted i-parameter ($i_x = i_y = i$) must be used.

(a) γ_{xy} graph $\equiv \varepsilon_x$ graph rotated about the z-axis through $\pi/2$.

(b) $x - z$ plane gradient $\dfrac{\partial w}{\partial x} = \dfrac{z_0 - z}{ni_x^2} u \sim u.$

(c) $y - z$ plane gradient $\dfrac{\partial w}{\partial y} = \dfrac{z_0 - z}{ni_y^2} v \sim v.$

(d) $\dfrac{\partial^2 w}{\partial x^2} = \dfrac{z_0 - z}{ni_x^2} \varepsilon_x \sim \varepsilon_x,$ $\dfrac{\partial^2 w}{\partial y^2} = \dfrac{z_0 - z}{ni_y^2} \varepsilon_y \sim \varepsilon_y,$ $\dfrac{\partial^2 w}{\partial x \partial y} = \dfrac{z_0 - z}{n} \dfrac{\gamma_{xy}}{i_x^2 + i_y^2} \sim \gamma_{xy}.$

F

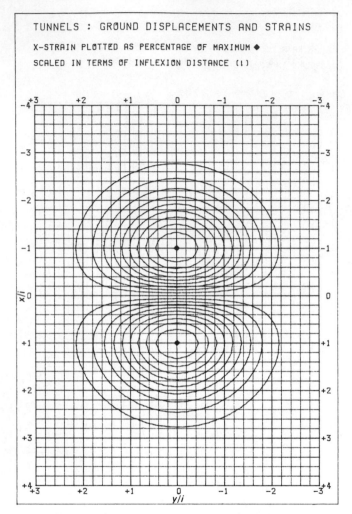

Figure 6.8 10% incremental ground strain (ε_x)-curves, x-axis.

(These second derivatives of w are approximations to the components of curvature of the settled surface, given small displacement and gradient, and to the components of bending strain referred to the initial pre-settlement horizontal plane, under conditions of small strain and small gradient.)

Sufficient parameters have probably now been covered for building foundations, but not for buried pipes. In particular, horizontal pipes also need the following:

(e)
$$\frac{\partial u}{\partial y} = \frac{i_x^2 \gamma_{xy}}{i_x^2 + i_y^2} \sim \gamma_{xy}, \quad \frac{\partial v}{\partial x} = \frac{i_y^2 \gamma_{xy}}{i_x^2 + i_y^2} \sim \gamma_{xy}.$$

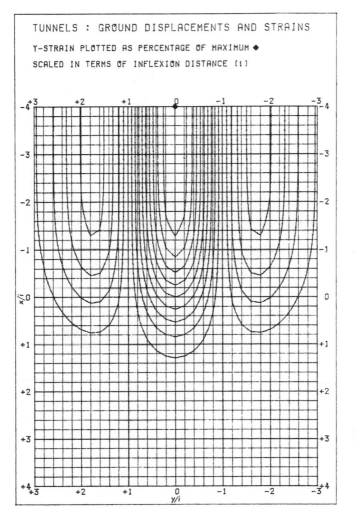

TUNNELS : GRØUND DISPLACEMENTS AND STRAINS

Y-STRAIN PLØTTED AS PERCENTAGE ØF MAXIMUM ◆

SCALED IN TERMS ØF INFLEXIØN DISTANCE (ι)

Figure 6.9 10% incremental ground strain (ε_y) curves, y-axis.

(f)
$$\frac{\partial^2 u}{\partial x^2} = \left(\frac{x^2}{i_x^2} - 1\right)\frac{u}{i_x^2}, \quad \frac{\partial^2 u}{\partial y^2} = \left(\frac{y^2}{i_y^2} - 1\right)\frac{u}{i_y^2}, \quad \frac{\partial^2 u}{\partial x \partial y} = -\frac{y}{i_y^2}\varepsilon_x ;$$

$$\frac{\partial^2 v}{\partial x^2} = -\frac{y}{i_x^2}\varepsilon_x, \quad \frac{\partial^2 v}{\partial y^2} = \left(\frac{y^2}{i_y^2} - 3\right)\frac{v}{i_y^2}, \quad \frac{\partial^2 v}{\partial x \partial y} = \left(\frac{y^2}{i_y^2} - 1\right)\frac{u}{i_x^2}.$$

It is noted in Reeves *et al.* (1983) that there is an argument for graphically displaying gradient magnitude $\sqrt{(\partial w/\partial x)^2 + (\partial w/\partial y)^2}$ and principal lateral strains/bending strains of the horizontal plane, but these curves depend on the ratio i_x/i_y, unlike contours for the above parameters.

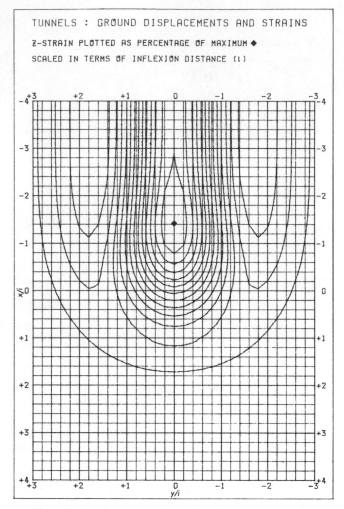

Figure 6.10 10% incremental ground strain (ε_z) curves, z-axis.

Provided that values are available for the essential input parameters, the percentages of the reference values for each displacement and strain parameter are then assessed by interpolation between curves on the graphs. Numerical values for the displacement and strain parameters are calculated by defining the reference displacements and strains corresponding to *absolute maxima* defined as follows:

$$w_m = \frac{1}{\sqrt{2\pi}} \frac{V}{i_y} \quad \text{at} \, (-\infty, 0) \tag{6.7}$$

$$v_m = -w_m \frac{ni_y}{z_0 - z} \exp[-\tfrac{1}{2}] \quad \text{at} \, (-\infty, 1) \tag{6.8}$$

$$u_m = -w_m \frac{1}{\sqrt{2\pi}\, z_0} \frac{ni_x}{z} \quad \text{at } (0,0) \tag{6.9}$$

$$\varepsilon_{zm} = w_m \frac{n}{z_0 - z} \left[\frac{\exp(-1)}{\sqrt{\pi}} + \frac{1 + \mathrm{erf}(1)}{2} \right] \quad \text{at } (-\sqrt{2}, 0) \tag{6.10}$$

$$\varepsilon_{ym} = -w_m \frac{n}{z_0 - z} \quad \text{at } (-\infty, 0) \tag{6.11}$$

$$\varepsilon_{xm} = w_m \frac{n}{z_0 - z} \frac{\exp(-\frac{1}{2})}{\sqrt{2\pi}} \quad \text{at } (1,0) \tag{6.12}$$

$$\gamma_{xym} = \varepsilon_{xm} \frac{i_x^2 + i_y^2}{i_x i_y} \quad \text{at } (0,1) \tag{6.13}$$

Input parameters $i_x, i_y, z_0 - z$ will often be specified in *metres* and V in *metres*[3] *per metre*. In that case the reference displacements expressed in *millimetres* are:

$$w_m = 339\frac{V}{i_y}, \quad v_m = -0.607 w_m \frac{ni_y}{z_0 - z}, \quad u_m = -0.399 w_m \frac{ni_x}{z_0 - z}.$$

Similarly, if w_m is in *millimetres*, then the reference strains in *microstrains* are:

$$\varepsilon_{zm} = 1.129 \varepsilon_{ym}, \quad \varepsilon_{ym} = -1000 w_m \frac{n}{z_0 - z}, \quad \varepsilon_{xm} = 0.242 \varepsilon_{ym},$$

$$\gamma_{xym} = \varepsilon_{xm} \frac{i_x^2 + i_y^2}{i_x i_y}.$$

6.2.3 *Ground movements and structures—use of design curves for preliminary assessment*

Two matters must be emphasized initially. First, the ground-structure interaction is one of ground shear against the (assumed) stiffer contact surface of the foundation or buried pipe structure combined with an induced bending, the latter simple or torsional. It is the ground displacements, not the strains, that are important although, as stated earlier, the calculated ground strains provide upper-bound values for the structural strains and so are used here. Second, assumed limiting levels of structural deformation must be known, and these will be specified in terms of the type of structure at risk (Attewell, 1978; Norgrove *et al.*, 1979). Although criteria suggested by Burland and Wroth (1975), and based on Skempton and Macdonald (1956) and others (see section 6.6 in this chapter), have been proposed for tunnelling-induced ground deformation, caution is required. Self-weight settlements and deformations develop long-term and much of their potential effect can be accommodated during the actual construction. Tunnelling-induced movements are imposed on a structure very quickly, and so the self-weight damage criteria will have no in-built conservatism with respect to dynamic movements.

6.2.3.1 *Application of design curves—use of overlays.* When a structure at risk has been identified, its foundation (in the case of a building) should be dimensionally scaled on a

transparent overlay and then superimposed on each of the design-curve graphs in turn. The distribution of the 10% *design curves* on the area of the structure is then noted and interpreted with respect to an adopted (usually empirical) damage criterion.

If the tunnel passes beneath a heavily built-up area it is recommended that the relevant portion of the national or state survey plan be photographed with national grid or other orientation feature marked. The scale of the plan would be noted and a further calculation performed to scale its dimensions by the factor i for further printing. Finally, a design-curve overlay transparency would be prepared from the print. Suppose that the scale of the survey plan is $1:S$. The design curve graphs, as originally drawn to scale by the microcomputer, reproduce a linear dimension scaled to i in units of $g = 25$ mm linear measure. If p is a dimension of a building measured (say in mm) on the plan, then the corresponding required length of the building on the photographic print and overlay transparency is gpS/i. It is noted that this procedure is valid only for foundations in the same horizontal plane.

Example 2(a): Building foundation assessment. A sewer tunnel, having an excavated radius 1.105 m, will be driven in a mainly granular soil at an axis depth of 10.5 m. The ground loss is estimated to be 5% (i.e. 0.192 m³/m) and it is assumed that this loss is fully transferred to ground surface as settlement. A building, having a raft foundation of rectangular area at a mean depth of 0.5 m, is located with respect to the tunnel centre line as shown in Fig. 6.11. Prediction of the effects on the building could proceed as follows:

(a) $\overline{z_0 - z} = 10.5\,\text{m} - 0.5\,\text{m} = 10\,\text{m}$.
(b) Let $i = 0.28(z_0 - z) - 0.1$ metres, after O'Reilly and New (1982). Thus $i = 2.7$ m.
    Let the n parameter be unity.

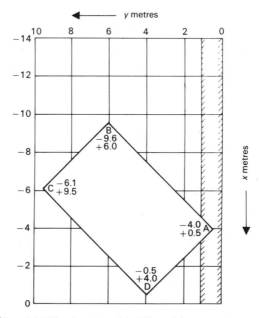

Figure 6.11 Plan location of building with respect to tunnel.

(Although the above equation is derived from U.K. tunnelling case histories, a check against a batch of international case histories suggests that it might be more generally relevant.)

(c) The scaled coordinates of the foundation are:

$x(m)$	x/i	$y(m)$	y/i
-4.0	-1.481	$+0.5$	$+0.185$
-9.6	-3.555	$+6.0$	$+2.222$
-6.1	-2.259	$+0.5$	$+3.518$
-0.5	-0.185	$+4.0$	$+1.481$

(d) The scaled foundation is redrawn in Fig. 6.12 for superimposition on the design curve graphs.

(e) *Lateral distortion and tensile failure*: as the tunnel face advances most of the foundation will eventually be subjected to virtually the whole deformation field on one side of the tunnel centre line. Tensile effects are most forcefully impressed on the foundation in two main regions: $x > 0$, $y^2 < i_y^2$ and $x < 0$, $y^2 > i_y^2$. However, if $i_x = i_y$, wherever $\varepsilon_x \geq \varepsilon_y$ the more tensile principal strain in the horizontal plane is less than the overall maximum tensile ε_y, $\varepsilon_{ytm} = -0.446\,\varepsilon_{ym}$. The region ahead of the tunnel face $x > 0$, $y^2 < i_y^2$ is thus less dangerous than

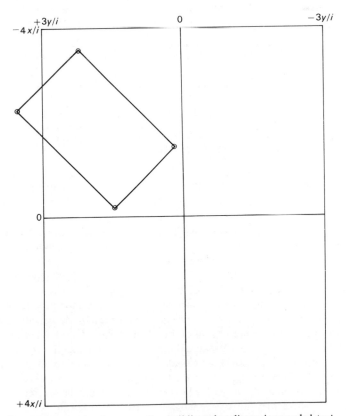

Figure 6.12 Overlay incorporating building plan dimensions scaled to i.

the region of permanent deformation $x < 0$, $y^2 > i_y^2$ transverse to the tunnel centre line. In this present examples, $\varepsilon_{ytm} \simeq 0.127\%$.

A factor of some importance is the area of the foundation that is required to sustain a large tensile strain field. Superimposing Fig. 6.12 on Fig. 6.9 shows that high permanent tensile strains would be experienced over a large area of the foundation, centred approximately on an x-axis just to the tunnel side of corner B. From Fig. 6.6, $-90\%v_m \simeq 0.0065$ m displacement occurs over a horizontal distance of about 4.86 m, suggesting horizontal (y) distortion (see Fig. 6.51) of approximately 0.096%. At a superficial damage threshold of 0.05% tensile strain the foundation would be vulnerable, even if a certain amount of interfacial slippage is allowed for.

(f) *Mid-planar bending* can be analysed in a manner similar to horizontal strain. However, whereas in the case of horizontal strain, it is the tensile directional component that is important in a brittle material foundation, with bending the positive (convex up) and negative (concave up) components are of equal concern. Noting again the equivalence between ground curvature and direct horizontal strain curves, and superimposing Fig. 6.12 overlay on Fig. 6.9, it is seen that the more negative principal bending strain at corner A exceeds

$$80\%\left(\frac{\partial^2 w}{\partial y^2}\right)_m = 80\%\frac{w_m}{i^2} \simeq 0.00389 \text{ m}^{-1},$$

and will eventually exceed $90\%(\partial^2 w/\partial y^2)_m$ as the tunnel face advances a short distance further. This bending strain is clearly much more significant than the smaller, but near maximum and stabilized, positive principal bending strain already present at corner B.

(g) If i_x is taken equal to i_y, the maximum ground slope, similar to the maximum horizontal displacement, is bounded absolutely in any region by the values

$$\left(\frac{\partial w}{\partial x}\right)_m = \frac{z_0 - z}{ni_x^2}u_m \simeq 0.42\% \quad \text{and} \quad \left(\frac{\partial w}{\partial y}\right)_m = \frac{z_0 - z}{ni_y^2}v_m \simeq 0.64\%.$$

As with tensile failure, the average value may be more significant. From Fig. 6.7 approximately 80% w_m ($\simeq 0.0227$ m) is developed over a distance of about 5m, giving a y-direction *angular distortion* (see Fig. 6.50) of approximately 0.454% (1 in 220). Threshold values of angular distortion depend upon the actual construction of the building (Attewell, 1978; Norgrove et al., 1979; see also Table 6.11). However, this predicted ground slope is sufficiently serve to create some concern for a building where the foundation is likely to follow the slope profile.

Example 2(b): *Buried pipeline assessment.* The initial assessment of induced longitudinal pipe strain generally follows the above lines, predicted bending strains being additive to direct axial strains (the latter being the only strains experienced by the pipe neutral axis). For superimposition on the design curve graphs the pipe can usually be represented by a line, since its scaled (to i) diameter will be small. As earlier, it is assumed that the pipe deforms conformably with the predicted ground deformation that would be experienced without the presence of the pipe, the design curves effectively representing upper bound predictions of pipe deformation. Any slippage at or near the pipe-soil interface implies that the actual direct and bending strains in the pipe might well be substantially less than those calculated on the basis of no interfacial slippage. Furthermore, the concept of an 'interface' may in many cases be invalid, since old cast iron pipes (for which assessments are likely to be made) may be severely corroded on the outside and so may have mechanical properties, over the zone of corrosion, intermediate between those of the uncorroded pipe material and those of the pipe bedding. Depending upon the type of pipe joint and its age, there may be a facility for rotation and direct translation at the joints, so absorbing some of the potential strain in the pipe material.

For the example calculation we conservatively assume locked joints and an upper-bound strain condition. Consider a 2ft (0.61 m) diameter pipeline buried at a depth of 1.5 m and lying parallel to—but 3.13 m from—the centre line of a tunnel, axis depth 13.5 m. The ground volume loss at the tunnel is estimated at 2.5%, or 0.103 m³/m, and the i-value is 6.26 m. It is assumed that n is close to unity.

An overlay incorporating a line distant $y/i = 3.13/6.26 = 0.5$ from and parallel to the tunnel centre line ($y/i = 0$) may be prepared (Fig. 6.13) but this is really unnecessary since for this simple structural configuration all the information can be readily drawn by direct inspection of the design curve graphs. From Fig. 6.8 the maximum tensile strain along the line of the pipe $|y/i| = 0.5$ is 85% ε_{xm}, approximately 110 $\mu\varepsilon$. Maximum vertical and horizontal axial bending strains occur at the same point, since $\partial^2 w/\partial x^2 \sim \varepsilon_x$ and $\partial^2 v/\partial x^2 = -(y/i^2)\varepsilon_x$. Consequently, maximum vertical axial bending $= 85\%$ $(\partial^2 w/\partial x^2)_m \simeq 32\ \mu\varepsilon/$m and maximum horizontal axial bending is approximately 9 $\mu\varepsilon/$m. The maximum resultant bending is approximately $\sqrt{32^2 + 9^2} \simeq 33\ \mu\varepsilon/$m. Thus,

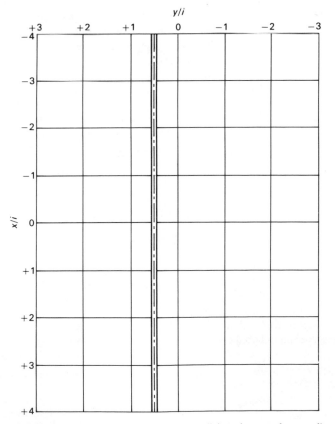

Figure 6.13 Example of pipeline lying parallel to the tunnel centre line.

assuming constant shear in a pipe of outside diameter d, the maximum axial tensile strain is approximately $\varepsilon_x + (d/2)[(\partial^2 w/\partial x^2) + (\partial^2 v/\partial x^2)] \simeq 120$ microstrain.

Analyses of the above type would be applied to old pipes of traditional brittle material construction (cast iron or clay). Take, for example, an old pit-cast grey iron pipe of pre-BS 78 vintage, corroded on the outside, and unable to translate or rotate at the joints. Subject to all the assumptions discussed earlier, all of the potential strain calculated above would be experienced by the pipe material. With an allowable pipe stress of 0.25 times the material ultimate tensile strength, an allowable strain of 400 $\mu\varepsilon$ would probably be applicable. It follows that the predicted upper-bound axial strain in the example pipe is less than $\frac{1}{3}$ of its allowable strain. Had the predicted strains been much closer to the allowable strain, a criterion of failure based on the biaxial state of strain would have been applied, using the fact that for all pipe orientations the mean transverse strain is everywhere equal, but of opposite sign, to the axial strain under the assumption of no volume change.

No information can be derived from the theory of elasticity if the material has passed beyond its elastic limit at any point. When permanent deformations occur, the material is said to have 'failed'. Failure does *not* imply rupture. As a guide to design, for brittle materials such as grey cast iron the Rankine maximum principle stress theory should be used. According to this theory, failure will occur when the maximum principal stress in the complex system reaches the value of the maximum stress at failure in simple tension. For ductile materials, two failure criteria could be used most conveniently: the Tresca maximum shear stress theory and the Huber–Hencky–von Mises distortional energy criterion. Examples of the application of these theories are not given in this chapter, but reference may be made to any standard textbooks on material mechanics.

Many of the major problems with underground services occur in urban areas when tunnel alignments and buried water and gas distribution mains are parallel to main roads—the example considered above. However, if a horizontal buried pipe is at an angle θ with the tunnel advance direction (x-axis)—defined by direction cosines $l, m, 0$—the following expressions apply for point values and mean values over a straight-line segment of length L between points 1 and 2:

point axial strain $= \varepsilon_x l^2 + \gamma_{xy} lm + \varepsilon_y m^2$,

mean axial strain $= [(u_2 - u_1)l + (v_2 - v_1)m]/L$;

point vertical axial bending $= \dfrac{\partial^2 w}{\partial x^2} l^2 + 2\dfrac{\partial^2 w}{\partial x \partial y} lm + \dfrac{\partial^2 w}{\partial y^2} m^2$,

mean vertical axial bending

$$= \left[\left\{\left(\frac{\partial w}{\partial x}\right)_2 - \left(\frac{\partial w}{\partial x}\right)_1\right\}l + \left\{\left(\frac{\partial w}{\partial y}\right)_2 - \left(\frac{\partial w}{\partial y}\right)_1\right\}m\right]\bigg/ L;$$

point horizontal axial bending

$$= \frac{\partial^2 v}{\partial x^2} l^3 + \left(2\frac{\partial^2 v}{\partial x \partial y} - \frac{\partial^2 u}{\partial x^2}\right)l^2 m - \left(2\frac{\partial^2 u}{\partial x \partial y} - \frac{\partial^2 v}{\partial y^2}\right)lm^2 - \frac{\partial^2 u}{\partial y^2} m^3,$$

mean horizontal axial bending

$$= [(\varepsilon_{y2} - \varepsilon_{y1} - \varepsilon_{x2} + \varepsilon_{x1})\sin 2\theta + (\gamma_{xy2} - \gamma_{xy1})\cos 2\theta]/2L.$$

It is re-stressed that this section in general, and the above equations in particular, represent what are, in effect, upper-bound solutions which assume that even a brittle pipe will follow the contours of the settlement trough. For specific solution of the buried pipe problem the attention of the reader is directed to section 6.4 of this chapter.

6.2.4 Numerical methods

The finite element method of analysis has direct application to the problem of tunnelling ground movements and the effect of those movements on structures. As shown earlier, any foundation or buried pipe within the zone of influence of the tunnel excavation will be subjected to a cycle of deformation. Accordingly, any finite element modelling must take account of the 3-dimensional character of the movements. It is not satisfactory to model two-dimensionally on a plane-strain basis solely for the permanent transverse deformations.

Finite element modelling cannot be considered in any detail in this chapter. Three-dimensional f.e. programs are available commercially (e.g. PAFEC, in the U.K.) and many have been purchased or rented by organizations such as universities for use on suitably powerful mainframe computers. Civil engineering consultants often have their own in-house facility. There will be varying degrees of refinement on offer—mesh generation, plasticity, viscoelasticity and so on. It will be realized, however, that these programs are non-specific, and further programming may be needed to apply them to a particular ground engineering problem. For this reason it is often desirable to develop a program for the special problem of tunnelling ground movements and ground-structure interaction. In outlining the application of such a program, some of the problems and some of the methods of avoiding them are mentioned.

The relative dimensions of the structures in the ground create difficulties. A typical tunnel may be 2.5 m excavated diameter at an axis depth of, say, 13.5 m. Pipes likely to be affected by ground movements could be 0.5 m diameter at a depth of 1.5 m, but with a wall thickness of only a few millimetres. In most cases it is impractical to mesh the pipe(s) wall(s) in continuity with and as part of, the general mesh which also incorporates the tunnel. The same restriction would apply to column footings.

Two methods, each with some disadvantages, have been used in an attempt to overcome this restriction. First, and in the case of buried pipes, the finite element mesh in the vicinity of a pipe or pipes has been constructed so that a suite of nodal points creates a definable boundary line around the pipe(s) at a distance of between 2.5 and 3 pipe diameters from it (them). The pipe(s) is (are) then removed from the mesh, the zone vacated by the pipe(s) and surrounding soil re-meshed to a scale compatible with that for the rest of the ground, and the program run for the tunnel in a pipe-less (and foundation-less) ground. Note is made of the orthogonal displacements at the nodes forming the above boundary. The limited zone(s) of ground around the pipe(s) is (are) then expanded to a scale suitable for meshing the ground and the thickness of the pipe wall cross-section. The boundary node displacements for this limited zone(s), as previously determined in the absence of the pipes, are then applied. Output data with respect to total strain are then retrievable for the pipe(s) with respect to position in advance of or behind the tunnel face. By noting the relative amplitude of strain, at pipe soffit and pipe invert, along the pipe length, it is then possible to resolve the bending components and direct components of strain which compose the total pipe strain at the different positions along the pipe, although clearly it is the total applied strain that will form the basis of any criticality assessment.

The second method again involves isolating a volume or volumes of ground around a pipe(s), but this time defining the movements u, v, w at the selected boundary nodes of the finite element mesh by the use of equations 6.1, 6.2, 6.3 in section 6.2 of this chapter. A computer program written in BASIC for an Exidy Sorcerer microcomputer has been used by the present authors for this purpose. These displacements are then entered, as above, into the 3-D f.e. program for solution of the pipe strains.

It follows that the two-stage finite element approach, and/or the analytical-numerical hybrid approach, can be adopted for the solution of isolated foundation strains.

6.3 Consolidation settlement

Having predicted ground losses at the tunnel, for clay soils perhaps on the basis of an overload factor, and then having decided on the degree to which those losses are transferred to form a settlement volume, there is still a need to predict additional consolidation effects. Although consolidation begins with settlement trough formation, it is its superposition on the shorter-term transverse settlement profile formed immediately after passage of the tunnel face that requires investigation. Deepening and widening of a ground loss settlement trough can obviously change the response of a building or in-ground structure over the longer term of consolidation.

Two of the mechanisms promoting consolidation of the ground above a tunnel have been noted earlier. There is the phenomenon of direct gravitational drainage, under transient drawdown conditions, into the excavated void. Drainage may continue until the tunnel is sealed (lined, caulked and contact grouted). There is also the reduction of pore-water pressure effect in the soil around the tunnel. Soil dilates into the tunnel void until restrained by the lining, and so encourages pore-water migration towards the disturbed ground.

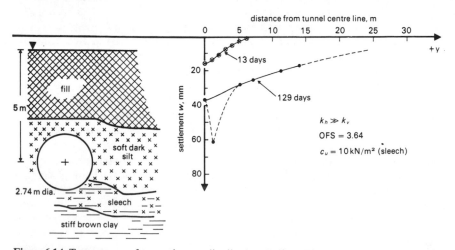

Figure 6.14 Transverse surface settlement distribution: Belfast, King George VI playing fields (after Glossop and Farmer, 1977, 1979).

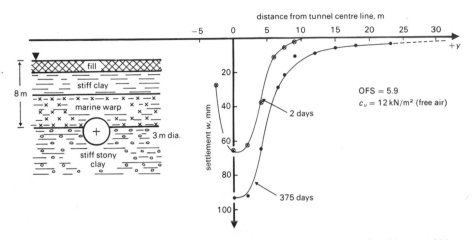

Figure 6.15 Transverse surface settlement distribution: Grimsby Array A (after Glossop, 1980; Glossop and O'Reilly, 1982).

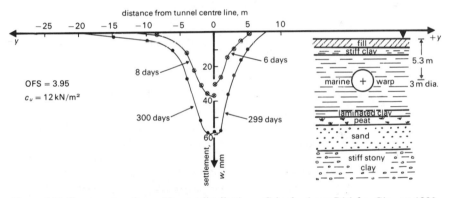

Figure 6.16 Transverse surface settlement distributions: Grimsby Array B1 (after Glossop, 1980; Glossop and O'Reilly, 1982).

There is little authoritative guidance that can be given on this important subject at the present time, since case history evidence is sparse. Some general comments may, however, be helpful for design purposes.

6.3.1 *Total maximum settlement prediction from simple overload factor*

Hurrell (1983) has examined several case histories and, for cohesive soils below the water table, has *tentatively* suggested an empirical relation for the total (ground loss plus longer-term consolidation) maximum settlement w_{mt} (mm) above the tunnel centre line:

$$w_{mt} = (2w_m)A. \text{ OFS} \tag{6.14}$$

where

w_m is the maximum ground loss settlement (mm),

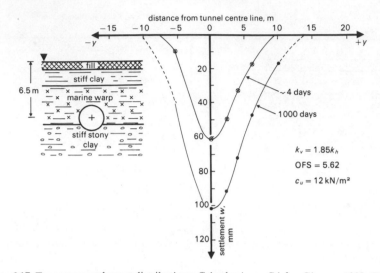

Figure 6.17 Transverse settlement distributions: Grimsby Array C (after Glossop, 1980; Glossop and O'Reilly, 1982).

A is a consolidation settlement coefficient to be determined,

and OFS is the simple overload factor $= \dfrac{\sigma_{z0} - \sigma_i}{c_u} = \dfrac{\gamma z_0 - \sigma_i}{c_u}$.

Coefficient A is evaluated below (Table 6.2) for the cases shown in Figs. 6.14–6.18.

Figure 6.19 suggests that the lower the ground loss settlement (that is, the stiffer the ground), the higher, proportionately, is the consolidation settlement contribution to the total settlement. The form of the relation between A and w_m can be used, albeit approximately, to predict the total maximum settlement, w_{mt}, for other case histories. This can be done by reading off the A value appropriate to a measured ground loss settlement and then inserting it in equation 6.14, or by incorporating the A–w_m relation directly into equation 6.14:

$$w_{mt} = 0.78 \, \text{OFS}(w_m - 0.01 \, w_m^2) \tag{6.15}$$

It is likely that this equation will be modified when more consolidation settlement data become available.

Table 6.2. Some settlement case history data for the evaluation of the A coefficient.

Belfast (Glossop and Farmer, 1977)	Grimsby Array B1 (Glossop, 1980)	Grimsby Array C (Glossop, 1980)	Willington Quay (Attewell et al., 1978)
$w_m = 17$ mm	$w_m = 36$ mm	$w_m = 55$ mm	$w_m = 25$ mm
OFS $= 3.64$	OFS $= 3.95$	OFS $= 5.62$	OFS $= 5.9$
$w_{mt} \simeq 40$ mm	$w_{mt} \simeq 70$ mm	$w_{mt} = 103$ mm	$w_{mt} = 85$ mm
From eq. 6.14	From eq. 6.14	From eq. 6.14	From eq. 6.14
$A = 0.32$	$A = 0.25$	$A = 0.17$	$A = 0.29$

6.3.2 *Total maximum settlement prediction using a compression index*

This method, used by Attewell *et al.* (1978) for analysis on a Northumbrian Water Authority interceptor sewer at Willington Quay, north-east England (see Fig. 6.18), requires measurement of the long-term pore-water pressure change by piezometer installed above the tunnel crown. Maximum consolidation settlement w_{mc} is related to piezometric pressure change Δp by the equation

$$\Delta V = \varepsilon_z = \frac{\Delta H}{H} = \frac{\Delta e}{1+e_0} = \frac{C_c}{1+e_0}\log\left(\frac{p_0 + \Delta p}{p_0}\right)$$

where

ΔV = volumetric strain = vertical strain ε_z for *one-dimensional* consolidation
ΔH = change in thickness of soil element having initial total thickness H,
Δe = corresponding change in void ratio from an initial value of e_0.
p_0 = initial vertical pressure at the tunnel crown before measurable soil consolidation occurs,
Δp = change in ground pressure above the tunnel crown as a result of drawdown and soil consolidation.

and

C_c = compression index.

It is noted that the thickness, H, of the consolidating layer must be estimated from settlement measurements in the ground. Effective stress normal to ground surface then increases as a result of draw-down and ΔH is equivalent to consolidation settlement w_{mc}.

In the case history quoted by Attewell *et al.* (1978) values of 1, 0.3, and 4 metres were assigned to e_0, C_c and H, respectively, for the soft silty alluvial clay. A p_0 value of $205\,\text{kN/m}^2$ was calculated from the tunnel depth to crown and the soil bulk unit weight. A recorded reduction in piezometric pressure Δp of $22\,\text{kN/m}^2$ was

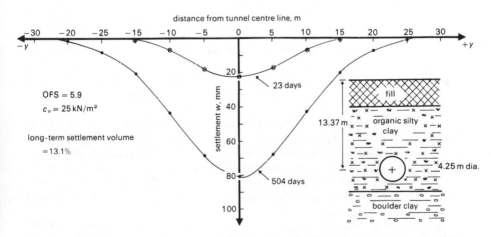

Figure 6.18 Transverse surface settlement distributions: Willington Quay (after Glossop, 1977).

Table 6.3 A guide to values of compression index C_c of saturated soils (Lee *et al.*, 1983, Table 5.1, p. 190).

| Soil type | Index properties | | | Source (see Lee *et al.* Table for full references) |
	Liquid Limit LL%	Plasticity Index PI%	C_c	
Normally consolidated estuarine silty clay (undisturbed)	100 +	High	1 to 1.4	Lee *et al.* (1983)
Marine sediment, B.C., Canada	130	74	2.3	Finn *et al.* (1971)
Re-moulded marine silty clay, Kyushu, Japan	70	43	1.1	Lee *et al.* (1983)
Deep-water brown marine clay	100 to 200	High	0.5 to 1	Noorany and Gizieski (1970)
Undisturbed organic silty clay, Delaware, U.S.A.	84	46	0.95	Schmidt and Gould (1968)
Undisturbed clay, New Orleans, U.S.A.	79	26	0.29	Lambe and Whitman (1969)
Stiff mottled clay	69	20	0.20	Lee *et al.* (1983)
Undisturbed Boston Blue Clay	41	20	0.35	Lambe and Whitman (1969)

accompanied by a measured settlement of 25 mm. The calculated settlement based on the above equation was 27 mm.

The Terzaghi and Peck (1967) equation

$$C_c \simeq 0.009 \, (\text{LL} - 10\%),$$

where LL is the liquid limit, could be used for a very approximate value of the compression index. Alternatively, Table 6.3, after Lee *et al.* (1983), could be used for guidance.

6.3.3 *Form of the transverse consolidation settlement profile*

The complete long-term behaviour of superadjacent structures cannot be assessed without some knowledge of the consolidation settlement profile. There is as yet no satisfactory method of predicting this.

Where the soil structure is such as to promote strong permeability anisotropy, the horizontal permeability k_h greatly exceeding the vertical permeability k_v, the long-term settlement trough width would seem to increase substantially beyond the short-term ground-loss trough. As a very tentative approximation, the final trough (curve) can be constructed by dropping ordinates, equal to the maximum consolidation settlement, from all points on the ground loss transverse trough (curve) and extrapolating beyond the span of the latter.

Where the soil permeability tends to isotropic, the width of the settlement trough will still tend to increase in the longer term but to a much lesser extent since the drainage direction is now dominantly vertical (see, for example, the Willington Quay case in Fig. 6.18). Because the potential for structural damage is a function of differential settlement it is suggested that for structures or parts of structures within the span of the ground loss trough, this trough be considered deepened by consolidation settlement but not widened. Thus, in the earlier equations, w_{mt} would be entered for w_m and i_t would be the same as i, so leading to a likely upper bound pessimistic assessment of possible damage. For structures or parts of structures outside the span of the ground loss trough, any differential settlements could be treated as ignorably small.

6.4   Ground movements and buried pipelines

This subject has been mentioned somewhat cursorily in section 6.2. Rapid appraisals of possible pipeline vulnerability can be performed using the curves in Figs. 6.5 to 6.10 inclusive. The aim of this present section is to promote more serious examination of the problem.

It has been estimated that the U.K. gas and water distribution system consists of 500 000 km of buried pipeline (at the year 1977, excluding service connections) of which over 90% is grey iron pipework (Department of Energy, 1977; National Water Council, 1977; Collins et al, 1973). Similar usage of grey iron distribution pipework is reported in a survey of the 100 largest cities in the U.S.A. (Sears, 1968). In terms of engineering structural design, failure can be by excessive pipe yielding, joint leakage or pipe fracture. It has been established that the primary causes of fracture in grey iron distribution mains are corrosion and system disturbance due to ground movement (for example, Department of Energy, 1977; Roberts and Regan, 1974, 1977; Needham and Howe, 1979). The main cause of fracture associated with ground movement is flexural failure in longitudinal bending, that is, beam bending, not ring bending. The differential ground movement necessary to fracture pipes arises from traffic loading,

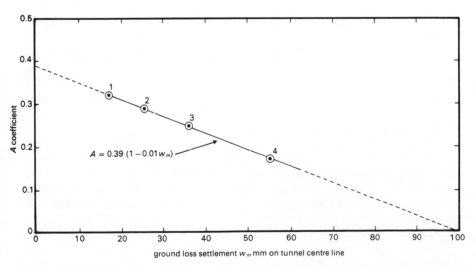

Figure 6.19 Variation of consolidation coefficient 'A' with short-term ground loss settlement. For data points, ref. Figs. 6.14–6.18 inclusive. 1. Belfast, OFS = 3.64; 2, Willington Quay, OFS = 5.9; 3, Grimsby B, OFS = 3.95; 4, Grimsby C, OFS = 5.62.

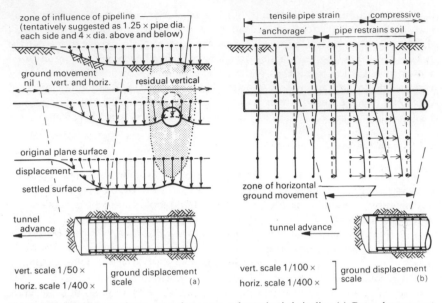

Figure 6.20 Typical restraint to ground movement due to buried pipeline. (a) Ground movement lateral to pipeline. (b) Ground movement longitudinal to pipeline.

ground temperature and moisture changes, and ground movement associated with adjacent excavation such as trenching or tunnelling.

6.4.1. *Movements lateral to a buried pipeline or beam*

'Lateral' in the context of this work is used to denote those ground and pipe movements that are at right angles to the long axis of the pipe or beam. 'Longitudinal' (axial) movements parallel to the pipeline are considered in section 6.4.4. Throughout this section reference is made to a pipeline, but the approach is equally applicable to any system that can be treated as a beam embedded in a elastic solid (for example a strip footing) where the ground movement can be taken as constant across the width. Furthermore, although the discussion is directed specifically to tunnels, the approach is applicable to ground movements in general (without necessarily referring to source) and their effects on buried pipes or beams.

The ground movements predicted in section 6.2, or any other specified movement profile, will be modified in the vicinity of a buried pipeline due to the pipe stiffness being many times greater than the soil stiffness. Where ground movement is lateral (that is, non-axial) to a buried pipeline, bending in the pipeline will result. The effect is illustrated in Fig. 6.20. Bending is associated with differential slope on the ground movement profile. Generally it is necessary to limit this bending stress; occasionally, limitations on pipeline joint rotation may be important.

6.4.2. *Subgrade–reaction analysis*

6.4.2.1. *Modulus of subgrade–reaction and limitations.* The subgrade–reaction foundation model, first introduced by Winkler (1867), is characterized by the assumption that the pressure in the foundation is proportional at every point to the deflection occurring at that point and is independent of pressures or deflections produced elsewhere in the foundation. An exact theoretical solution can be found for the bending under a concentrated load of an infinite beam resting on an isotropic elastic solid. The solution of Biot (1937) has been extended by Vesic (1961) to give the full solution for bending moment, shear force and deflection. Vesic has investigated in detail the reliability of the modulus of subgrade–reaction approach and the magnitude of error induced by its application compared with an elastic continuum. The Winkler foundation model is practically satisfied for beams of infinite length and the limits of application are given. For an infinite beam,

$$k_\infty d = K_\infty = 0.65 \sqrt[12]{\frac{E_s d^4}{E_p I_p}} \left[\frac{E_s}{1 - v_s^2} \right] \tag{6.16}$$

where
k is the *coefficient* of subgrade reaction (dimensions FL^{-3}), K is the *modulus* of subgrade–reaction $= kd$ (dimensions FL^{-2}),
d is the beam width (e.g. pipe outside diameter),
$E_p I_p$ is the flexural rigidity of the beam, and
E_s, v_s are elastic properties of the foundation.
For a pipe, the moment of inertia about the centre is

$$I_p = \frac{\pi}{64} [d^4 - (d - 2t)^4]$$

where t is the pipe wall thickness.

Table 6.4 Recommended procedures for analysis

Size of beam	Free length (L) beyond loaded length...(c)	Recommended procedures	
		For rough estimates	For refined analysis
Long	$\lambda L > 2.5$	Conventional analysis assuming infinite beam and K from equation 6.16...(a)	
Moderately long	$1 < \lambda L < 2.5$	Conventional analysis assuming finite beam and K from equation 6.16...(a)	
Moderately short	$0.4 < \lambda L < 1$	Conventional analysis (a)	(b)
Short	$\lambda L < 0.4$	Treat as perfectly rigid[a] i.e. contact pressure linear	(b)

(a) After Vesic (1961).
(b) Use of the equations of Mindlin (1936) or finite-element methods.

(c) $\lambda = \sqrt[4]{\dfrac{K}{4E_p I_p}}.$

The recommended procedure for analysis is shown in Table 6.4 and, according to Vesic, has been substantiated by large-scale model tests.

For a beam below the surface of an elastic solid, the foundation modulus is the sum of the moduli on each side of the beam. At infinite depth, and for Poisson's ratio of 0.5, the effective modulus,

$$K_{\text{eff}} = 2K_\infty$$

where K_∞ is given by equation 6.16. It is suggested that for practical purposes the presence of a free boundary (ground surface) above the pipeline can be ignored and is conservative. Thus for all cases the effective modulus can be taken as twice the value given by equation 6.16.

6.4.2.2 *Basic theory*. In the foundation model, the pressure (p) in the foundation is proportional at every point to the deflection occurring at that point, that is,

$$p = kw_p$$

where k is the *coefficient* of subgrade reaction (FL^{-3}), and w_p is the deflection of the pipe.

Initially, the pipeline is in equilibrium with soil pressure p_1 and p_2, respectively below and above the pipeline, as illustrated in Fig. 6.21(a). The foundation coefficient is equivalent to $k = k_1 + k_2$. The ground movement profile, $w = f(y)$, is the predicted ground movement that occurs remote from the pipeline or if the pipeline were not present. The effect of this movement is shown in Fig. 6.21(c) and causes pipeline movement w_p. Adjacent to the pipeline, soil and pipeline movement are equal, since it is assumed that the soil and pipe remain in contact. The increase in soil pressure is

$$p = k(w_p - w).$$

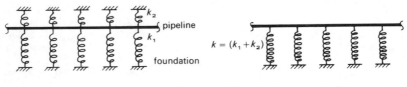

(a) initial condition which is equivalent to (b)

(b)   loading pressure $= kw$
        loading intensity $= q = kwd = Kw$

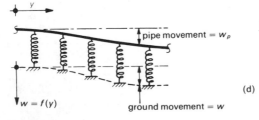

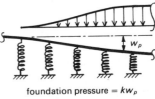

foundation pressure $= kw_p$

(d)

(c) effect of ground movement which is equivalent to (d)

Figure 6.21 Beam on elastic foundation model. Note: representation of foundation reaction as discrete springs is diagrammatic only—reaction in foundation model is continuous.

This is equivalent to a foundation pressure kw_p acting upwards plus an applied loading pressure kw acting downwards, that is

$$p = kw_p + (-kw)$$

as shown in Fig. 6.21(d). Putting the foundation modulus $K = kd$, where d is the pipe outside diameter or beam width, the applied loading intensity is

$$q = Kw.$$

The equation of the elastic line for the pipeline is

$$\frac{\partial^4 w_p}{\partial y^4} + 4\lambda^4 w_p = 4\lambda^4 \cdot f(y) \tag{6.17}$$

where w_p is the pipeline movement,

$f(y) = w$ is the predicted ground movement profile,

λ is the damping factor (dimensions L^{-1})

and

$$\lambda = \sqrt[4]{\frac{K}{4E_p I_p}} \quad \text{(see Table 6.4)} \tag{6.18}$$

The parameter $1/\lambda$ will be called the soil-pipe stiffness. Thus, the procedure to find the deflection, shear force and bending moment in the pipeline due to the settlement, w, of the foundation, is to apply a loading to a beam on an elastic foundation where the loading intensity is $q = Kw$. Hetenyi (1946) gives a comprehensive treatment of the theory of beams on elastic foundations together with solutions that are applicable to this problem.

6.4.2.3 *General method of analysis.* The soil/pipeline model is linear elastic and the assumptions are:

(i) The Winkler foundation model applies.
(ii) The soil is already pre-compressed and always remains in contact with the pipe.
(iii) The pipe material is linear elastic, homogeneous, and isotropic.
(iv) The pipeline is homogeneous (that is, a plain pipeline with rigid joints) and continuous for a distance of $3/\lambda$ beyond the point at which the slope of the ground movement profile is negligible.
(v) The soil around the pipe is linear elastic and homogeneous.

Since the pipeline is assumed continuous for at least $3/\lambda$ past the point of load variation, analysis is for an infinite beam on an elastic foundation. The following should be noted:

(i) w is the ground movement profile at *pipeline level* and it is generally convenient to put $w_m = 100$ and pro rata for the required w_m value at the end of the calculation.
(ii) λ is derived from equation 6.18 and has dimensions of L^{-1}.
(iii) $K = (K_1 + K_2)$ and has dimensions FL^{-2}. Generally take $(K_1 + K_2) = 2K_\infty$ where K_∞ is derived from equation 6.16.

(iv) In equations 6.16 and 6.18, E_p and E_s are appropriate *secant values corresponding to the range of stress and are values applicable adjacent to the pipeline, not for the soil as a whole.*

(v) A range of E_s, and where appropriate I_p, should be substituted to test sensitivity to chosen values.

(vi) Possible ground movement profiles, in addition to those described in section 6.2, should be considered to test sensitivity to smoothness of profile.

(vii) The increase in soil pressure below the pipeline is $K_1(w_p - w)/d$, and the decrease above it is $K_2(w_p - w)/d$. These expressions involve a small difference which makes the calculation of soil pressure approximate. This is not a defect in the method, but merely a reflection of the physical reality of the problem.

(viii) The method also applies to horizontal soil movement lateral to a pipeline which would have to be combined with the vertical bending to give the maximum resultant bending.

6.4.2.4 *Solutions for a rigidly-jointed pipeline transverse to a tunnel drive.* The maximum soil movement lateral to a pipeline occurs when the pipe is transverse to the tunnel drive and the settlement trough is fully developed. The pipeline is in the plane $x = $ constant and $4i$ or more behind the tunnel face, where i is the trough width parameter. *This lateral soil movement is wholly vertical.* For a pipeline with rigid joints extending $3/\lambda$ into undisturbed ground ($1/\lambda$ being defined from equation 6.18), dimensionless curves are shown in Fig. 6.22 which relate the pipe curvature $\partial^2 w_p/\partial y^2$ to transverse distance y/i and soil-pipe stiffness $1/\lambda$. The bending moment is given by

$$M = E_p I_p \frac{\partial^2 w_p}{\partial y^2} \tag{6.19}$$

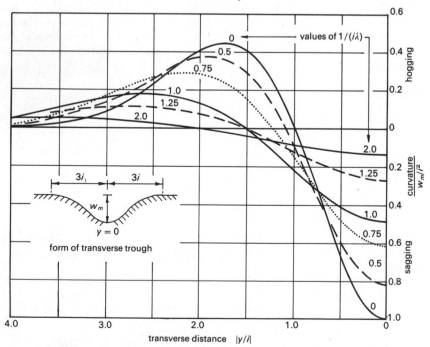

Figure 6.22 Pipe curvature diagram variation with soil-pipe stiffness.

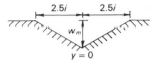

Figure 6.23 Worst transverse discontinuous slope profile.

and the extreme fibre bending strain is

$$\varepsilon_y = \frac{d}{2} \frac{\partial^2 w_p}{\partial y^2} \tag{6.20}$$

The assumed ground movement profile is the normal probability form. For $1/\lambda = 0$, the trough is assumed to be smooth; for other values of stiffness the settlement curve was divided into 12 linear elements of length $0.5\, y/i$. Interpolation on Fig. 6.22 is facilitated by Figs. 6.24 and 6.25 which give the maximum sagging and hogging curvatures as a function of soil-pipe stiffness and the position of zero and maximum moments. The maximum sagging moment is always at $y = 0$. Sensitivity to the assumed settlement trough is shown in Fig. 6.24. The area of the settlement trough shown in Fig. 6.23 is the same as the normal probability form, that is, both troughs reflect the same settlement volume, V. For practical purposes there is no difference in maximum bending moment for $1/\lambda > 0.5i$. For $1/\lambda < 0.5i$ the maximum sagging curvature for Fig. 6.23 is

$$\frac{\partial^2 w_p}{\partial y^2} = \frac{\lambda i}{2.5} \frac{w_m}{i^2}.$$

For the settlement trough shown in Fig. 6.22, the relationship between maximum sagging curvature and soil-pipe stiffness shown in Fig. 6.24 can be read as a reduction factor associated with soil-pipe stiffness. For negligible stiffness ($1/\lambda = 0$), the curvature factor given by Fig. 6.24 is 1.0, reducing to 0.8 at $1/\lambda = 0.5i$, 0.45 at $1/\lambda = 1.0i$, and so on. The maximum hogging curvatures are reduced by a similar amount. This reduction

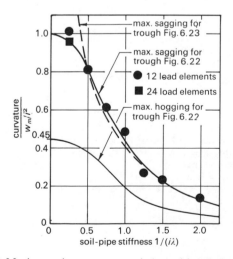

Figure 6.24 Maximum pipe curvature variation, with soil-pipe stiffness.

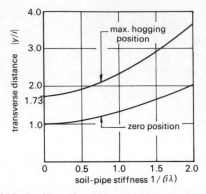

Figure 6.25 Position of maximum and zero bending moment.

factor for bending curvature due to soil-pipe stiffness is greatest where the 'span of the beam' is least. The fully-developed transverse settlement trough is the plane in the 3-dimensional ground movement field which has the least 'span'. Thus, irrespective of the orientation of the pipeline relative to the tunnel drive, the factor (bending curvature for actual soil-pipe stiffness)/(bending curvature for negligible soil-pipe stiffness) cannot be less than the value given by Fig. 6.24.

Example 3: Calculation of bending strain for pipeline.

Tunnel excavated diameter, $2R = 1.5$ metres in granular soil at tunnel level.
Volume of ground loss/unit advance, $V = 5\%$ of face area $= 0.0884 \text{m}^3/\text{m}$.
Tunnel depth, ground level to axis level, $z_0 = 5.0$ metres.
Pipeline depth, ground level to pipe axis, $z = 1.5$ metres.
                                                Thus $z_0 - z = 3.5$ metres.
Settlement trough width parameter $i = 0.43(z_0 - z) + 1.1$ metres $= 2.6$ metres.
(based on predominantly cohesive soil above tunnel level).
Maximum ground displacement at pipeline level (i.e. at $z_0 - z = 3.5$ m) is

$$w_m = \frac{V}{\sqrt{2\pi} i} = \frac{0.0884}{\sqrt{2\pi} \times 2.6} = 0.0136 \text{ metres (13.6 mm)}.$$

Pipe outside diameter, $d = 0.50$ metres.
Pipe wall thickness, $t = 0.018$ metres.

Pipe moment of inertia, $I_p = \dfrac{\pi}{64}[d^4 - (d - 2t)^4]$

$$= \frac{\pi}{64}(0.50^4 - 0.464^4) = 7.93 \times 10^{-4} \text{m}^4.$$

Pipe secant modulus, $E_p = 70 \text{ GN/m}^2 = 7 \times 10^{10} \text{ N/m}^2$.
Soil secant modulus, $E_s = 10 \text{ MN/m}^2 = 10 \times 10^6 \text{ N/m}^2$.
(value adjacent to pipeline)
Soil Poisson's ratio, $v_s = 0.5$.
Modulus of subgrade-reaction, K_∞ from equation 6.16:

$$K_\infty = 0.65 \sqrt[12]{\frac{E_s d^4}{E_p I_p}} \left[\frac{E_s}{1 - v_s^2}\right]$$

$$= 0.65 \sqrt[12]{\frac{10 \times 10^6 \times 0.50^4}{7 \times 10^{10} \times 7.93 \times 10^{-4}}} \cdot \frac{E_s}{0.75} = 0.596 E_s$$

$$= 5.96 \times 10^6 \text{ N/m}^2.$$

Effective K value $= 2 \times K_\infty = 11.9 \times 10^6 \, \text{N/m}^2$.
(due to depth of burial etc)
Damping factor, λ from equation 6.18:

$$\lambda = \sqrt[4]{\frac{K}{4E_p I_p}} = \sqrt[4]{\frac{11.9 \times 10^6}{4 \times 7 \times 10^{10} \times 7.93 \times 10^{-4}}} = 0.481 \, \text{m}^{-1}.$$

Soil-pipe stiffness parameter $1/\lambda = 2.078$ metres

$$= \frac{2.078}{2.6} i = 0.8i.$$

For the pipeline transverse to the tunnel drive there are permanent bending curvatures given by Figs. 6.22 and 6.24.
Above the tunnel centreline:

$$\frac{\partial^2 w_p}{\partial y^2} = 0.55 \frac{w_m}{i^2} = \frac{0.55 \times 0.0136}{2.6^2} = 1.107 \times 10^{-3} \, \text{metres}^{-1}.$$

The extreme fibre bending strain $\varepsilon_y = \dfrac{d \, \partial^2 w_p}{2 \, \partial y^2} = \dfrac{0.50}{2} \times 1.107 \times 10^{-3}$

$$= 2.77 \times 10^{-4} = 277 \, microstrain.$$

At a transverse distance of approximately $2.2i = 5.7$ metres there is a maximum hogging curvature of

$$\frac{\partial^2 w_p}{\partial y^2} = 0.27 \frac{w_m}{i^2} = 5.43 \times 10^{-4} \, \text{metres}^{-1}.$$

The extreme fibre bending strain $\varepsilon_y = \dfrac{d \, \partial^2 w_p}{2 \, \partial y^2} = 136 \, microstrain.$

From Fig. 6.25, the transverse distance to zero bending moment is approximately $1.2i = 3.1$ metres each side of the tunnel centreline. Although the analysis has been performed for a rigidly-jointed pipeline, perfectly flexible joints in the pipeline at 6.2 metre centres would have no effect on the bending strains if these joints were in the most adverse position at $|y| = 3.1$ metres, 9.3 metres, and so on. A spacing encountered in practice of 5.5 metres (18 feet) would also have only a small reducing effect if the joints happened to be in the worst position.

The effect of soil drainage can be allowed for by substituting drained soil parameters in the expression for K. Taking $E'_s = \frac{2}{3}(1 + v'_s)E_s$ and $v'_s = 0.3$, the effective K' value $= 0.486 E'_s = 4.21 \times 10^6 \, \text{N/m}^2$ and $\lambda' = 0.441$; $1/\lambda' = 2.266$ metres $= 0.87i$. The bending strains are reduced from 277 $\mu\varepsilon$ to 257 $\mu\varepsilon$ and from 136 $\mu\varepsilon$ to 121 $\mu\varepsilon$. Soil and pipe material creep would give similar small reductions in pipe strain with time.

6.4.2.5 *Solutions for a rigidly-jointed pipeline parallel to a tunnel drive.* For a pipeline parallel to the tunnel, the maximum lateral soil movement occurs when the pipeline is directly above the tunnel drive. The pipeline is in the plane $y = 0$ and the lateral soil movement is wholly vertical. In a similar manner to section 6.4.2.4, Figs. 6.26, 6.28 and 6.29 relate the pipe curvature to distance ahead of the tunnel face, x/i, and soil-pipe stiffness. The term 'distance ahead of tunnel face' is interchangeable with 'distance ahead of point of maximum slope on parallel settlement trough' without affecting the validity of the diagrams. For the cumulative normal probability form of ground movement shown in Fig. 6.26, the sagging bending curvature at x/i behind the tunnel

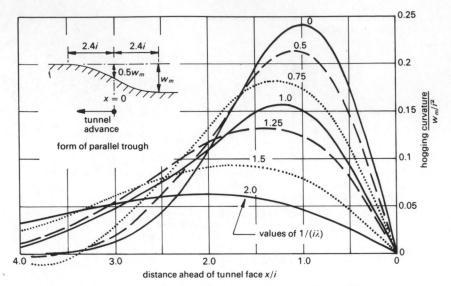

Figure 6.26 Pipe curvature diagram variation with soil-pipe stiffness.

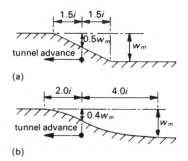

Figure 6.27 Parallel settlement troughs. (*a*) Worst parallel discontinuous slope profile. (*b*) Case-history data for tunnels in firm/stiff clay.

face is numerically equal to the hogging curvature at x/i ahead of the face. Sensitivity to the assumed settlement trough is shown in Fig. 6.28. The area displaced by the settlement trough in Fig. 6.27(a) is the same as that displaced by the cumulative normal probability form and reflects the same total load on the pipeline. Figure 6.27(b) is based on case history data for bolted segmentally-lined tunnels in firm to stiff cohesive soil where the forward part of the settlement trough is approximately 80% of the form shown in Fig. 6.26 and post-grouting deformation makes a significant contribution, extending the development of the trough to $4i$ behind the face.

For practical purposes, the maximum hogging bending moment associated with Fig. (6.27(b) is merely 80% of the value for Fig. 6.26. For $1/\lambda > 1.0i$ there is no practical difference between the maximum bending moment for Figs. 6.27(a) and 6.26 movement profiles. For $1/\lambda < 1.0i$ the maximum curvature for Figure 6.27(a) is

$$\frac{\partial^2 w_p}{\partial x^2} = \frac{\lambda i}{6} \frac{w_m}{i^2}.$$

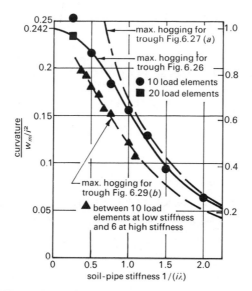

Figure 6.28 Maximum pipe curvature variation with soil-pipe stiffness.

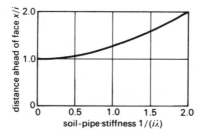

Figure 6.29 Position of maximum bending moment.

In a similar manner to section 6.4.2.4, the relationship between maximum curvature and soil-pipe stiffness shown in Fig. 6.28 for the ground movement profile indicated in Figure 6.26 can be read as a reduction factor associated with soil-pipe stiffness. Reduction factors are shown on the right hand scale taking the relative curvature for $1/\lambda = 0$ as 1.0. The curvature factors are 1.0 at $1/\lambda = 0$, 0.89 at $1/\lambda = 0.5i$, 0.65 at $1/\lambda = 1.0i$ and so on. As mentioned in section 6.4.2.4, the reduction is less than that for pipelines transverse to the tunnel drive; also, for pipelines parallel to—but offset from—the tunnel drive (in the plane $|y| = \text{constant} > 0$) the amount of reduction due to soil-pipe stiffness cannot be more than that indicated by Fig. 6.28.

6.4.2.6. *Solutions for pipeline deflection and soil pressure.* The change in soil pressure at the soil-pipe interface due to the settlement of the ground is:

$$\text{increase below pipeline,} \quad p_1 = K_1(w_p - w)/d \qquad (6.21)$$

$$\text{decrease above pipeline,} \quad p_2 = K_2(w_p - w)/d \qquad (6.22)$$

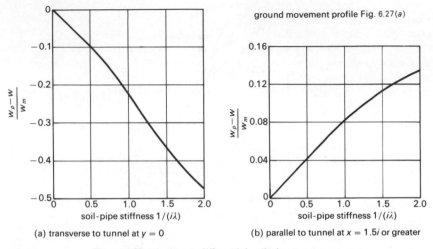

Figure 6.30 Maximum differential soil-pipe movement.

where K_1 and K_2 are the foundation modulus values below and above the pipeline (dimensions FL^{-2}) as discussed in section 6.4.2.1, w_p and w are the respective pipe and ground movements, and d is the pipe outside diameter. The positions of maxima, minima and zero change in soil pressure coincide with the maxima, minima and zero points on the pipe curvature diagram, for example Fig. 6.22. For the rigidly-jointed pipeline transverse to the tunnel drive, as described in section 6.4.2.4, the maximum numerical value of $(w_p - w)$ occurs directly above the tunnel at $y = 0$. Figure 6.30(a) shows the variation of dimensionless movement $(w_p - w)/w_m$ with dimensionless soil stiffness, $1/i\lambda$, for the ground movement profile shown in Fig. 6.23. The negative value of $(w_p - w)$ indicates that at $y = 0$ the pipe restrains ground movement. At $y = 0$, for $1/\lambda < 0.5i$:

$$w_p = w_m\left(1 - \frac{1}{5i\lambda}\right) \rightarrow w_m \quad \text{as} \quad \frac{1}{i\lambda} \rightarrow 0.$$

For the pipeline parallel to the tunnel drive, as described in section 6.4.2.5, and taking the ground movement profile shown in Fig. 6.27(a), the maximum value of $(w_p - w)$ is shown in Fig. 6.30(b). This maximum occurs at $x = 1.5i$ or more ahead of the tunnel face position; thus $w = 0$ at this position for the ground movement profile adopted. The value of $(w_p - w)$ at $(x/i) = -n$ is $-(w_p - w)$ at $(x/i) = n$, where n is any number. Calculation of the change in soil pressure and pipeline movement is carried out so that the assumptions concerning soil secant modulus and pipeline remaining in contact with the soil can be checked. Knowledge of pipeline movement is required where a *network of connected pipes* is to be analysed and the *load-deformation equations are required at each connection*.

6.4.2.7 *Effect of non-uniformity in foundation to pipeline.* In the preceding sections, the assumption of homogeneity in the model gives zero residual bending stresses in pipelines parallel to the tunnel drive when the tunnel face is sufficiently far advanced. Also, for an assumed smooth transverse settlement trough, theoretical bending stresses in small diameter pipelines are small since the bending is proportional to $w_m/i^2 \times d/2$. Non-uniformity in the soil or pipe stiffness could be taken as proportional to the effect of an applied lateral displacement, where this displacement is directly proportional to the maximum ground movement. The implication is that the relationship between maximum bending stress and soil-pipe stiffness is similar for a non-uniformity and an applied lateral displacement. (It is not suggested that a non-uniformity necessarily does apply significant concentrated forces to the pipe.) On this basis, a displacement of $\delta = \text{constant} \times w_m$ causes an extreme fibre bending strain in a buried pipeline of

$$\varepsilon_{\max} = (0.45 \text{ to } 0.71) \times \left(\frac{E_s}{E_p I_p}\right)^{1/2} d \cdot \delta \qquad (6.23)$$

where

$v_s = 0.5$,

pipe diameter, $d = 0.075$ to 0.75 metres,
pipe diameter/wall thickness ratio, $(d/t) = 5$ to 50,
pipe modulus/soil modulus ratio, $(E_p/E_s) = 20$ to 200 000,
and the factor 0.45 is for a small stiff pipe in weak soil and 0.71 is for a large flexible pipe in stiff soil.

At constant d/t ratio, I_p is proportional to d^4, so equation 6.23 implies a pipe strain approximately *inversely* proportional to diameter. For commercially-available pipes, d/t is generally constant or increases with increasing diameter, which slightly reduces this effect. Since non-uniformity in the pipeline and foundation cannot be ignored, it is tentatively suggested that a displacement lateral to the pipe, given by $\delta = 0.05w_m$, be applied to the pipeline. This should be considered as an alternative to the lateral soil movement considered in preceding sections. Equation 6.23 reduces to approximately

$$\varepsilon_{\max} = 0.05\frac{d}{2}w_m\left[\frac{E_s}{E_p I_p}\right]^{0.5} \qquad (6.24)$$

Equation 6.24 is comparable with a relationship that has been tentatively suggested for predicting the effect of ground movements caused by deep trenching on adjacent buried pipelines. The empirical formula for trenching is

$$\varepsilon_{\max} = 0.058\frac{d}{2}w_m\left[\frac{E_s}{E_p I_p}\right]^{0.4} \qquad (6.25)$$

where E_s and E_p are in kN/m^2, I_p is in metres^4 and d and w_m are in metres. The formula is the result of linear-elastic modelling coupled with the results of measurements on full-scale trench construction. Equation 6.25 is said to have a typical range of possible error of $\pm 25\%$ and is applicable to pipelines up to 300 mm (12 inches) diameter.

6.4.3 Ultimate lateral soil pressure

An estimate of the maximum restraint to differential soil-pipe movement can be obtained by equating the ultimate soil resistance with the bearing capacity of a strip footing. The general form for a strip footing given by Terzaghi and Peck (1967) is:

Ultimate bearing capacity $= q_u = cN_c + \gamma z_p N_q + 0.5\gamma BN_\gamma$

For a pipe and lateral soil movement:

z_p is the depth from the ground surface to the pipe axis level,
$B = d$, which is the pipe outside diameter,
γ is the soil unit weight,
$q_u = p_u$ where $p_u \times d$ is the ultimate load on the pipeline/unit length,
c is the soil cohesion, and
N_c, N_q are bearing capacity factors with respect to cohesion and surcharge, and N_γ
accounts for the influence of the weight of the soil.
The bearing capacity factors are dimensionless and depend on the angle of internal friction ϕ and the failure mechanism. The failure mechanism depends on the depth/width ratio, z_p/d, whether the soil is dense or loose (or stiff or soft) and the direction of loading relative to the ground surface.

For vertical ground movement, and where the pipe is pushing downwards into the soil, $w_p > w$, the usual bearing capacity factors for footings apply. Typical values are:

ϕ	z_p/d	N_c	N_q	N_γ	ϕ	z_p/d	N_c	N_q	N_γ
0	1.5	6.7	1.0	0	35°	1.5		35	40
0	10	7.5	1.0	0	35°	10	(45)		

Where the ground movement is horizontal and lateral to a pipe, the bearing capacity factors given by Brinch Hansen (1961) may be used. Model tests by Audibert and Nyman (1977) substantiate the use for pipes. Typical values are:

ϕ	z_p/d	N_c	N_q	N_γ	ϕ	z_p/d	N_c	N_q	N_γ
0	1.5	5	0	–	35°	1.5	(30)	10	–
0	10	7	0	–	35°	10	(60)	20	–

For the case of vertical ground movement and $w_p < w$, soil failure occurs above the pipe. Tests on model pipes with granular soils have been carried out by Matyas and Davis (1983). For $\phi = 35°$, the suggested relation between the pipe load and z_p/d is equivalent to $N_q = 1.3$ at $z_p/d = 1.5$, $N_q = 5.0$ at $z_p/d = 10$. These values would be increased if the pipe were buried under a stiff road pavement. For this case, the bearing capacity factors will be nearer to those suggested previously for horizontal ground movement. These apply to the general case of a $c - \phi$ soil.

6.4.4 Movements longitudinal to a buried pipeline

Where ground movement is longitudinal to a buried pipeline, axial tensile and compressive forces are induced in the pipeline and the effect is illustrated in Fig. 6.20.

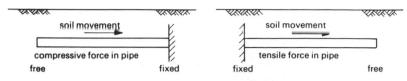

Figure 6.31 Effect of fixity on force in pipeline.

Load is transferred to the pipeline by the shear stress at the soil-pipe interface and the mechanism is analogous to the transfer of load to the ground by friction piles as described by Cooke (1975). For the case illustrated in Fig. 6.20, there can only be a load induced in the pipeline if there is a shear stress between the soil and the pipe. This requires that the pipeline restrains the soil movement in the zone of ground movement and the soil restrains the pipeline movement in the zone of pipeline 'anchorage'. The fixity conditions of the pipeline are crucial in determining whether tensile or compressive forces are induced, as illustrated in Fig. 6.31.

Where the main pipeline stiffness is such that there is significant local restraint to longitudinal ground movement, this can induce large bending stresses in small diameter relatively flexible branches off the main pipeline.

The horizontal ground movement associated with tunnelling in soil can be predicted by the methods described in section 6.2. There is little case-history data, but the data available does not conflict with the expressions given. For the fully-developed transverse settlement trough, the ratio of horizontal to vertical movement,

$$\frac{v}{w} = \frac{n}{z_0 - z} y.$$

For cohesive soils, $n = 1.0$, which implies ground movements towards a 'sink' source located along the tunnel axis. For granular soils $n < 1.0$, which implies ground movement towards a point below tunnel axis level, vertical movement predominating. Both of these relations for cohesive and granular soils have been demonstrated in model tests and have a simple physical basis.

The following discussion and analysis is for assessing the effect of horizontal ground movement on buried pipelines but a similar approach could be used for a strip footing where the ground movement can be taken as constant across the width. Tensile forces in the pipeline are associated with differential movement on the ground movement profile. (Compare lateral soil movement where bending is associated with differential slope.) Generally, it is necessary to limit the axial tensile stress; occasionally, limitations on pipeline joint extension may be important.

6.4.5 Elastic analysis

6.4.5.1 Basic analysis. Poulos and Davis (1980) give the results of a linear elastic analysis for a pile subjected to longitudinal soil movement. The problem and notation is illustrated in Fig. 6.32.

The assumptions are:
1. The soil is a homogeneous elastic half-space.
2. The soil-pile interface does not slip.
3. The pile is incompressible and does not fail in tension or compression.

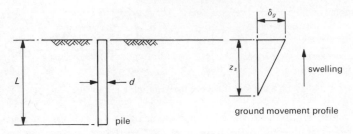

Figure 6.32 Pile in swelling soil.

On this basis the pile movement to soil movement ratio, δ_p/δ_g, is related to z_s/L and L/d. The maximum load in the pile, $P_{max}/E_s d\delta_g$, is also related to z_s/L and L/d.

For the case of a pile subjected to axial loading, results are given for pile movement in terms of $PI_f/E_s d$, where I_f is an influence factor that depends, amongst other things, on L/d. The results for swelling soil and axial load can be combined to give the maximum load in a pipeline subjected to longitudinal soil movement associated with tunnelling. The load, calculated on the basis of an incompressible pipe, over-estimates tensile forces.

6.4.5.2 *Method for pipeline parallel or transverse to tunnel drive.* For a pipeline parallel to and directly above the tunnel drive, the horizontal ground movement is given in section 6.2 as

$$u = \frac{-n}{z_0 - z} w_m \frac{i}{\sqrt{2\pi}} \exp\left[\frac{-x^2}{2i^2}\right] \text{when } y = 0.$$

The maximum value of u occurs at $y = 0$, $x = 0$ and

$$u_m = \frac{-n}{z_0 - z} \frac{i}{\sqrt{2\pi}} w_m \qquad (6.26)$$

Thus,

$$u = u_m \exp\left[\frac{-x^2}{2i^2}\right] \qquad (6.27)$$

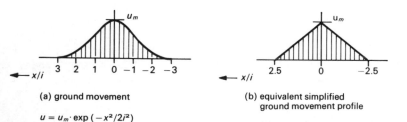

(a) ground movement

(b) equivalent simplified ground movement profile

$u = u_m \cdot \exp(-x^2/2i^2)$

Figure 6.33 Horizontal ground movement parallel to and directly above the tunnel drive.

The negative value of u indicates ground movement in the direction of decreasing x. The distribution of horizontal ground movement along the tunnel line is shown in Fig. 6.33(a). Since it is the differential ground movement that leads to load in the pipeline, the movement profile in Fig. 6.33(a) can be approximated by Fig. 6.33(b), where the integral from $x = \infty$ to 0 for (a) and (b) is the same.

Using the simplified movement profile in Fig. 6.33(b), and neglecting the free boundary above the pipeline at the ground surface (which over-estimates the load in the pipeline), the effect of horizontal movement on the pipeline is simplified to the arrangement shown in Fig. 6.34.

From the results of Poulos and Davis (1980), the maximum strain in the pipeline is

$$\varepsilon_m = \frac{1.273}{K^*}\left(\frac{i}{d}\right)\frac{u_m}{i} \times I_1$$

where I_1 depends on z_s/L which is equal to 0.25, and L/d which is equal to $10(i/d)$.

Thus, the maximum pipe strain depends on the pipe stiffness factor K^*, the factor i/d and the maximum ground movement factor u_m/i.

For the pipeline transverse to the tunnel drive, in the plane $x = $ constant and $4i$ or more behind the tunnel face, where i is the trough width parameter, the horizontal ground movement is given in section 6.2 as

$$v = \frac{-n}{z_0 - z}yw_m\exp\left[\frac{-y^2}{2i^2}\right]$$

The maximum value of v occurs at $y/i = 1$

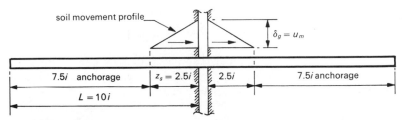

Figure 6.34 Simplified arrangement of analysis of the effect of ground movement longitudinal to a buried pipeline—pipeline parallel to tunnel drive. Notation after Poulos and Davis (1980):

Pipe outside diameter $= d$ Soil Poisson's ratio $= v_s = 0.3$
Pipe elastic modulus $= E_p$ Soil elastic modulus $= E_s$
Area of pipe wall $= A_p = \pi(d - t)t$ Depth of 'swelling' soil $= z_s$

Pipe area ratio $= R_A = \dfrac{A_p}{\pi d^2/4}$ Pipe length $= L$

$$= \frac{4(d - t)t}{d^2}$$

Pipe stiffness factor $K^* = (E_p/E_s)R_A$ and is a measure of the relative compressibility of the pipe and the soil. The more relatively compressible the pipe the smaller is the value of K^*.

G

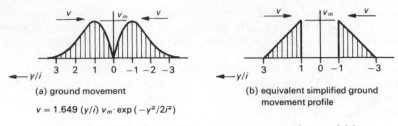

(a) ground movement

$v = 1.649 \, (y/i) \, v_m \cdot \exp(-y^2/2i^2)$

(b) equivalent simplified ground movement profile

Figure 6.35 Horizontal ground movement transverse to the tunnel drive.

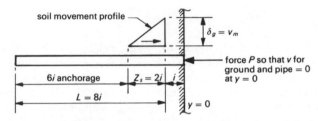

Figure 6.36 Simplified arrangement for analysis of the effect of ground movement longitudinal to buried pipeline—pipeline transverse to tunnel drive. Notation as in Fig. 6.34.

and

$$v_m = \frac{-n}{z_0 - z} i \exp[-\tfrac{1}{2}] w_m \qquad (6.28)$$

Thus

$$v = 1.649(y/i)v_m \exp\left[\frac{-y^2}{2i^2}\right] \qquad (6.29)$$

The distribution of horizontal ground movement transverse to the tunnel drive is shown in Fig. 6.35(a) and in a similar way to movement parallel to the tunnel the approximate equivalent movement is shown in Fig. 6.35(b).

Using the simplified movement profile in Fig. 6.35(b) and neglecting the free boundary above the pipeline, the effect of horizontal movement on the pipeline is simplified to the arrangement shown in Fig. 6.36.

Assuming simple linear load distributions in the pipeline due to 'swelling' soil and axial load, P, the loadings can be superimposed to give the resulting maximum compressive and tensile strains.

6.4.5.3 *Solutions for rigidly-jointed pipeline transverse to tunnel drive.* The maximum soil movement longitudinal to a pipeline occurs when the pipe is transverse to the tunnel drive and the settlement trough is fully developed. For a pipeline with rigid (non-extending) joints anchored $6i$ into undisturbed ground (where i is the trough width parameter as defined earlier) dimensionless curves are shown in Fig. 6.37. These curves relate the maximum tensile pipe strain, $\varepsilon_{ymtensile}$, to the diameter/trough width ratio, d/i, for various values of pipe stiffness factor K^*. The maximum load in the

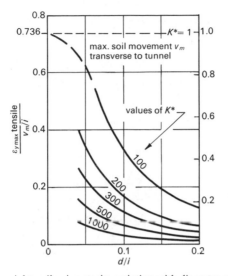

Figure 6.37 Maximum axial tensile pipe strain variation with diameter and stiffness factor K^*.

pipe has been calculated on the basis that the pipe is incompressible. This significantly over-estimates the load for values of K^* less than 1000. In Fig. 6.37 the maximum pipe strain is inversely proportional to K^*. For a pipe with negligible stiffness, $K^* = 1$, the pipe strain is given by the maximum ground strain in section 6.2, that is,

$$\varepsilon_{ym} = 0.736\frac{v_m}{i}$$

at a transverse distance of $y = 1.73i$. The maximum tensile pipe strain occurs at a transverse distance of between $1.73i$ and $3i$ and, for the purpose of calculating tensile bending plus axial strain, should be taken as coincident with the position of maximum hogging moment (see section 6.4.2.4). The variation of maximum compressive strain with d/i and K^* is shown in Fig. 6.38. This maximum occurs at $y = 0$ and has a limiting value for $K^* = 1$ of

$$\varepsilon_{ym} = 1.65\frac{v_m}{i} \quad \text{at} \quad y = 0.$$

The reduction in pipe strain associated with pipe stiffness is greatest where the pipeline is transverse to the tunnel drive, and the reduction compared with the strain for $K^* = 1$ is shown on the right-hand scale of Figs. 6.37 and 6.38. Irrespective of the orientation of the pipeline, the factor (pipe strain for actual pipe stiffness factor, K^*)/(pipe strain for negligible stiffness) cannot be less than the value given by Figs. 6.37 and 6.38.

Example 4: *Calculation of axial strain for pipeline.* The pipeline is transverse to the tunnel drive and pipe and tunnel details are the same as in Example 3. The trough width parameter $i = 2.6$ metres and the maximum settlement, $w_m = 0.0136$ metres.

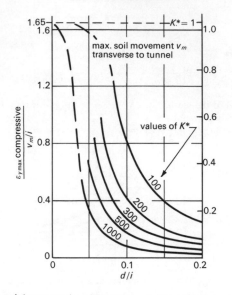

Figure 6.38 Maximum axial compressive pipe strain variation with diameter and stiffness factor K^*

The maximum horizontal ground movement is given by equation 6.28 as

$$v_m = \frac{-n}{z_0 - z} i \exp[-\tfrac{1}{2}] w_m.$$

For cohesive soil above the tunnel, $n = 1$.
Thus,

$$v_m = -\frac{1}{3.5} \times 2.6 \times \exp[-\tfrac{1}{2}] \times w_m$$

$$= -0.45 w_m = 0.0061 \text{ metres (6.1 mm) towards the tunnel centreline.}$$

The pipe stiffness factor, $K^* = \dfrac{E_p}{E_s} R_A$.

The pipe area ratio, $R_A = \dfrac{4(d-t)t}{d^2}$.

Thus $K^* = \dfrac{7 \times 10^{10}}{10 \times 10^6} \times \dfrac{4(0.5 - 0.018)0.018}{0.5^2} = 972.$

The ratio $d/i = \dfrac{0.5}{2.6} = 0.192.$

The permanent tensile strain is given by Fig. 6.37 for $K^* = 100$ as

$$\varepsilon_{ym} = .135 \frac{v_m}{i} = 0.135 \times \frac{0.0061}{2.6}.$$

For $K^* = 972$, the strain is

$$\varepsilon_{ym} = \frac{100}{972} \times 0.135 \times \frac{0.0061}{2.6} = 3.26 \times 10^{-5}$$

$$= 33 \; microstrain \; tensile.$$

(This is additional to the tensile bending strain of 136 microstrain in example 3, that is, total $= 169$ microstrain)

The compressive strain above the tunnel centreline is given by Fig. 6.38 as

$$\varepsilon_{ym} = \frac{100}{972} \times 0.26 \times \frac{0.0061}{2.6} = 6.28 \times 10^{-5}$$

$$= 63 \; microstrain \; compressive.$$

(This is additional to the tensile bending strain of 277 microstrain in example 3, that is, total $= 214$ microstrain tensile at the bottom of the pipe; 340 $\mu\varepsilon$ compressive at the top of the pipe.)

Since the axial load in the pipe due to horizontal soil movement is sensitive to soil-pipe slip and joint movement, horizontal ground movement cannot be relied on to *reduce* tensile bending strains associated with ground movement lateral to the pipeline. For example, assuming perfectly telescopic joints at a transverse distance of $y = 2.6$ m, 7.8 m and so on (a spacing of 5.2 metres $= 17$ feet), from section 6.4.5.2 the maximum tensile strain in the pipe is

$$\varepsilon_{ym} = \frac{1.273}{K^*} \left(\frac{i}{d}\right) \frac{v_m}{i} I_f$$

where I_f depends on $z_s/L = 1.0$ and $L/d = 5.2/0.5 = 10.4$. The value of I_f is given by Poulos and Davis (1980) as 1.8. Thus

$$\varepsilon_{ym} = \frac{1.273}{972} \times \frac{2.6}{0.5} \times \frac{0.0061}{2.6} \times 1.8 = 2.87 \times 10^{-5}$$

$$= 29 \; microstrain \; tensile.$$

This value is virtually the same as the previously-calculated value for rigid joints of 33 microstrain. With a telescopic joint directly above the tunnel centre-line the previously-calculated compressive strain of 63 microstrain is reduced to zero.

To allow for the effect of soil drainage, drained soil parameters can be substituted in the expression for K^*. Taking $E_s' = \frac{2}{3}(1 + v_s')E_s$, $E_s'/E_s = 0.87$. Thus, the tensile strain is reduced from 33 microstrain to $0.87 \times 33 = 29$ microstrain. (Poisson's ratio has a minor effect and $v_s = 0.3$ has been used throughout.)

Long-duration pile loading tests have indicated that at loads above about one-third of the ultimate, the settlement increases with time, long after consolidation should have finished. This time-dependent pile settlement is primarily the result of shear creep. For a pipeline transverse to the tunnel drive, shear creep is expected to give a significant reduction in pipe strain with time when the shear stress adjacent to the pipeline is similarly more than one-third the ultimate value (see section 6.4.5.5). In effect, this is equivalent to a decrease in E_s' with time, the decrease being greater as the shear stress approaches the ultimate value.

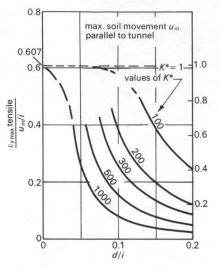

Figure 6.39 Maximum axial tensile pipe strain variation with diameter and stiffness factor K^*.

6.4.5.4. *Solutions for rigidly-jointed pipeline parallel to tunnel drive.* For a pipeline parallel to the tunnel, the maximum longitudinal soil movement occurs when the pipeline is directly above the tunnel drive. For a rigidly-jointed pipeline anchored distance $7.5i$ into undisturbed ground, Fig. 6.39 relates the maximum tensile pipe strain to the ratio d/i for values of pipe stiffness factor K^*. The maximum pipe strain is inversely proportional to K^*. For a pipe with negligible stiffness, $K^* = 1$, the limiting value for strain is

$$\varepsilon_{xm} = 0.607\frac{u_m}{i} \text{ at } x = i \text{ ahead of the tunnel face.}$$

The maximum tensile pipe strain occurs between a distance of i and $2.5i$ ahead of the tunnel face, and, for the purpose of calculating bending plus axial strain, should be taken as coincident with the maximum hogging moment (see section 6.4.2.5). The compressive pipe strain at x/i behind the tunnel face is numerically equal to the tensile strain at x/i ahead of the face and the strain is zero at $x = 0$. The reduction in pipe strain associated with pipe stiffness is shown on the right-hand scale of Fig. 6.39 taking the relative strain for $K^* = 1$ as 1.0. For pipelines parallel to, but offset from, the tunnel drive, the reduction in pipe strain due to pipe stiffness cannot be more than that indicated by Fig. 6.39.

6.4.5.5 *Solutions for pipe displacement and soil-pipe shear stress.* For the rigidly-jointed pipeline transverse to the tunnel drive, as described in section 6.4.5.3, the maximum horizontal pipe movement occurs where the axial pipe strain is zero at a transverse distance between i and $3i$ depending on the stiffness factor, K^*, and the pipe diameter. Figure 6.40(a) shows the variation of dimensionless maximum pipe movement v_p/v_m with pipe stiffness and diameter/trough width ratio. The pipeline movement, v_p, was calculated by integrating the pipe strain diagram starting from $v_p = 0$ at $y = 0$. For the pipeline parallel to the tunnel drive, as described in section 6.4.5, the maximum pipe movement is shown in Fig. 6.40(b). This maximum occurs where the

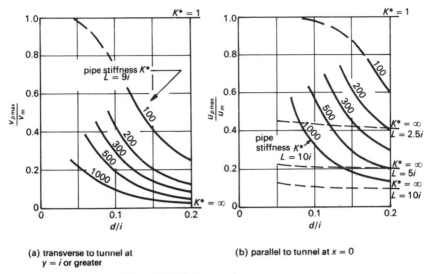

Figure 6.40 Maximum pipe movement.

axial pipe strain is zero at $x = 0$. The integration was started at the free end of the pipeline ($x = 10i$) on the assumption that the movement of the free end is zero for a very flexible pipe ($K^* = 100$) and the movement is half the incompressible pipe movement for a pipe of stiffness $K^* = 1000$. Figure 6.40 generally over-estimates the pipe movement which will under-estimate the bending induced in flexible branches off the main pipeline. For this purpose, the main pipeline movement could be assumed to be not less than the movement of an incompressible pipe. This is zero for the transverse case and is given in Fig. 6.40(b) for various anchorage conditions in the parallel case.

The maximum soil-pipe shear stress occurs where the differential soil-pipe movement is a maximum. This is coincident with the point of maximum pipe movement and zero axial strain (that is, change from tensile to compressive strain). Considering an element of pipe length ΔL, subjected to axial stress σ, and soil-pipe shear stress, τ, then for equilibrium

$$\tau \pi d \Delta L = \Delta \sigma A_p.$$

Substituting for the pipe stiffness factor,

$$K^* = \frac{E_p}{E_s} \frac{A_p}{\pi d^2 / 4},$$

then

$$\frac{\tau}{E_s} = \frac{K^* d}{4} \frac{\Delta \varepsilon}{\Delta L} = f\left(\frac{u_m}{i} \text{ or } \frac{v_m}{i}, \frac{i}{d}\right).$$

The relation between maximum soil-pipe shear stress and the pipe diameter/trough width ratio is shown in Fig. 6.41 for a pipeline transverse or parallel to the tunnel drive. The results of a finite-element analysis are given in Fig. 6.41(b) and the simplified analysis used here over-estimates the shear stress by up to 30%.

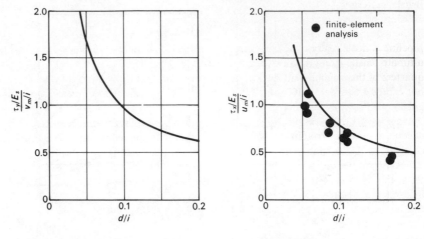

(a) transverse to tunnel   (b) parallel to tunnel

Figure 6.41 Maximum soil-pipe shear stress assuming full bond at soil-pipe interface.

6.4.6 *Soil-pipe slip*

With increasing magnitude of ground movement, the soil-pipe shear stress in the zone of ground movement reaches the limiting soil-pipe shear strength τ_a. Provided that the pipeline extends a minimum distance of $2.5i$ beyond the zone of ground movement, the shear stress in the anchorage length is less than τ_a. At full slip between the soil and the pipe, the the shear force along the pipeline can be integrated to give the maximum load in the pipe. Taking τ_a as constant, then at full slip the maximum tensile pipe strains are

$$\text{Transverse to the tunnel:} \quad \varepsilon_{y\ max} = \frac{5}{K^*} \frac{i\,\tau_a}{d\,E_s} \tag{6.30}$$

$$\text{Parallel to the tunnel:} \quad \varepsilon_{x\ max} = \frac{10}{K^*} \frac{i\,\tau_a}{d\,E_s} \tag{6.31}$$

where K^* = pipe stiffness factor,
$\quad\quad i/d$ = settlement trough width parameter/pipe diameter,
and $\quad \tau_a/E_s$ = soil-pipe shear strength/soil modulus.

Equations 6.30 and 6.31 give the maximum pipe strain when τ_a/E_s is less than the value of τ/E_s from Fig. 6.41.

6.4.7 *Application of theoretical analysis to practical problems*

6.4.7.1 *Introduction.* The principal effect to be evaluated when considering ground movements and buried pipelines is longitudinal bending of the pipeline associated with lateral soil movement. Axial tensile force in the pipe, which results from longitudinal soil movement, is additive but of secondary importance. The possibility of pipe fracture being caused by ground movements associated with tunnelling is confined to relatively brittle materials—that is, grey iron pipelines. Even for small-diameter (less than 300 mm) grey iron pipelines, which constitute the greatest risk and also the major part of the gas and water distribution system of the U.K., the risk is low provided that the pipeline is not already overstressed (that is, stressed to more than one-quarter of the ultimate tensile strength), or excessively corroded (that is, reduced to less than three-quarters of the original wall thickness) and provided that the ground movement does not exceed the order of 100 mm. Although there are relatively few tunnels where the ground surface movement exceeds 100 mm, there must be many old grey iron pipelines that are locally overstressed or excessively corroded. It is these pipelines that fracture as a result of seasonal ground movements associated with soil temperature and moisture changes, or ground movements associated with traffic loading.

The maximum design stress for grey iron at working loads, which is 0.25 U.T.S., is set at this level so as to limit non-elastic strain, that is, permanent set. At stresses well below the tensile strength permanent set constitutes failure in normal engineering terms. The intention should be to limit the pipe stress to the maximum design stress, although for temporary loadings this stress could be increased from 0.25 U.T.S. to 0.30 U.T.S. An indication of the pipe strain 'locked-in' as a result of installation and traffic loading is given in section 6.4.7.6. For a 100 mm diameter grey iron pipeline these may total approximately 250 microstrain compared with an elastic strain equivalent to the design stress of 400–500 microstrain.

6.4.7.2 *Soil properties.* The application of the methods of analysis in sections 6.4.2 and 6.4.5 requires knowledge of appropriate values for the soil deformation parameters E_u, E_s' and v_s'. It is recognized that if the analysis is linear-elastic then the appropriate elastic parameters are not simply a function of the soil but depend on the method of analysis, the means of evaluating the parameters and the type of construction process being modelled (Marsland, 1980). The elastic parameters are obtained from:

Table 6.5 Typical soil modulus values

Soil type (*Standard penetration N-value or undrained shear strength* (kN/m^2) *is given*)	$E_s(MN/m^2)$
Loose sand and gravel $N = 4$ 10	4–8
Medium dense sand and gravel $N = 10–30$	8–24
Dense sand and gravel $N = 30–50 +$	24–50 +
Soft clay $c_u = 20–50$	1–3
Firm clay $c_u = 50–75$	3–10
Stiff clay $c_u = 75–150$	10–20
Very stiff clay $c_u = 150 +$	20–40 +

Table 6.6 Typical pipe material properties for short-term static loading in direct tension (gradually applied, non-repeated loading without creep and at 20°C)[1][3]

Material	Ultimate tensile stress (N/mm²)	Lower yield stress or proof stress (N/mm²)	Typical max. design stress for working loads (N/mm²)	Design stress/ultimate stress	Design stress/yield or proof stress	Secant elastic modulus to max. design stress (GN/m²)	Elastic strain equivalent to design stress (microstrain)
Grey iron							
Pit cast before year 1914	110 to 140	70 to 90	27 to 35			67 to 87	400
Vertically cast to BS 78:1917	145 $\Big\}$ 20 N±/mm²	95 $\Big\}$ 0.1% proof stress	36	0.25	0.38	90 $\Big\}$ Ref[4]	400
Vertically cast to BS 78:1965	155	100	38			93	400
Spun cast to BS 1211:1958 over 16 in. dia. grade 12;	185	120	46			103	450
3 in. to 16 in. dia. grade 14	215	140	54			108	500
Ductile iron							
Spun cast to BS 4772 grade 420/12 material	420 min.	0.2% proof stress 300 min.	155	0.37	0.52	165	940
Mild steel to BS 534:1966 grade 320 material	320 min.	yield or 0.2% proof stress 195 min.	generally 80–115 (occasionally up to 140)	depending on duty 0.25–0.36 (occasionally up to 0.44)	0.4–0.6 (occasionally up to 0.72)	210	380–550
grade 410 material	410 min.	235 min.	95–140				450–660

Plastic[2]

UPVC to BS 3505	45 min.	—	20	0.5	—	—	2.8	7000
MDPE	30 at 50mm/min.	19 upper yield	7	0.4	—	0.7	10000
HDPE type 1	19 at strain rate 125 mm/min.	—	8	0.4	—	—	—
HDPE type 2	32 at strain rate 125mm/min.	14 lower yield / 22 upper yield	8	0.25	0.57	0.9	9000

(1) For iron and steel pipes, creep is not significant at 20°C at the maximum design stress indicated. These stresses are therefore also used for long-term loading.

(2) After 50 years, plastic properties are UPVC at 10°C – UTS = 25 N/mm², max. design stress = 12 N/mm², creep modulus = 1.4 GN/m²

MDPE at 20°C – U.T.S. = 8 N/mm², max. design stress = 5 N/mm², creep modulus at 3 N/mm² = 0.13 GN/m²

HDPE type 1 at 20°C – U.T.S. = 6.5 N/mm², " = 5 N/mm², " = 0.13 GN/m²

HDPE type 2 at 20°C – U.T.S. = 9.5 N/mm², " = 5 N/mm², " = 0.13 GN/m²

HDPE type 1 at 10°C – U.T.S. = 9.5 N/mm²

HDPE type 2 at 10°C – U.T.S. = 12.5 N/mm².

(3) For all the above materials the maximum design tensile bending stress is not less than the maximum design direct tensile stress.

(4) Secant elastic modulus for grey iron is for the elastic component of stress/strain curve, not stress/total strain curve.

(5) HDPE is high density polyethylene

MDPE is medium density polyethylene

(a) laboratory or *in-situ* testing

(b) loading tests and back-figuring the relevant parameters

(c) empirical correlations based on previous experience.

In the absence of more precise data it is suggested that the values of soil modulus in Table 6.5 are appropriate. These values have been suggested as suitable for use in an empirical formula for predicting pipe bending strain resulting from adjacent trench excavation. The formula is the result of linear-elastic modelling coupled with the result of field experiments. The magnitudes of the ground movements adjacent to the pipe are similar to those caused by tunnelling, and the data base/formula covers 0.075 metres < pipe diameter < 0.300 metres and $150 \, \text{kN.m}^2 < E_p I_p < 16\,000 \, \text{kN.m}^2$.

For granular soils an average value of Poisson's ratio v_s is 0.30. For clay soils, v_s is 0.50 for undrained conditions and the drained value is 0.1 to 0.45 (say 0.3 average). For the purpose of calculating relative values of pipe strain the drained modulus for clay soils can be taken as $E_s' = \frac{2}{3}(1 + v_s')E_s$, where E_s is from Table 6.5. The soil parameters should be based on the material adjacent to the pipe, which is a remoulded or disturbed soil, and not on the soil mass as a whole.

6.4.7.3 *Pipeline properties.* The stress/strain properties of distribution pipe material in common use are summarized in Table 6.6. It is worth mentioning that grey cast iron is not truly elastic even from zero stress. The elastic component of the stress/strain behaviour is non-linear and the path on unloading is different from that on reloading. In the design of grey iron pipelines for normal service conditions these effects can safely be ignored if the working stress does not exceed 0.25 times the ultimate tensile strength. Angus (1976) discusses in detail the engineering properties of cast irons. Elastic beam formulae, coupled with the ultimate tensile strength in direct tension, under-estimate the true rupture strength of a pipe in longitudinal bending. Roark (1965) gives data on rupture factors for various loading cases. The maximum size of service connection that may be drilled into a main is one-quarter of the main diameter. This causes a reduction of 50% in the beam bending strength if the hole is in the worst position, that is, this reduction allows for the stress concentration factor at rupture *and* the reduced section area (Boden, 1956; Andrews, 1972).

Lead/yarn joints are semi-rigid and under *ideal* circumstances can accommodate small rotations and axial slip. O'Rourke and Trautmann (1980) have summarized some test data on this type of joint. Further detailed information on the moment-rotation performance of lead caulked joints taken from non-corrosive soils is given by Harris and O'Rourke (1983). Rubber ring push-in joints and bolted-gland joints have little resistance to pull-out or moment under *ideal* conditions but this could be affected by pipe corrosion. A description of the joint types that have been used in the U.K. gas distribution system is contained in Department of Energy (1977).

Some typical pipe dimensions are given in Table 6.7. By substituting the values of Tables 6.6 and 6.7 in equation 6.24, and for comparative purposes taking $E_s = 10 \, \text{MN/m}^2$ and $w_m = 100 \, \text{mm}$, a measure of the sensitivity of a pipeline to system disturbance is obtained. This is given in Table 6.8 in terms of resistance to overstress. The relative position of ductile iron and plastic would change in favour of plastic if resistance to fracture were to be considered.

6.4.7.4 *Lateral ground movement.* The manner in which the various parameters interact is summarized in Table 6.9. Frequently it can be assumed that the pipeline is transverse

Table 6.7 Typical pipe dimensions

Nominal Diameter		Vertically cast grey iron BS 78 class B mean values		Spun cast grey iron BS 1211 class B mean values		Spun cast ductile iron BS 4772 class K9 mean values		Mild steel BS 534 : 1966 minimum values		UPVC BS 3505 class B mean values		HDPE and MDPE Minimum values			
													SDR = 41	SDR = 17	SDR = 11
mm	(inches)	d (mm)	t (mm)	d (mm)	t (mm)	d (mm)	t (mm)	d (mm)	t (mm)	d (mm)	t (mm)	d (mm)	t (mm)	d (mm)	t (mm)
75	3	96	9.7	96	7.4	98	6.0	89	3.2	89	3.1	90		5.1	8.2*
100	4	122	9.9	122	7.6	118	6.1	114	3.6	114	3.6	125		7.1	11.4*
150	6	177	10.9	177	8.4	170	6.3	168	3.6	168	4.7	180	4.4	10.2*	16.4
200	8	232	11.9	232	9.1	222	6.4	219	4.0	219	5.5	250	6.1	14.3*	22.8
250	10	286	13.2	286	9.9	274	6.8	273	4.0	273	6.8	280	6.9	15.9	25.5
300	12	334	14.5	334	10.9	326	7.2	324	4.5	324	8.1	315	7.7	18.0*	28.7
375	15	413	16.0	413	11.9	—	—	—	—	—	—	400	9.8	22.7	
450	18	492	17.5	492	13.2	480	8.6	457	6.3	457	11.4	450	11.0	25.5	
575	21	572	19.0	572	14.2	—	—	—	—	—	—	560	13.7	31.7	
600	24	650	20.3	650	15.2	635	9.9	610	6.3	610	15.3	630	15.4	35.7	
675	27	729	21.6	729	17.3	—	—	—	—	—	—	710	17.4	40.2	

*Dimensions of MDPE to British Gas Standard BGC/PS/PL2
HDPE is high density polyethylene, MDPE is medium density polyethylene.

Table 6.8 Resistance to pipe overstress associated with ground movements

Nominal Pipe Diameter mm inches		Pipe stress/Max. design stress					
		Vertically cast grey iron	Spun cast grey iron	Mild steel grade 410	UPVC short-term	HDPE SDR = 17 short-term	Ductile iron
100	4	3.42	2.77	2.16	1.76	1.71	1.33
200	8	2.16	1.77	1.45	1.02	0.85	0.92
300	12	1.61	1.34	1.16	0.69	0.68	0.71
600	24	0.96	0.91	0.68	0.37	0.34	0.43

Ground movement = 100 mm; soil modulus = $10 \, MN/m^2$.
For metal pipes the stress is reduced for pipes above 200 mm diameter if flexible joints are at 3.66 metre centres or above 300 mm diameter at 5.5 metre centres—joints are assumed to be in the least helpful position. For plastic pipes, flexible joints have little effect.

or parallel to and directly above the tunnel and use can be made of the solutions in sections 6.4.2.4 to 6.4.2.6. Where it is necessary to refine the analysis for the true position, the ground movement profile can be numerically differentiated twice to give the curvature, and hence pipe strain, neglecting flexural rigidity. This is only possible where the ground movement profile has the smooth form defined by the equations given in section 6.2 and the soil-pipe stiffness is sufficient to smooth out irregularities in the true ground movement. The numerical procedure and the limitations of numerical differentiation are given by Hartree (1958).

Where the tunnel passes beneath a network of connected pipes, the restraining effect of small diameter or very flexible branches off the main pipeline can be neglected when considering the bending of the main. The restraint to movement of a small diameter pipe because of the connection to a more rigid pipe or structure cannot generally be ignored. An approximate solution to this problem can be found by 'cutting' the joint between the branch and the main and then applying the forces necessary to restore compatibility of displacements and rotations at the joint.

Where a pipeline has flexible joints, it is necessary to assess the rotation at the joints, since excessive rotation can cause locked joints with associated high stress levels. It is sufficient to assume that the pipe between the joints is perfectly rigid, and the joints then lie on the ground movement profile with the pipe straight between the joints.

Example 5: Calculation of joint rotation. The pipeline is transverse to the tunnel drive and the pipe and tunnel details are the same as in Example 3. A joint is immediately above the tunnel drive and joints are at a transverse distance of 3.66 metres, 7.32 metres, and so on.

At $y = 0$, $w = 13.6$ mm.
At $y = 3.66$ m, $w = 5.05$ mm.
Differential $w = 8.55$ mm in 3.66 m.

A similar differential movement occurs between $y = 0$ and -3.66 m
The rotation at the joint immediately above the tunnel is

$$\theta = \frac{2 \times 8.55 \times 10^{-3}}{3.66} = 4.7 \times 10^{-3} \, rad = 0.27 \, degrees.$$

Table 6.9 Effect of tunnel, pipe and soil parameters on pipe bending stress and strain

Variation in parameter	Effect on pipe bending stress and strain	
Smooth ground movement profile or irregular slope as Fig. 6.23 and $1/\lambda > 0.5i$, or Fig. 6.27(a) and $1/\lambda > 1.0i$.	Irregular slope to ground movement profile and $1/\lambda < 0.5i$ transverse or $1/\lambda < 1.0i$ parallel to tunnel	Non-uniformity in foundation to pipe
Increasing tunnel depth	Increases trough width → decreases maximum settlement → decreases stress and strain	
Increasing ground loss at tunnel face	Increases settlement trough volume → increases maximum settlement → increases stress and strain	
Increasing tunnel diameter or face area	Increases ground loss Possible increase in trough width $\Big\}$ → increases maximum settlement → increases stress and strain	
Increasing pipe diameter at $d/t =$ constant or d/t increasing within the limits for commercially-available pipes	Increases $1/\lambda$ Increases d $\Big\}$ → increases stress and strain Increases $1/\lambda$ Increases d $\Big\}$ → practically no effect on stress and strain for $d/t =$ constant, but increase in stress and strain for increasing d/t	Decreases $d/\sqrt{I_p}$ → decreases stress and strain
Increasing pipe modulus E_p	Increase $1/\lambda$ → decreases strain; increases E_p/λ → increases stress	Increases $d/\sqrt{E_p}$ → decreases strain Increases $d \cdot \sqrt{E_p}$ → increases stress
Increasing depth of pipe below ground surface	Increases effective soil modulus, increases maximum settlement at pipe level $\Big\}$ → increases stress and strain and decreases trough width	

(*Contd.*)

Table 6.9 (Contd.)

Variation in parameter	Effect on pipe bending stress and strain		
	Smooth ground movement profile or irregular slope as Fig. 6.23 and $1/\lambda > 0.5i$ and $1/\lambda > 1.0i$.	Irregular slope to ground movement profile and $1/\lambda < 0.5i$ transverse or $1/\lambda < 1.0i$ parallel to tunnel	Non-uniformity in foundation to pipe
Increasing soil modulus E_s	Decreases $1/\lambda \rightarrow$ increases stress and strain		Increases $d \cdot \sqrt{E_s} \rightarrow$ increases stress and strain
Yielding of soil	Decreases pipe stress and strain compared with linear-elastic analysis		
Yielding of pipe	Decreases pipe stress and increases pipe *total* strain compared with linear-elastic analysis		
Pipe position oblique to tunnel drive compared with transverse case	Decreases stress and strain (Increases strain as a proportion of zero flexural rigidity case)		No effect
Pipe position parallel to and offset from tunnel drive compared with parallel to and directly above tunnel drive	Decreases stress and strain (Increases strain as a proportion of zero flexural rigidity case)		Decreases maximum settlement; decreases stress and strain
Flexible joints or non-continuous compared with rigid joints and continuous	Generally has little effect for narrow settlement troughs (which are associated with higher bending stresses and strains) if the joints are in the most adverse position for transverse or any position for parallel. Theoretically can increase the stress and strain for non-practical joint spacings.		Decreases stress and strain

The solutions given in sections 6.4.2.4, 6.4.2.5 and 6.4.2.7 have been tested against the small amount of published data concerned with the effect of tunnelling on buried pipelines. Agreement between the analysis and the measurements or observations is good; the data are however, very sparse—see Howe et al. (1980) and O'Rourke and Trautmann (1982). The solutions given in section 6.4.2.7 which allow for the effect of non-uniformity in the soil-pipe system are virtually the same as an empirical formula which 'fits' all the reliable U.K. data on pipe strains induced by ground movement associated with trenching. Since the formula for trenching shows no variation with the form of movement and relates to maximum movement only, there is reason to suppose that the relationship derived for tunnelling is also reliable.

In general, it should be expected that the solutions in section 6.4.2 over-estimate pipe bending moment since all the simplifications have been on the safe side where necessary. Given an accurate specification for the ground movements, it is expected that section 6.4.2 should predict pipe stress to within $\pm 50\%$ or better, if the correct assumptions are made about joint stiffness and position. It should be remembered that it is a good prediction of ground movement if the maximum settlement can be estimated to within 50% without having measurements relating to a similar tunnel constructed under similar conditions in similar ground.

6.4.7.5 *Longitudinal ground movement.* In considering the effect of longitudinal ground movements on pipelines it is necessary to estimate the possible pulling effect on branches off the main pipeline. Consider, for example, the main pipeline parallel to and directly above the tunnel line and a branch off the main at a position directly above the tunnel face position. Horizontal movement of the branch is restrained due to the connection to the main pipeline. This may induce significant horizontal bending stress in the branch although a flexible joint greatly reduces the effect. If the main pipeline is offset from the tunnel centreline, axial pulling of the branch is possible due to movement of the main pipeline horizontally towards the tunnel centreline. If there is freedom of movement at the joint, excessive joint extension can cause leakage. Alternatively, a rigid joint may induce tensile stress in the branch pipeline.

Where a pipeline has joints that are free to accommodate longitudinal movement, it is necessary to assess the joint slip since excessive movement can cause leakage. For this assessment the pipeline may be assumed to be incompressible and the movement calculated as in Poulos and Davis (1980). Alternatively, the slip at the joint is not more than the differential ground movement between mid-points of adjacent pipes.

6.4.7.6 *Other considerations.* When assessing the effect of tunnelling on buried pipelines it is useful to have an indication of the pipe stresses that arise from other causes. For gas distribution mains the effect of internal pressure on pipe stress levels is usually insignificant and the principal causes of stress are externally applied. In some water mains the internal pressure, which causes uniform ring tension, may be large enough to contribute significantly to failure. For 100 mm diameter spun grey iron pipelines a summary of published data on measured longitudinal bending strains is given in Table 6.10.

The transient load attributable to traffic can be increased by a damaged or uneven road surface. The dynamic load is generally in the range 1.5 to 2.5 times the static load, but can reach 4.0 times the static load. Longitudinal bending stresses reduce with increasing pipe diameter (that is, increasing beam strength) and, on the basis of section

Table 6.10 Typical longitudinal bending strain in 100 mm spun grey iron pipelines

	Pipe strain (microstrain) for different standards of pipeline construction		
Cause of pipe strain	Very good (granular bedding/ densely compacted backfill)	Average (trimmed trench bottom/ compacted backfill)	Very poor (uneven trench bottom/loose clay backfill)
'Locked-in' due to main laying and trench reinstatement	25	50	150 +
'Locked-in' after consolidation of backfill due to traffic loading etc. (up to 1 year after installation)	25	50	150 +
Total 'locked-in' due to installation	50	100	300 +
Transient static load due to traffic on smooth road surface	50	100	250 +

6.4.2.7, are approximately proportional to the value of $d(1/E_pI_p)^{0.5}$ which at d/t and E_p constant implies strains inversely proportional to diameter. For comparative purposes, Table 6.10 is for $d = 0.122$ metres and $E_pI_p = 485$ kN·m^2.

6.4.8 *Design to accommodate ground movement*

6.4.8.1 *Measures to protect old pipe distribution systems.* When tunnelling beneath established pipeline networks, concern is likely to focus on grey iron pipes— particularly small-diameter pipes which are known to be sensitive to disturbance. This is partly due to failure being related to fracture only and not including the possibility of failure due to non-recoverable or excessive deformation in less brittle materials. Reference to Table 6.8 shows that steel pipelines are not very much better than grey iron in terms of failure due to permanent deformation (at least at the minimum wall thickness given in Table 6.7). It is interesting to note that the performance of small-diameter steel distribution pipework when subjected to earthquake movements is no better than that of grey iron (see, for example, Mikaoka, 1978).

The effect of tunnelling on pipelines can be reduced by controlling the ground loss at the tunnel and reducing consolidation due to lowering of the water table (see earlier in this chapter). Factors to be considered include adequate ground investigation before and during tunnelling, tunnel face support and erection of the primary lining and grouting of overbreak without delay, driving the tunnel as quickly as possible and continuously, and as far as possible making the tunnel watertight.

Ground surface movements can be measured quite easily, and besides providing a data base for future tunnelling operations, measurements can provide an advanced warning of possible problems on critical sections. Pipe-joint leakage can be monitored before and during tunnelling and repairs made as necessary. Similarly, pipe strains can be measured but the technique requires considerable experience to obtain reliable results. Generally it is necessary to make an assessment of the effect of the tunnelling works *before* the work has started so that re-laying or polyethylene sliplining of critical pipelines can be carried out before tunnelling begins. Renovation of old pipelines by thermoplastic pipe insertion has been extensively used in the gas industry and 'size-for-size' replacement of cast iron with polyethylene is now available for 75 mm and 100 mm diameter pipes. The method involves a machine being drawn through the iron main, expanding and fragmenting the pipe as it progresses. A PVC replacement sleeve is simultaneously drawn through. It supports the ground and protects the new polyethylene main. The polyethylene main is installed inside the sleeve by conventional means. A description of 'size-for-size' replacement with polyethylene is given by Freeman (1983). Conventional polyethylene sliplining is described by Reed (1978) and by Whipp and Glennie (1982).

6.4.8.2 *Design of new pipelines to accommodate ground movement.* When subjected to the moderate ground movements associated with tunnelling, all modern pipeline materials perform well when compared with grey iron. It is emphasized that flexible pipe materials require at least as good bedding and backfilling as was used when the grey iron distribution system was installed. In particular, an uneven machine-dug trench bottom is not adequate if the pipeline has to carry significant external loads. Flexible pipes also rely on adequate compaction of backfill around the pipe. Where ground movement is lateral to a pipeline, ductile iron and high or medium-density polyethylene at the diameter-to-wall thickness ratio commonly used are known to perform well. Steel pipelines, if required to accommodate ground movement, can be sized accordingly. Flexible joints are beneficial, but, if high security against joint pull-out is required, welded or anchored flexible joints should be used. Where there is a large change in pipeline flexural stiffness, a 'pinned-joint' should be used if possible since this reduces bending strains by about 70%. This applies where a pipeline is connected to a structure and differential movement is expected, or where a small branch is attached to a relatively stiff main pipeline. Apart from the preceding case, flexible joints on small diameter pipelines do not have much beneficial effect at the spacings commonly used, that is, 3.66 metres or more.

Where ground movement is horizontal and longitudinal to a pipeline the effect is reduced if the pipeline can accommodate axial movement at the joints. Smooth-surfaced pipelines or loose polyethylene sleeving (applied as a corrosion protection measure), or pipelines installed within sleeves, promote soil-pipe slip which reduces the pulling forces associated with horizontal ground movement.

6.5 Ground movements and buildings

The Winkler foundation approach to the modelling of a beam on an (assumed) elastic subgrade is obviously as relevant to the problem of a structural ground beam as to the problem of buried pipe deformation. In the present chapter it is clearly not possible to discuss buildings and their foundations in general. Rather, it is proposed to use just one example as a guide to procedure. This takes the form of a multi-bay portal framed structure supported on pad footings which themselves rest on the surface of a Winkler model subgrade. The problem, studied by Couldery (1982), is considered in two-dimensions only with respect to a terminal transverse settlement trough.

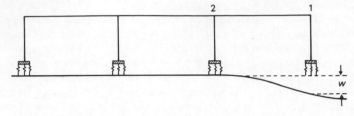

Figure 6.42 Settlement beneath a single footing of a multi-bay structure.

Under the influence of tunnelling-induced ground settlement each footing may settle differentially, thereby inducing rotation (sway) in structural members. Simply, from Fig. 6.42, footing settlement $w_f(\leq w)$ creates a net stress $k(w - w_f)$ in the subgrade, resulting in structural member 1-2 deflection. Letting force P_s cause deflection d_s in member 1-2 having length L, then

$$P_s = \frac{12EI}{L^3} d_s$$

Equating force P_s to the force in the subgrade,

$$\frac{12EI}{L^3} w_f = ka(w - w_f)$$

where a is the area of the footing.

Thus,

$$\left(\frac{12EI}{L^3} + ka \right) w_f = kaw.$$

Deflection w_f, equivalent to sway in the structural member, is therefore calculated by applying force kaw to the footing. In general, for differential settlements at the footings a load of $ka_i w_i$ is applied to each footing, where w_i is the predicted unloaded ground settlement at footing i having soil base-contact area a_i.

Analysis of a multi-bay, multi-storey framed structure involves the definition of structural, foundation and ground nodes and solution of the structural stiffness matrix **K** in

$$\mathbf{Kd} = \mathbf{F} \qquad (6.32)$$

where **F** is the force vector and **d** is the displacement vector at a node. Reduction out of the equation of those forces and displacements known to be zero requires the solution of the set of equations

$$\sum_j (k_{ij} d_j) = F_i \qquad (6.33)$$

where i is the number of non-zero forces and j is the number of non-zero displacements.

In practice, the structural stiffness matrix is formed by merging the individual stiffness equations of the structure's component members.

Force F is taken to include active and passive point and distributed loads. The degrees of freedom may be divided into exclusive classes and labelled (subscripted in the matrices) s (structure), f (foundation) and g (ground). This system is shown diagrammatically in Fig. 6.43.

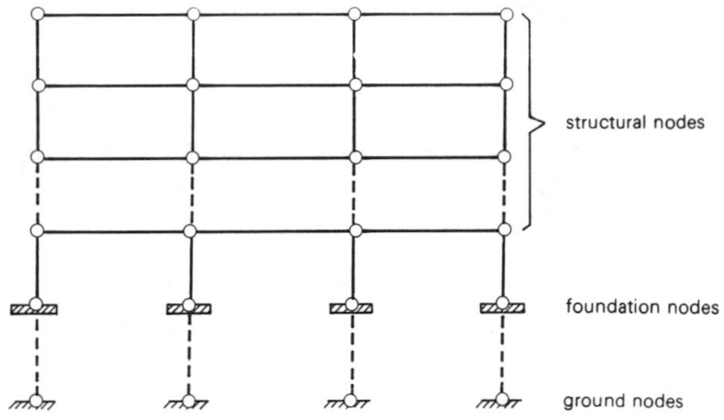

Figure 6.43 Designation of structural foundation and ground nodes.

The force vector $\mathbf{F}_g$ for the ground nodes is unknown. However, the ground deflection vector $\mathbf{d}_g$ can be defined because it is assumed to correspond to the deflections that the ground nodes would have undergone as a result of tunnelling, had the structure not been present. This means that, in terms of vertical deflection (settlement), $\mathbf{d}_g$ represents the predicted w (function of x, y) values. The ground nodes and the foundation nodes are coupled according to the Winkler—or any other chosen—sub-grade model relationship.

Considering the coupling of *structure and foundation only*, the stiffness equation for the nodes is:

$$\begin{bmatrix} \mathbf{K}_{ff} & \mathbf{K}_{fs} \\ \mathbf{K}_{sf} & \mathbf{K}_{ss} \end{bmatrix} \begin{bmatrix} \mathbf{d}_f \\ \mathbf{d}_s \end{bmatrix} = \begin{bmatrix} \mathbf{F}_f \\ \mathbf{F}_s \end{bmatrix} \tag{6.34}$$

This equation is composed by merging the member stiffness equations relating each coupled pair of nodes in the structure-foundation system. The vertical stiffness matrix for a two-node (1, 2) member would be:

$$\begin{bmatrix} \frac{AE}{L} & 0 & 0 & \frac{-AE}{L} & 0 & 0 \\ & \frac{12EI}{L^3} & \frac{-6EI}{L^2} & 0 & \frac{-12EI}{L^3} & \frac{-6EI}{L^2} \\ & & \frac{4EI}{L} & 0 & \frac{6EI}{L^2} & \frac{2EI}{L} \\ & \text{SYMMETRIC} & & \frac{AE}{L} & 0 & 0 \\ & & & & \frac{12EI}{L^3} & \frac{6EI}{L^2} \\ & & & & & \frac{4EI}{L} \end{bmatrix} \begin{bmatrix} v_{s1} \\ w_{s1} \\ R_{s1} \\ v_{s2} \\ w_{s2} \\ R_{s2} \end{bmatrix} = \begin{bmatrix} P_{y1} \\ P_{z1} \\ M_1 \\ P_{y2} \\ P_{z2} \\ M_2 \end{bmatrix} \tag{6.35}$$

where, considering the yz plane,

v_{s1}, w_{s1}, R_{s1} are the y, z and rotational displacements at node 1,
v_{s2}, w_{s2}, R_{s2} are the y, z and rotational displacements at node 2.

P_{y1}, P_{z1}, M_1 are the associated forces at node 1,
P_{y2}, P_{z2}, M_2 are the associated forces at node 2, and
A, E, I, L represent the area, elastic modulus, second moment of area, and length of the member.

Now consider the coupling between the foundation nodes and the ground nodes. The stiffness equation for this system may be written

$$\mathbf{Nd} = \mathbf{F} \tag{6.36}$$

where $\mathbf{N}$ represents the stiffness matrix for the ground-foundation interaction. The stiffness equation is:

$$\begin{bmatrix} \mathbf{N}_{gg} & \mathbf{N}_{gf} \\ \mathbf{N}_{fg} & \mathbf{N}_{ff} \end{bmatrix} \begin{bmatrix} \mathbf{d}_g \\ \mathbf{d}_f \end{bmatrix} = \begin{bmatrix} \mathbf{F}_g \\ \mathbf{F}_f \end{bmatrix} \tag{6.37}$$

Since the ground-foundation system is in equilibrium in its displaced position, there are no net forces across the ground-foundation interface. Thus

$$\mathbf{F}_g + \mathbf{F}_f = 0$$

that is,

$$\mathbf{F}_g = -\mathbf{F}_f \tag{6.38}$$

For calculation of the terms in the stiffness matrix $\mathbf{N}$ consider the force and displacement system between the ith foundation node f_i and its corresponding ground node g_i. The force system between them comprises a vertical force and two moments, and the displacement system comprises a vertical settlement and two rotations:

$$\mathbf{F}_{gi} = -\mathbf{F}_{fi} = \begin{bmatrix} P_i \\ M_{ix} \\ M_{iy} \end{bmatrix}, \qquad \mathbf{d}_i = \begin{bmatrix} w_i \\ \theta_{ix} \\ \theta_{iy} \end{bmatrix} \tag{6.39}$$

where θ_i (radians) defines the two rotational components (x, y) of the displacement vector. Comparatively small horizontal x- and y- direction forces and displacement are ignored.

Vertical force P_i may be written

$$P_i = ka_i(w_{fi} - w_{gi}) \tag{6.40}$$

where, as per earlier notation,

$a_i \equiv a$,   the foundation area at the ith position
              (or the horizontal projection of the foundation if complex),
$w_{gi} \equiv w$, the ground settlement without the presence of the structure, and
$w_{fi} \equiv w_f$, the foundation-subgrade interface settlement.

For the derivation of *moment M_i* we have

$$M_{ix} = k \int_{a_i} (w_{fi} - w_{gi}) x \, dx \, dy \tag{6.41}$$

and

$$M_{iy} = k \int_{a_i} (w_{fi} - w_{gi}) y \, dx \, dy \tag{6.42}$$

where the node is at the origin.

In terms of the angular displacements,

$$w_i = \theta_{ix} x + \theta_{iy} y \tag{6.43}$$

and so

$$M_{ix} = k\{(\theta_{fix} - \theta_{gix})J_{ixx} + (\theta_{fiy} - \theta_{giy})J_{ixy}\} \tag{6.44}$$

and

$$M_{iy} = k\{(\theta_{fix} - \theta_{gix})J_{ixy} + (\theta_{fiy} - \theta_{giy})J_{iyy}\} \tag{6.45}$$

where

$$J_{ixx} = \int_{a_i} x^2 \, da, \quad J_{ixy} = \int_{a_i} xy \, da, \quad J_{iyy} = \int_{a_i} y^2 \, da. \tag{6.46}$$

Expressing

$$\theta_i = \begin{bmatrix} \theta_{ix} \\ \theta_{iy} \end{bmatrix}, \quad J_i = \begin{bmatrix} J_{ixx} & J_{ixy} \\ J_{ixy} & J_{iyy} \end{bmatrix}, \quad M_i = \begin{bmatrix} M_{ix} \\ M_{iy} \end{bmatrix} \tag{6.47}$$

we have

$$k \begin{bmatrix} a_i & -a_i \\ -a_i & a_i \end{bmatrix} \begin{bmatrix} w_{fi} \\ w_{gi} \end{bmatrix} = \begin{bmatrix} P_{fi} \\ P_{gi} \end{bmatrix} \tag{6.48}$$

and

$$k \begin{bmatrix} J_i & -J_i \\ -J_i & J_i \end{bmatrix} \begin{bmatrix} \theta_{fi} \\ \theta_{gi} \end{bmatrix} = \begin{bmatrix} M_{fi} \\ M_{gi} \end{bmatrix}$$

Thus, the ground-foundation stiffness equation

$$\mathbf{Nd} = \mathbf{F}$$

can be expressed as

$$k \begin{bmatrix} \mathbf{a} & 0 & -\mathbf{a} & 0 \\ 0 & \mathbf{J} & 0 & -\mathbf{J} \\ -\mathbf{a} & 0 & \mathbf{a} & 0 \\ 0 & -\mathbf{J} & 0 & \mathbf{J} \end{bmatrix} \begin{bmatrix} \mathbf{w}_f \\ \boldsymbol{\theta}_f \\ \mathbf{w}_g \\ \boldsymbol{\theta}_g \end{bmatrix} = \begin{bmatrix} \mathbf{P}_f \\ \mathbf{M}_f \\ \mathbf{P}_g \\ \mathbf{M}_g \end{bmatrix} \tag{6.49}$$

where

$$\mathbf{a} = \begin{bmatrix} a_1 & & & 0 \\ & a_2 & & \\ & & \cdot & \\ & & & \cdot \\ 0 & & & a_n \end{bmatrix}; \quad \mathbf{J} = \begin{bmatrix} J_1 & & & 0 \\ & J_2 & & \\ & & \cdot & \\ & & & \cdot \\ 0 & & & J_n \end{bmatrix}; \tag{6.50}$$

$$\mathbf{w}_f = [w_{f1} \, w_{f2} \ldots w_{fn}]^T,$$
$$\boldsymbol{\theta}_f = [\theta_{f1} \, \theta_{f2} \ldots \theta_{fn}]^T, \tag{6.51}$$

and similarly for $\mathbf{w}_g, \boldsymbol{\theta}_g, \mathbf{P}_f, \mathbf{M}_f, \mathbf{M}_g$.

Noting the form of the stiffness equation and from earlier equations that

$$N_{ff} = N_{gg} = k \begin{bmatrix} a & 0 \\ 0 & J \end{bmatrix} \qquad (6.52)$$

and

$$N_{fg} = N_{gf} = -k \begin{bmatrix} a & 0 \\ 0 & J \end{bmatrix} \qquad (6.53)$$

the equations for the foundation-ground system and the structure-foundation system can be merged to yield the full structural system stiffness matrix:

$$\begin{bmatrix} N_{gg} & N_{gf} & 0 \\ N_{fg} & N_{ff}+K_{ff} & K_{fs} \\ 0 & K_{sf} & K_{ss} \end{bmatrix} \begin{bmatrix} d_g \\ d_f \\ d_s \end{bmatrix} = \begin{bmatrix} F_g \\ F_f \\ F_s \end{bmatrix} \qquad (6.54)$$

This equation is soluble since d_g, F_s are known, and there are three unknowns: d_s, d_f and $F_g (= F_f)$. There are two possible approaches to the solution. The full system stiffness matrix can be calculated, inputting d_g, F_s and solving for d_f, d_s, F_f. Alternatively, the predicted (unloaded) ground settlements d_g may be converted to equivalent loads (as shown earlier), these loads added to the force vector F_s, the ground deflection vector d_g set to zero, and then the system solved for d_f, d_s, F_s. Since the first method requires the solution of a set of equations of size close to the sum of the order of d_g plus d_f plus d_s, the second method, which requires the solution of a set of equations of order d_f plus d_s only, is to be preferred. For this equivalent load method the following truncation of equation 6.54 must be solved:

$$\begin{bmatrix} N_{ff}+K_{ff} & K_{fs} \\ K_{sf} & K_{ss} \end{bmatrix} \begin{bmatrix} d_f \\ d_s \end{bmatrix} = \begin{bmatrix} F_f \\ F_s \end{bmatrix} \qquad (6.55)$$

Stiffness matrices $K_{ff}, K_{fs}, K_{sf}, K_{ss}$ can be calculated from input data on structural member lengths, areas, moduli of elasticity, and second moments of area. Matrix N_{ff} is defined by equations 6.52 and 6.50. Since it is diagonal, only diagonal terms need to be added to the basic stiffness matrix. This may be accomplished by adding linear and rotational spring elements to the foundation nodes, the stiffness of a linear spring being given by

$$S_{li} = k a_i \qquad \text{(units of N/m)} \qquad (6.56)$$

and the stiffness of the rotational spring being given by

$$S_{\theta i} = k J_i \qquad \text{(units of Nm/radian)} \qquad (6.57)$$

To summarize, the method of ground-building analysis is:

(a) Assuming the building not to be there, predict the tunnelling-induced ground settlement profile beneath the building. Obviously one is not restricted to a normal probability form of transverse settlement distribution, but such a distribution does have the backing of case-history data.

(b) Calculate an equivalent force for each foundation element, where the force at node $i = k a_i w_i$.

(c) Use a structural analysis computer program to solve the structural system stiffness equation. At each foundation node the force-deflection relation is defined by linear stiffness ka_i and a rotational spring stiffness kJ_i. There is assumed restraint on any horizontal deflection at foundation nodes.

(d) Analyse the output from the structural system stiffness equation to determine the structural member loads when affected by the tunnelling-induced ground settlement.

(e) Interpret these loadings in terms of possible structural damage. In the absence of empirically-based damage criteria related to 'fast' ground settlement, standard self-weight settlement damage criteria—optimistic in the present setting—must be used (see section 6.6.).

6.6 Building damage: practical appraisal

Time and financial constraints, and the sheer number of buildings adjacent to a tunnel route may dictate that rapid assessments of structural vulnerability be performed at an early site investigation stage. Two methods of data presentation for such a rapid assessment have been used by one of the present writers (Attewell, 1981) for an urban tunnelling contract in England. Only permanent (transverse) settlements and derivatives are shown.

In the first of these methods, surface settlement, transverse displacement and transverse strain profiles are computed and plotted on a transverse vertical cross-section to a scale of 1 : 200 (see, for example, Fig. 6.44). The position of the building foundation within the settlement trough is then readily noted and preliminary inferences with respect to possible structural deformation can be drawn. These inferences are based on one or more damage criteria related to the building as a whole but take no account of individual structural detail. During the property surveys both before and after tunnelling, attention would be directed to such points as zones of stress concentration

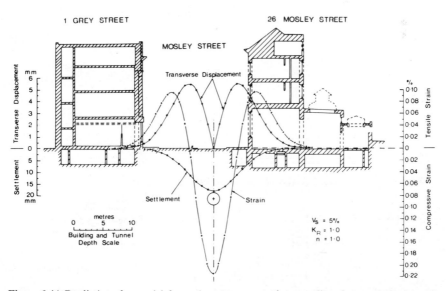

Figure 6.44 Prediction of ground deformation transverse to the centre line of a tunnel driven in soil beneath a major street in Newcastle upon Tyne (Section C on Figure 6.48).

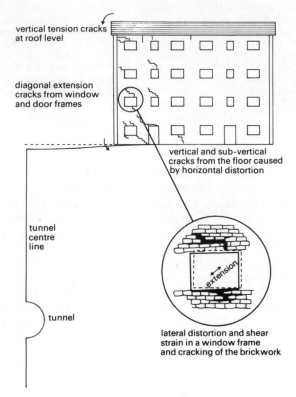

Figure 6.45 Fractures that could develop in a load-bearing brick wall affected by hogging settlement.

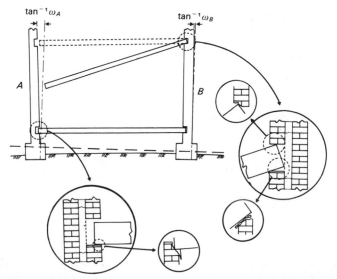

Figure 6.46 Rotation of bearing walls and consequent instability of joists.

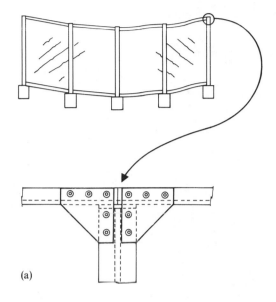

(a)

Figure 6.47 (a) Frame structure with exaggerated sagging deformation. The columns are braced at the base, shear deformation being induced in the brittle panel infilling as a result of differential settlement and the cracking being caused by diagonal tensile strain. Full bending moments are shown for the two right-hand side top and floor cross-beams, with obvious points of contraflexure. The left-hand side cross beams have rotated and do not support on bending moments. There will also be horizontal in-plane and out-of-plane ground distortion superimposed in the system.

and potential weakness at the corners of windows and doors (see Fig. 6.45) and note, for example, that sub-vertical crack propagation will follow the joints in weak lime mortar or possibly pass through the bricks where the mortar is a strong Portland cement mix, the position of floor joists with respect to the tunnel advance direction (Fig. 6.46), the detailed construction of frame buildings (as indicated in Fig. 6.47 of the panel infilling, stiffening of joints at heads of column, restriction on column rotation at footings) and so on. These matters are described in more detail in Attewell (1978), to which reference should be made.

In the second (often additional) method, an example of which is shown in Fig. 6.48, the most important ground deformation parameters are quantified along key lines drawn on a 1:500 scale plan of an area. Since the predicted normal probability transverse settlement profile is assumed to be symmetrical about the tunnel centre line, the settlement, horizontal displacement and ground strain and slope predictions located on Fig. 6.48 are also reflected on the other side of the centre line.

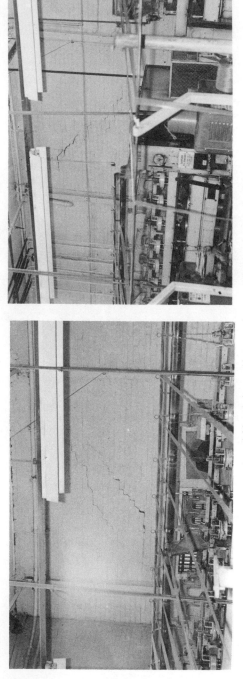

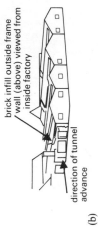

tunnel driven below and between the two wall areas photographed above

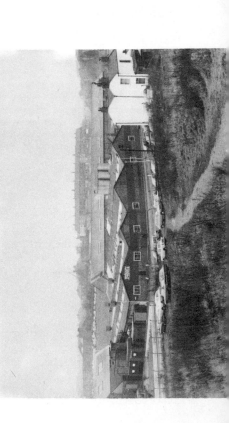

(b)

brick infill outside frame wall (above) viewed from inside factory

direction of tunnel advance

(b) Indirect tensile failure in panel infilling.

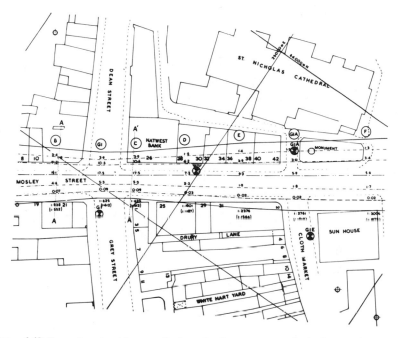

Figure 6.48 Example extract from a site investigation report: method of presenting predicted ground deformation information.

○ Transverse section reference.

❂ Exploratory borehole.

—Upper and lower lines: practical limit of settlement at $\sqrt{2\pi i}$.

— — — Settlement at $\sqrt{3}i$ from tunnel centre line (mm).

·· —·· — Settlement at distance i from tunnel centre line (mm).

— · — · Settlement (maximum) on tunnel centre line (mm).

— ·· —·· Maximum lateral displacement (mm) at distance $\sqrt{3}i$.

— — — Maximum tensile strain (%) at distance $\sqrt{3}i$.

1 : x Average slope of settlement profile.

(1 : x) Maximum slope of settlement profile.

Building response to settlement depends upon the length of the structure. Fig. 6.49 indicates that a short building may 'ride' a settlement trough, rigid body tilt releasing the structure from much internal distortion. For an overall assessment, notwithstanding the earlier analyses in this chapter, it is usually assumed at the site investigation stage that the building foundation will deform conformably with the ground. Three structural damage criteria, based on deflection ratio, angular distortion and horizontal distortion (see Figs. 6.50 and 6.51) have been discussed in Attewell (1978) and in which some detailed consideration is given to actual styles of failure. Additional references for further reading are Skempton and MacDonald (1956), Meyerhof (1956), Polshin and Tokar (1957), Bjerrum (1963), Grant et al. (1974), Leonards (1975), Burland and Wroth (1975), O'Rourke et al. (1976) and Wahls (1981). It is recommended that the latter reference be acquired first and the values for the damage criteria appropriate to framed

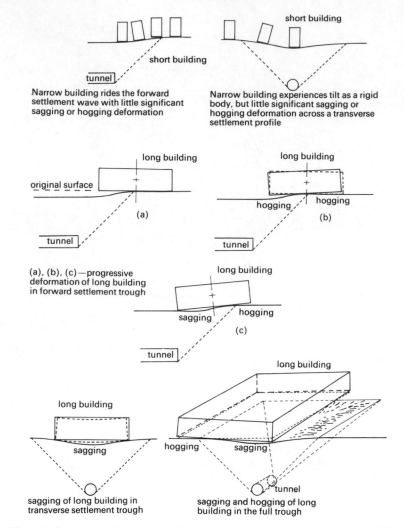

Figure 6.49 Some idealized modes of behaviour—narrow buildings and long buildings.

structures and load bearing walls assessed with those discussed in Attewell (1978). Some damage criteria are listed in Table 6.11. Wahls has pointed out that allowable differential settlements in a structure tend to be significantly smaller when the settlement pattern is concave downwards (hogging mode), in contrast to the concave upwards sagging mode. Leonards has also made an important point relevant to framed structures of the general type considered at greater length in section 6.5. He notes that whereas it is the usual practice in damage prediction and assessment to remove the rigid body tilt element of the rotation when adopting an angular distortion criterion, tilting does in fact contribute to the stress and strain in a framed structure.

Obviously, different criteria are required for different types of building (see Table

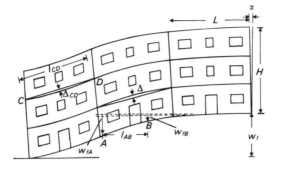

α is the rigid body rotation (tilt)

w_{fA}, w_{fB} are the foundation settlements at points

A and B respectively

Angular distortion $\omega = (w_{fA} - w_{fB})/l_{AB}$

Δ_{CB} is the relative deflection between points C and D

Deflection ratio $= \Delta_{CD}/l_{CD}$

Figure 6.50 Definition of angular distortion and deflection ratio. The building response to settlement is, of course, exaggerated. Δ_{CD} is a relative sagging deflection. Δ (unsubscripted) is a relative hogging deflection.

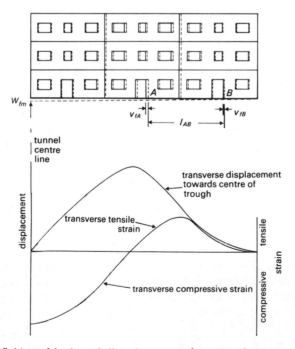

Figure 6.51 Definition of horizontal distortion. v_{fA} and v_{fB} are the transverse (horizontal) displacements of points A and B, respectively, (related to the foundation), towards the centre of the settlement trough. l is the distance between A, and B. Horizontal distortion, ε_{yft} (foundation tensile strain) is $\varepsilon_{yft} = (v_{fA} - v_{fB})/l_{AB}$.

Table 6.11 Criteria for building damage (self-weight settlement)
(a) Angular distortion

Structure	Limit		Notes	Reference
Load bearing walls or panel walls in frame structures	1/300	(0.003)	Cracking likely	Skempton and McDonald (1956)
"	1/150	(0.007)	Structural damage probable	"
"	1/500	(0.002)	Design criterion against cracking	"
Frames with diagonals	1/600	(0.002)	Danger	Bjerrum (1963)
Buildings generally	1/500	(0.002)	Safe limit for no cracking	"
Panel walls	1/300	(0.003)	First cracking	"
"	1/150	(0.007)	Considerable cracking of panel and brick walls	"
Buildings generally	1/150	(0.007)	Danger of structural damage	"
Flexible brick walls $L/H > 4$	1/150	(0.007)	Safe limit	"

Structure	Sand and hard clay	Plastic clay	Reference
Column foundations, steel and reinforced concrete structures	0.002 (1/500)	0.002 (1/500)	Polshin and Tokar (1957)
Column foundations for end rows of columns with brick cladding	0.007 (1/150)	0.001 (1/1000)	"
Column foundations for structures where auxiliary strain does not arise during non-uniform settlement of foundations	0.005 (1/200)	0.005 (1/200)	

(b) *Deflection ratio*

Structure	Limit		Notes	Reference
	Sand and hard clay	*Plastic clay*		
Plane load bearing brick walls	0.003 (1/3333)	0.004 (1/250)	For multi-storey buildings at L/H ≤ 3	Polshin and Tokar (1957)
,,	0.005 (1/2000)	0.0007 (1/1500)	For multi-storey buildings at L/H ≥ 5	,,
,,	0.001	(1/1000)		Grant *et al.* (1974)

(c) *Horizontal distortion*

Structure	Limit	Notes	Reference
Load-bearing walls/continuous brick cladding	1/2000 (0.0005)	Onset of cracking	Burland and Wroth (1975)

6.11). As an overall comment, it can be said that for open-frame structures a limiting angular distortion of 1/250 could be adopted. For infill frame structures and panel walls, an angular distortion of about 1/500 should be taken as an allowable limit, 1/300 would be expected to induce some cracking in architectural elements, and 1/150 could cause structural damage. In the case of load-bearing walls and continuous brick cladding, the tolerable and allowable differential settlement would tend to be smaller than those values adopted for frame structures and would depend upon the length/height ratio of the wall. A general limiting angular distortion would be about 1/1000. In the case of a deflection ratio criterion, a limiting value of about 1/3300 has been suggested for loadbearing walls, irrespective of foundation soil.

Acknowledgements

The following colleagues have worked on the ground movements problem with P.B. Attewell at Durham University: John Woodman, Malcolm Reeves, Mitsuo Tsutsumi, Peter Couldery, Chris Golightly and Mick Hurrell. The views expressed by the authors do not necessarily represent the views or policies of the Institutions that employ them.

Notation

A Consolidation settlement coefficient

A Area of structural member

A_p Area of pipe wall

a Area of pad footing

B Plate width

C_a Pipe-lead (joint) adhesive strength

C_c Compression index

c Soil cohesion

c_u Undrained shear strength of soil

d Beam width

d Outside diameter of pipe

d_s Deflection in structural member

$\mathbf{d}$ Displacement vector

E Elastic modulus

E_p Elastic secant modulus of pipe

E_s Elastic secant modulus of soil

E_u Undrained elastic secant modulus of soil

e Void ratio

e_0 — Initial void ratio

Δe — Change in void ratio

$\mathbf{F}$ — Force vector

$\mathbf{F}_f$ — Force vector for foundation nodes

$\mathbf{F}_g$ — Force vector for ground nodes

$\mathbf{F}_s$ — Force vector for structural nodes

H — Initial thickness of soil element

ΔH — Change in soil thickness

I — Moment of inertia

I_f — Influence factor

I_p — Moment of inertia of pipe

I_1 — Influence factor

I_2 — Influence factor

i — Parameter defining the form and span of the settlement trough on the assumption that the semi-transverse (y-axis) settlement profile can be described by a normal probability equation

i_t — Value of i composited by ground-loss and consolidation settlement

i_x — x-axis i-parameter

i_y — y-axis i-parameter

K — Modulus of subgrade reaction (FL^{-2})

K_{eff} — Effective modulus of subgrade reaction

K_r — Empirically-determined constant

K_1 — Modulus of subgrade reaction below a pipeline

K_2 — Modulus of subgrade reaction above a pipeline

K_∞ — Value of the modulus of subgrade reaction for an infinitely long beam

K^* — Pipe stiffness factor

$\mathbf{K}$ — Structural stiffness matrix

k — Coefficient of subgrade reaction (FL^{-3})

k_h — Horizontal permeability

k_v — Vertical permeability

L — Length of pipe between points 1 and 2

L Free length of beam

L Length of structural member

L Length of pipe

l Direction cosine

M Bending moment

m Direction cosine

$\mathbf{N}$ Stiffness matrix for ground-structure interaction

N_c Bearing capacity factor with respect to cohesion

N_q Bearing capacity factor with respect to surcharge

N_γ Bearing capacity factor which accounts for influence of soil weight

n Power of $\overline{z_0 - z}$ to which i_x, i_y, i are proportional

P Force

$P_{\max}$ Maximum load in pipeline

P_s Force on structure

p Normal pressure

p Piezometric pressure

p_0 Initial vertical pressure at tunnel crown before measurable soil consolidation occurs

Δp Change in ground pressure above tunnel crown as a result of drawdown and soil consolidation

q Intensity of pressure applied to pipe

q_u Ultimate bearing capacity

R Excavated radius of circular cross-section tunnel

R Rotational displacement

R_A Pipe area ratio

S_{li} Stiffness of linear spring

$S_{\theta i}$ Stiffness of rotational spring

t Pipe wall thickness

u Ground displacement in x-coordinate direction

u_m Maximum value of u

u_p Horizontal movement in pipeline (x-coordinate direction)

$u_{p\max}$ Maximum horizontal movement in pipeline (x-coordinate direction)

V Volume of settlement trough per unit distance of tunnel advance and attributable to ground losses at the tunnel

V_s Volume of surface settlement trough

ΔV Volumetric strain

v Ground displacement in y-coordinate direction

v_m Maximum value of v

v_p Horizontal movement in pipeline (y-coordinate direction)

$v_{p\max}$ Maximum horizontal movement in pipeline (y-coordinate direction)

w Ground displacement in z-coordinate direction

w_f Settlement of footing

w_m Maximum value of w

w_{m_c} Maximum consolidation settlement

w_{m_t} Total (ground loss plus consolidation) maximum settlement (above tunnel centre line)

w_p Vertical deflection of pipe

x Cartesian coordinate of any point in the ground deformation field

x_f Face or final tunnel position ($y = 0$)

x_i Initial or tunnel start point ($y = 0$)

y Cartesian coordinate of any point in the ground deformation field

z Cartesian coordinate of any point in the ground deformation field

z_0 Depth of effective source of ground loss (taken as tunnel axis)

z_p Depth of pipe axis from ground surface

γ Soil unit weight

γ_{xy} Ground shear strain in xy (horizontal) plane

γ_{xym} Maximum value of γ_{xy}

δ Displacement

δ_g Soil movement (pile in swelling soil analogy)

δ_p Pile movement (pile in swelling soil analogy)

ε_x Ground strain in x-coordinate direction

ε_{xm} Maximum value of ε_x

ε_{max} Extreme fibre bending strain in buried pipeline

ε_y Ground strain in y-coordinate direction

ε_{ym} Maximum value of ε_y

ε_{yt} Maximum tensile strain in y-direction

ε_z Ground strain in z-coordinate direction

ε_{zm} Maximum value of ε_z

θ Angle formed between linear horizontal buried pipe and the tunnel advance direction (x-axis)

θ Rotation at pipe joint

θ Rotation between foundation and ground node

λ Damping factor (units of L^{-1}). λ^{-1} defines the soil-pipe stiffness

ν_s Poisson's ratio of soil

σ Axial stress in pipe

σ_i Internal support pressure at the tunnel

σ_{z0} Vertical ground pressure at tunnel axis level

τ Soil-pipe shear stress

τ_a Limiting soil-pipe shear strength

τ_x Soil-pipe shear stress (pipe parallel to direction of tunnel drive)

τ_y Soil pipe shear stress (pipe transverse to tunnel drive)

ϕ Soil friction angle

References

Andrews, E.N. (1972) *Cast iron Pipelines—their Manufacture and Installation*. Stanton and Staveley, British Steel Corporation, Ilkeston, nr. Nottingham, England, Publication Ref. No. PJF 149/7.

Atkinson, J.H. and Potts, D.M. (1976) *Subsidence Above Shallow Circular Tunnels in Soft Ground*. Dept. of Engineering, University of Cambridge Report CUED/C—Soils/T.R. 27.

Angus, H.T. (1976) *Cast Iron: Physical and Engineering properties*. Butterworths, London.

Attewell, P.B. and Farmer, I.W. (1973) *Measurement and interpretation of ground movements during construction of a tunnel in laminated clay at Hebburn, County Durham*. Report to Transport and Road Research Laboratory, DoE, Research Contract No. ES/GW/842/68.

Attewell, P.B. (1978) Ground movements caused by tunnelling in soil. In *Proc. Conf. on Large Ground Movements and Structures, Cardiff, July* 1977, ed. J.D. Geddes, Pentech Press, London, pp. 812–948.

Attewell, P.B. (1981) Engineering contract, site investigation and surface movements in tunnelling works. In *Soft Ground Tunnelling*, eds. D. Reséndiz and M.P. Romo, A.A. Balkema, Rotterdam, pp. 5–12.

Attewell, P.B. and Woodman, J.P. (1982) Predicting the dynamics of ground settlement and its derivatives caused by tunnelling in soil, *Ground Engineering* **15** (8) 13–22, 36.

Audibert, J.M.E. and Nyman, K.J. (1977) Soil restraint against horizontal motion of pipes. *J. Geotechnical Engineering Division, ASCE*, **103** (GT10) October 1977, 1119–1142.

Biot, M.A. (1937) Bending of an infinite beam on an elastic foundation. *J. Applied Mechanics, Transactions, American Society of Mechanical Engineers* **59**, March 1937, pp. A1–A7.

Bjerrum, L. (1963) Allowable settlement of structures. In *Proc. European Conf. on Soil Mechanics and Foundation Engineering, Wiesbaden, Germany*, vIII, pp. 135–137.

Boden, W. (1956) *A Handbook on Cast Iron Pipes and Joints.* Stanton Ironworks Company Limited.

Burland, J.B. and Wroth, C.P. (1975) *Settlement of buildings and associated damage.* Building Research Establishment Current Paper CP 33/75.

Collins, H.H., Fuller, A.G. and Harrison, J.J. (1973) Corrosion characteristics and protection of buried ductile iron pipe. *12th World Gas Conference, Nice*, 1973. London International Gas Union, 1973.

Cooke, R.W. (1975) *The settlement of friction pile foundations.* Building Research Establishment Current Paper CP12/75. Watford, U.K.

Couldery, P.A.J. (1982) The effect of tunnelling settlement upon frame structures, Dissertation: M.Sc. Advanced Course in Engineering Geology, University of Durham (to be submitted).

Department of Energy (1977) *Report of the Inquiry into serious Gas Explosions.* H.M.S.O., London, June 1977.

Dobson, C., Cooper, I., Attewell, P.B. and Spencer, I.M. (1979) Settlement caused by driving a tunnel through fill. *Proc. Midland Geotech. Soc. Symp. on the Engineering Behaviour of Industrial and Urban Fill, Birmingham*, pp. E41–E50.

Freeman, L. (1983) New mains for old—will a gas system work for water? *Water Bulletin*, No. 65, National Water Council, London, July 1983, pp. 6–7.

Glossop, N.H. (1977) Soil deformations caused by soft ground tunnelling, Ph.D. thesis, University of Durham.

Glossop, N.H. (1980) *Ground deformation caused by tunnelling in soft ground at Grimsby.* Report to TRRL and DoE, University of Newcastle-upon-Tyne.

Glossop, N.H. and Farmer, I.W. (1977). *Ground deformation during construction of a tunnel in Belfast.* Rept. to the DoE for Northern Ireland and TRRL, No. R6/77, October 1977, by the University of Newcastle-upon-Tyne.

Glossop, N.H. and Farmer, I.W. (1979) Settlement associated with removal of compressed air pressure during tunnelling in alluvial clay, *Géotechnique* **29** (1) 67–72.

Glossop, N.H. and O'Reilly, M.P. (1982) Settlement caused by tunnelling through soft marine silty clay. *Tunnels and Tunnelling* **14** (9) 13–16.

Gordon, R.I.G. (1981) Ground movements associated with soft ground tunnelling and its effects on buried services, Dissertation: M.Sc. Advanced Course in Engineering Geology, University of Durham.

Grant, R., Christian, J.T. and Vanmarke, E.H. (1974) Differential settlement of buildings. *J. Geotech. Eng. Div., ASCE* **100**, (GT9) 973–991.

Hansen, B.J. (1961) *The ultimate resistance of rigid piles against transversal forces.* Geoteknisk Institut., Bull. No. 12, Copenhagen.

Harris, C.W. and O'Rourke, T.D. (1983) *Response of jointed cast iron pipelines to parallel trench construction.* Report to New York Gas Group, Geotechnical Engineering Report 83–5, School of Civil and Environmental Engineering, Cornell University, Ithaca, N.Y. (March 1983), 158 p.

Hartree, D.R. (1958) *Numerical Analysis.* 2nd edn. OUP, London.

Hetenyi, M. (1946) *Beams on Elastic Foundation—Theory with Applications in the Fields of Civil and Mechanical Engineering.* Ann Arbor: The University of Michigan Press.

Howe, M., Hunter, P. and Owen, R.C. (1980) Ground movements caused by deep excavations and tunnels and their effect on adjacent mains, *Proc. 2nd Conf. on Ground Movements and Structures, UWIST, Cardiff*, ed. Geddes, J.D., Pentech Press, pp. 812–840.

Hurrell, M.R. (1983) Personal communication.

Lee, I.K., White, W. and Ingles, O.G. (1983) *Geotechnical Engineering.* Pitman, Melbourne.

Leonards, G.A. (1975) Discussion on paper by R. Grant, J.T. Christian and E.H. Vanmarke 'Differential settlement of buildings', *J. Geotech. Eng. Div. ASCE.* **101**, (G77) 700–702.

Marsland, A. (1980) Discussion design parameters in geotechnical engineering. *Proc. VIIth Eur. Conf. S.M.F.E. Brighton*, Vol. 4, pp. 159–163.

Matyas, E.L. and Davis, J.B. (1983) Experimental study of earth loads on rigid pipes. *J. Geotechnical Engineering, ASCE*, **109**, (2) February 1983, 202–209.

Meyerhof, G.G. (1956) Discussion on paper by A.W. Skempton and D.H. MacDonald: 'The allowable settlements of buildings', *Proc. Inst. Civ. Engrs*, **5**, 774–775.

Mikaoka, T. (1978) Pipelines laid in difficult conditions including earthquake zones. *Proc. 12th Congress International Water Supply Association, Kyoto, October* 1978.

Mindlin, R.D. (1936) Forces at a point in the interior of a semi-infinite solid. *Physics*, **7**, 195.

National Water Council (1977) *Sewers and Water Mains—a national assessment*. Department of the Environment Standing Technical Committee. Report No. 4, National Water Council, June 1977.

Needham, D. and Howe, M. (1979) *Why pipes fail*. Communication 1103, Institution of Gas Engineers, London 1979.

Norgrove, W.B., Cooper, I.C. and Attewell, P.B. (1979) Site investigation procedures adopted for the Northumbrian Water Authority's Tyneside Sewerage Scheme, with special reference to settlement prediction when tunnelling through urban areas, *Proc. Tunnelling '79*, ed. M.P. Jones, Publ. I.M.M., London, pp. 79–104.

O'Reilly, M.P. and New, B.M. (1982) Settlement above tunnels in the United Kingdom—their magnitude and predication, *Proc. Tunnelling '82*, ed. M.P. Jones, Publ. I.M.M., London, pp. 137–181.

O'Rourke, T.D., Cording, E.J. and Boscardin, M. (1976) *Ground movements related to braced excavations and their influence on adjacent buildings*. Rept. to U.S. Dept. of Transportation, Washington, D.C. by University of Illinois.

O'Rourke, T.D. and Trautmann, C.H. (1980) *Analytical modelling of buried pipeline response to permanent earthquake displacements*. Geotechnical Engineering Report 80–4, School of Civil and Environmental Engineering, Cornell University, Ithaca, N.Y.

O'Rourke, T.D. and Trautmann, C.H. (1982) Buried pipeline response to tunnelling ground movements. *Europipe '82 Conf., Basle, Switzerland*, pp. 9–15.

Peck, R.B. (1969) Deep excavations and tunnelling in soft ground. State of the Art Report, *7th Int. Conf. on Soil Mechanics and Foundation Engineering, Mexico City*, pp. 225–290.

Polshin, D.E. and Tokar, R.A. (1957) Maximum allowable non-uniform settlement of structures, *Proc. 4th Int. Conf. on Soil Mechanics and Foundation Engrg, London*, vol. 1, pp. 402–405.

Poulos, H.G. and Davis, E.M. (1980) *Pile Foundation Analysis and Design*. Wiley, New York.

Reed, E.C. (1978) Renovation of old pipelines. *Proc. 12th Congress International Water Supply Association, Kyoto, October* 1978.

Reeves, M.J., Attewell, P.B. and Woodman, J.P. (1983) *Graphical method of estimating ground movements caused by tunnelling in soil*. Internal Report, Engineering Geology Laboratories, University of Durham, England.

Roark, R.J. (1965) *Formulas for Stress and Strain*. 4th edn., McGraw-Hill Book Co., N.Y.

Roberts, N.P. and Regan, T. (1974) *Causes of fractures in grey cast iron water mains*. Dept. of Civil Engineering, City University, and Metropolitan Water Board, London, March 1974.

Roberts, N.P. and Regan, T. (1977) *Fractures in Water Mains*. Internal Report to Thames Water Authority, Dept. of Civil Engineering, City University, London, July 1977.

Schmidt, B, (1969) Settlements and ground movements associated with tunnelling in soil, Ph..D. thesis, University of Illinois, 224p.

Sears, E.C. (1968) Comparison of the soil corrosion resistance of ductile iron pipe and grey cast iron pipe. *Materials Protection* 7, 33–36, October 1968.

Skempton, A.W. and MacDonald, D.H. (1956) Allowable settlement of buildings. *Proc. Inst. Civ. Engs* **5**, 727–768.

Széchy, K. (1973) *The Art of Tunnelling*. 2nd edn., Akadémiai Kiadó, Budapest.

Terzaghi, K. and Peck, R.B. (1967) *Soil Mechanics in Engineering Practice*. 2nd edn., Wiley, New York.

Tsutsumi, M. (1983) Tunnelling in Soil—Movements and Structures, Ph.D. thesis, University of Durham.

Vesic, A.B. (1961) Bending of beams resting on isotropic elastic solid. *J. Engineering Mechanics Division, ASCE*, **87** (EM2) 35–53, April 1961.

Wahls, H.E. (1981) Tolerable settlement of buildings, *J. Geotech. Eng. Div.*, *ASCE*, **107** (GT11) 1489–1504.

Whipp, S.H. and Glennie, E.B. (1982) *The Renovation of Sewers by Sliplining with Polyethylene, a Discussion Document*. External Report, Water Research Centre, January 1982.

Winkler, E. (1867) *Die Lehre von der Elastizität und Festigkeit (On Elasticity and Flexibility)*. Prague.

7 Mining subsidence

G. WALTON and A.E. COBB

7.1 Introduction

The extraction of minerals by underground mining inevitably induces a risk of
surface subsidence. Different methods of mining give rise to different risks of subsidence
and different styles of ground movement. Most underground mining takes place in
layered mineral deposits and employs partial or total extraction methods. This chapter

(a)

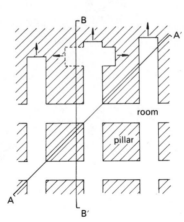

Plan

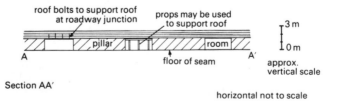

roof bolts to support roof
at roadway junction

props may be used
to support roof

Section AA'

horizontal not to scale

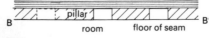

Section BB'

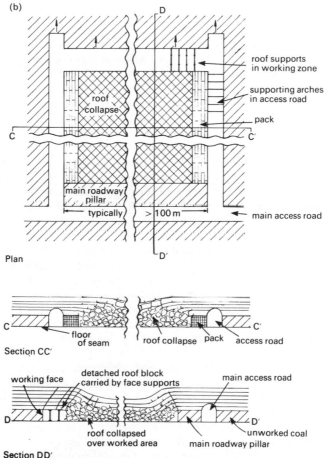

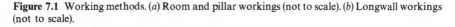

Figure 7.1 Working methods. (*a*) Room and pillar workings (not to scale). (*b*) Longwall workings (not to scale).

considers subsidence resulting from these mining methods; mineral extraction by solution, as in salt recovery (brine pumping), is not considered. Since coal-mining is in many parts of the world the most extensive and most researched form of mining, most of the examples come from this sector of the mining industry.

Subsidence damage due to underground coal-mining in Britain well exceeds £90 million per annum. The majority of this damage results from active longwall mining operations but mining subsidence may also result from old or active room and pillar workings. Substantial sums are spent annually investigating and treating old room and pillar workings so as to reduce the risk of surface subsidence.

Figure 7.1 illustrates diagrammatically the two basic extraction methods that may give rise to subsidence. Surface movements may also be caused by unstable mine entrances; such entrances include vertical shafts and both horizontal and inclined adits.

7.2 Partial extraction

This typically involves room and pillar mining in which pillars of mineral are left unworked to support the roof span during working, sometimes aided by timber or steel supports or by roof bolting (see Fig. 7.1(a)). Many patterns of working have been employed in different minerals over the last three or four centuries. Whilst the most common comprise regular square pillars, rectangular, herringbone and irregular pillars were often used, especially in coal and ironstone mines. Regional preferences to different systems were quite usual, particularly where steeply-dipping minerals required special layouts to permit underground transport. Room and pillar mining is thought to have developed rapidly during the sixteenth century; changes in mining methods up to 1850 are well described by Galloway (1969). Room and pillar coal-mining continued on a large scale in Britain for another century, but is now generally employed only in small-scale coal operations, chiefly licensed shallow mines. In other countries and in other minerals, room and pillar methods remain a major mining method. However, as minerals are worked at ever-increasing depths so the problem of roof support increases. Consequently, in deeper exploitable coal seams in the U.S.A. and elsewhere, longwall mining is becoming more common.

With few exceptions, subsidence from partial extraction workings in Britain is associated with mine takes which were abandoned many years ago. Room widths in coal-mines tend to vary between 2 and 5 m, with initial square pillars of 10–12 m side (although these may subsequently be reduced) and with extraction rates ranging from 30–70%. In deeper mines, it has been found necessary to reduce the rate of extraction in order to maintain roof support. Since roadways are generally of a constant size, pillar sizes are necessarily increased.

At the end of mining, props or arches may be withdrawn and pillars may be removed or reduced on retreat, so enlarging the width of rooms. Possible oxidation of pillars can prejudice stability, and flooding can encourage floor heave or the breakdown of roof strata. Complex ground movements can follow in which either the pillars may be forced into the mine floor, giving widespread surface lowering, or the integrity of roofs and pillars is affected. Eventual ground surface problems may thus arise due to (i) roof failure, (ii) pillar failure, (iii) floor heave, or (iv) a combination of these effects.

7.2.1. Roof failure

The failure of roofs into rooms occurs most commonly with the characteristic surface evidence of crown-holes or depressions which generally develop at roadway crossings as shown in Fig. 7.2(a) (section) and 7.2(b) (plan), although they may also occur at any point along a roadway. Roof collapse or overbreak is seen by numerous authors as a progressive phenomenon in which the void generated by mining travels vertically through the overlying strata until arrested by choking due to bulked roof debris (Wardell and Wood, 1965; Price et al.. 1969), by more massive roof strata of higher tensile strength, or by natural arching processes. The several constraints to void migration include:

(i) widths and heights of mine openings,

(ii) character and thickness of overlying bedrock strata, including any roof beam of remaining mineral,

(iii) character and thickness of overlying superficial deposits.

(a) Section

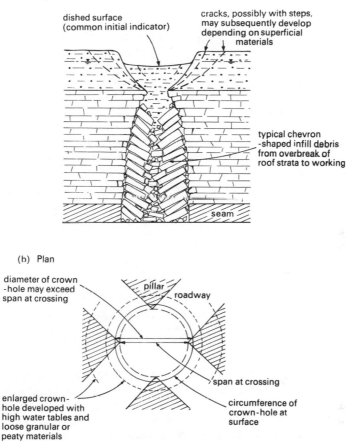

dished surface
(common initial indicator)

cracks, possibly with steps,
may subsequently develop
depending on superficial
materials

typical chevron
-shaped infill debris
from overbreak of
roof strata to working

seam

(b) Plan

diameter of crown
-hole may exceed
span at crossing

pillar

roadway

span at crossing

enlarged crown-
hole developed with
high water tables and
loose granular or
peaty materials

circumference of
crown-hole at
surface

Figure 7.2 Crown-hole development. (*a*) Section. (*b*) Plan.

Tincelin (1958), Price *et al.* (1969) and Piggott and Eynon (1977) developed simple relationships between the height of collapse, the thickness of extracted mineral and the bulking characteristics of the roof strata. Piggott and Eynon have considered the heights to which notional rectangular, wedge and conical collapses may rise for various bulking factors as shown in Fig. 7.3. The upper limits of void migration in bedrock with a flat or gently inclined seam may be up to 10 times the extracted thickness, although an upper limit of 8 (Taylor, 1975) is probably more reasonable. It is unusual to find many collapses reaching more than 6 times the room height. Voids that reach ground level typically form steep sided crown-holes. It is important to recognize that steeply inclined seams with strong water flows in the abandoned workings may inhibit choking; this potential for increased void migration is commonly encountered above drift mine entrances where supports have been removed or have failed.

Alternative assessments of the height to which ground movements may reach have been based on Voussoir beams and arching theories, for example Evans (1941);

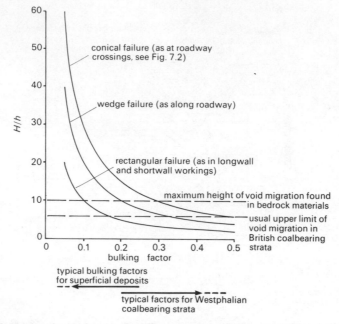

Figure 7.3 Heights of void migration, for different types of roof failure in relation to bulking factors. H/h = height of void migration as a proportion of the extracted thickness of seam (values equivalent to actual height of migration from seam of unit thickness). After Piggott and Eynon (1977).

Terzaghi (1946); Szechy (1967); and Beer and Meek (1982). Observations of old mine workings, both underground and exposed in surface mine workings, also give indications of void migration as noted in Jones and Davies (1919); Challinor (1976); and Walton and Taylor (1977; see especially Fig. 1). The latter authors have shown relationships to obtain between the height of overbreak and span or room width, as may be inferred from Fig. 4. Three times the room width appears to be an upper limit to the extent of most vertical overbreaks and 1.5 times the span is a commonly-found factor.

Either of the aforementioned approaches may be used to determine a zone of surface instability, but because of the difficulty in determining actual room widths, due partly to the unreliability of old mine plans, the zones are best determined using depth of bedrock cover and extracted coal/strata thickness criteria. Depth of cover should not include superficial deposits or made-ground such as embankments or spoil dumps, since very low bulking factors are appropriate for these materials (see Fig. 7.3). The writers have recently recorded a crown-hole that developed at the surface of an 18 m-high spoil dump due to mine workings with 1.5 m high openings, in an underlying seam with at least 12 m of intervening bedrock. Roof failures in gypsum and limestone mines with 3–3.5 m high workings have been observed to produce surface crown-holes above workings more than 40 m below ground level as a consequence of thick superficial deposits. Arrowsmith and Rankilor (1981) demonstrate that weak, superficial deposits may flow into voids that have reached rockhead, so giving rise to a variety of features which range from gentle dishing of the surface to inverted cone-like depressions of

Figure 7.4 Void migration in room and pillar workings and typical surface effects.

considerable lateral extent. Apart from increasing the height to which movements may extend, the inflow of saturated unconsolidated superficials need not directly overlie the bedrock overbreak, but can be offset according to the structural and hydrogeological setting (see for example the experience of Cameron, (1956) in Glasgow).

It should be noted that overbreak may be arrested by competent strata such as limestones or massive sandstone, provided that joining and bedding separation surfaces permit a rock beam to form which is capable of spanning the opening. Such competent strata should have a *thickness* equal to *twice the span width* to realistically allow for arching to develop. Massive strata, however, can be the locus of chimney-type collapses in which downward diverging joints give rise to abnormally high-level migrations.

Overbreak frequently gives rise to a chevron-shaped infill irrespective of whether the

Figure 7.5 Chevron-shaped collapse of roof strata (room and pillar workings).

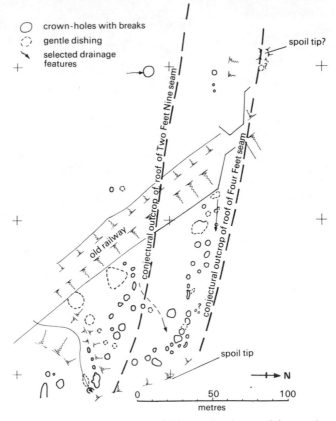

Figure 7.6 Plan showing crown-holes and subsidence related to partial extraction of two coal seams.

void migration is joint-controlled. Initial failures are therefore mainly in the centres of openings (Fig. 7.2, Fig. 7.5). The form of overbreak arches may vary as shown in Figs. 7.4 and 7.5. Examination of this infill in surface coal-mines shows it to be generally loose, and at best, poorly compacted. It is usual to expect that at least minor movements can occur with changes in physical conditions such as the raising or lowering of the water table.

Surface features that develop from void migration or overbreak are essentially irregular in form, and although not usually isolated, their distribution may be somewhat uneven within the depth constraints noted above. More collapses generally occur in the shallower rather than in deeper sections of the zone of instability. Frequently crown-holes are circular or sub-circular in plan; in section they commonly develop with steep-sided cracks at the limit and a central inverted cone-like depression. Sometimes they appear at the surface as gentle depressions of no more than a few tens of millimetres and may remain in this state without further movement for many years. The diameter of crown-holes is very variable and may range from one to many tens of metres, reflecting variations in bedrock and superficial deposits, although such large variations are unusual in a single area. Figure 7.6 is a plan of typical crown-holes mapped on a site in South Wales.

7.2.2. *Pillar failure*

Pillar failure due to imposed loads depends on a variety of factors including pillar height and width, the strength of the pillar materials, depth and seam dip as well as the character of immediate roof and floor to the mine workings. It is not the purpose of this section to examine these constraints in detail, but they are dealt with adequately by Hoek and Brown (1980), who outline both stress analysis and empirical design methods which may be used to check the safety of existing pillars. In deeper coal workings in

Table 7.1 Checklist for shallow mine workings

Item	Critical parameters etc.
Depth to workings and mined thickness	Thickness of overlying bed rock (not including superficials), generally adequate if equal to or greater than $10x$ the mined thickness (which must be determined).
Span width and overlying and underlying strata	Wide spans to uncollapsed working may be important if $2x$ span width exceeds $10x$ mined thickness. Critical span width depends on character of overlying material. Assessment relies on mine plans, therefore may be suspect. Mine plans should be fully researched and accuracy assessed by check drilling; mining method, geometry and current conditions are all important.
Pillar size	As for span width, accuracy of mine plans should be checked. In coal workings, pillar dimensions should not be less than 10% of depth. Acceptable pillar widths and coal seam thicknesses may be related empirically. Pillars under massive sandstone with open workings on dips steeper than about 15 degrees should be carefully assessed with respect to potential pillar failure.
Groundwater conditions	Check on levels and likelihood of changes. Chemical analysis essential if grouting is contemplated.

Britain, it has been empirically established that pillar widths should exceed 10 per cent of the working depth to ensure pillar stability, as shown by Orchard (1964; see especially Table 7.1).

Most pillar failures occur during working, with relatively few cases of failure after mining. Reported pillar failures include, for example, those at Stamphill, Cumbria, England (gypsum); Middleton, Derbyshire, England (limestone); Lick Creek, Illinois, U.S.A. (shales); and Coalbrook, South Africa (coal). All these mines experienced pillar failures which were attributable to the transfer of tributary area loads on to small internal pillars from larger ribside pillars during working. Failures of abandoned coal and limestone mines, whilst generally less well-known, have been reported by Bruhn *et al.* (1980) and Gray and Meyers (1970), both from Pennsylvania, and by Carter *et al.* (1980) from Bathgate, Scotland. The Pennsylvanian examples include suspected pillar punching or floor heave as well as pillar failures at overburden thicknesses ranging from less than 10 m to more than 130 m. This contrasts with data for crown-hole development in the same area, few of which were found with more than 30 m of cover. Figure 7.7 illustrates these findings. Bruhn *et al.* (*op. cit.*) attribute pillar failure and 'trough development' to weathering and pillar deterioration in areas with high extraction rates, but also recognize the importance of other factors, including drainage and further subjacent underground mining. Carter *et al.* (*op. cit.*) attribute the possible Bathgate pillar failures to high extraction rates (65–75 per cent), steep dip (*c.* 25 degrees), strong roof and floor, subjacent workings, and the age of the pillars (late eighteenth century) which showed both oxidation and spalling. The setting was one in which little support was available to pillars from overbreak debris or floor heave, because the debris would have been unlikely to remain in position with such a steep dip. At Bathgate, as in other places, it was not possible to be completely certain that primary pillar failure had occurred. It is sometimes possible to ascribe the initial failure mechanism to the collapse of a large arch spanning several pillars, or to an area of near-complete extraction as mentioned previously. Whatever the exact mechanism, the resulting surface subsidence takes on the character of a large trough with relatively steep sides often accompanied by boundary fissuring. The floor of the trough is frequently irregular, so reflecting the remains of collapsed or crushed pillars.

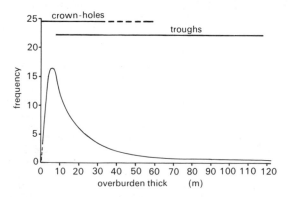

Figure 7.7 Observed surface movements in relation to overburden. Movements due to shallow mine workings include crown-holes and pillar failure with trough development. Based on data from Bruhn *et al.* (1980).

The potential for long-term pillar failure in old mines should not therefore be ignored, particular warning signs being strong roofs and floors allied with high extraction rates and moderate to steep seam dips. Little is known of the time element in pillar failure (or roof or floor movements) but such problems may increase with the age of pillars, owing to long-term creep movements, weathering and spalling. No guidelines can be given for time before failure, nor for critical depths, since several factors are involved. It is wise to assume that the potential for failure is independent of time.

7.2.3 Floor movements

The movement or heave of floor material into mine openings is a common phenomenon in room and pillar mining and has long been recognized, even in workings less than 50 m deep. Many underground coal-mines have weak or very weak seatearths or underclays which may in part approximate to engineering soils. Floor behaviour probably varies according to floor loading and to the mineralogical composition of the floor strata. Seatearths which include (highly active) expandable mixed-layer clay minerals are more likely to give problems (see for example Pearson and Wade, 1967). Such clays are not, however, essential for floor heave to occur. It is widely recognized in areas with beds of low activity minerals where softening has occurred following the ingress of mine-water.

The initial effects due to the absence of confining conditions are the bulging of the floor with a squeezing of floor material from beneath the pillars, outwards and upwards into the opening. Simplistically this may be seen as a bearing capacity failure in which the applied stress exceeds the strength of the footings in the floor strata. Simple calculations can be made for pillars of various shapes using conventional bearing capacity equations, known shear strengths for floor materials and average pillar stresses, assuming a uniform distribution of overburden loading as in the tributary theory (Duval, 1977). Floor heave may therefore give rise to either local or more widespread surface subsidence as the pillars are lowered or 'punched' into the floor. Associated disruption may occur to roof strata and to pillars which may also be spread and crushed. Although floor heave can occur during working, it commonly accelerates shortly after abandoned mines or sections of mines have been allowed to flood. There are, however, no overriding reasons why floor heave could not occur over a period of many years. The surface effects of floor heave are not well documented, but are apparently little different from those which result from pillar failure, namely the development of a trough with relatively steep sides.

7.2.4. Other considerations

External factors may be important in the assessment of the risk of subsidence due to room and pillar workings. These include:

(i) surface loadings such as embankments, spoil dumps, or foundations of structures
(ii) undermining by either total or partial extraction methods
(iii) seismic events due either to natural phenomena or induced by longwall caving, as recently in Stoke-on-Trent, England.

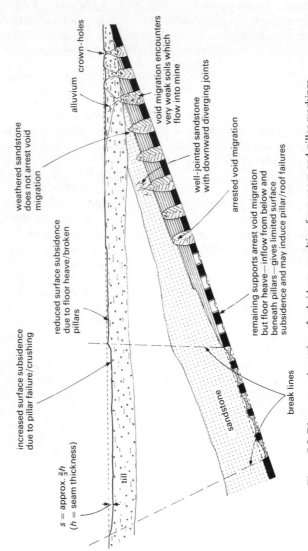

increased surface subsidence
due to pillar failure/crushing

s = approx. $\frac{2}{3}h$
(h = seam thickness)

till

reduced surface subsidence
due to floor heave/broken
pillars

weathered sandstone
does not arrest void
migration

alluvium   crown-holes

void migration encounters
very weak soils which
flow into mine

well-jointed sandstone
with downward diverging joints

arrested void migration

remaining supports arrest void migration
but floor heave — inflow from below and
beneath pillars — gives limited surface
subsidence and may induce pillar/roof failures

sandstone

break lines

Figure 7.8 Diagrammatic setting of subsidence resulting from room and pillar workings.

All these factors may act to increase or alter/concentrate strains in the roofs, the pillars or the floors of room and pillar workings, thereby increasing the possibility of surface movements.

Figure 7.8 shows diagrammatically the setting and character of various surface movements due to partial extraction as discussed above. Guidelines which can be used in the assessment of potential problems from these sources at the 'desk study' phase of investigations are listed in Table 7.1.

7.3 Total extraction

In this method of mining the mineral is recovered and the roof is allowed to collapse into the opening, thereby causing an inevitable movement of the ground surface towards the mining void. Total extraction is exemplified by the longwall mining technique (see Fig. 7.1) in which a single face of coal is worked with access from two parallel roadways, the working face being temporarily supported. The supports are then removed or advanced as the working area or panel is extended. The roof of the unsupported workings collapses behind the working face and between the gate roads (or ribsides) either gradually or periodically as the unsupported arch distance is exceeded. These movements, which generally extend up to 6 to 10 metres above the working level (albeit higher with increased extraction rates), are rapidly transmitted to the surface as a result of flexing and inter-bed slippage in the overlying strata, thus giving rise to eventual surface subsidence. Roof fracturing to extended heights in some situations may increase mass permeability and can disrupt overlying aquifers. Shadbolt et al. (1973) summarize surface effects as:

   (i) vertical subsidence (ground lowering), tilt (differential vertical subsidence)
  (ii) curvature (differential tilt)
 (iii) horizontal movement (displacement) and strain (differential horizontal movement).

Figure 7.9 shows the various effects diagrammatically. A number of important empirically-derived relationships should be noted, namely:

   (i) The limits of influence extend beyond the workings by an amount defined by the 'limit angle' or 'angle of draw' which in Britain is typically c. 35 degrees to the normal of the worked seam.
  (ii) For horizontal seams the subsidence and strain profiles are symmetrical about the panel centre line.
 (iii) Maximum ground lowering varies according to panel width and depth, but is no greater than 90% of the extracted thickness.
  (iv) The maximum subsidence lies on the normal to the panel centre line for panels less than, or equal to, those of critical width, (that is, panels with a width-to-depth ratio of 1.4). For wider or supercritical panels, (i.e. panels with a width-to-depth ratio greater than 1.4) the subsidence trough becomes flat-bottomed.
   (v) Ground tilt is zero in the centre of a panel. Maximum tilt for critical and supercritical panels lies at a distance of c. 0.15 times the depth inside the ribside. With subcritical panels, the position of maximum tilt gradually approaches the ribside crossing it at a width-to-depth ratio of c. 0.4.
  (vi) Tensile strains occur in the zone of convex upwards curvature, which lies in the area bounded by the limit of the angle of draw and the point of maximum tilt. These ground strains reach their maximum value over the ribsides of

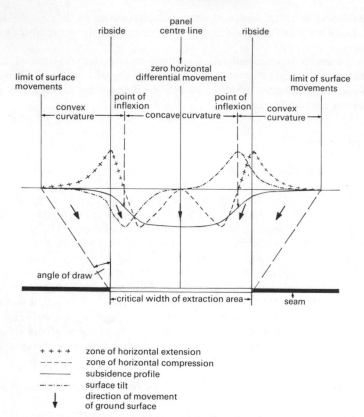

Figure 7.9 Basic subsidence effects and geometry resulting from longwall workings.

critical and supercritical width panels. For subcritical width panels, the maximum tensile strain lies outside the ribsides, although only just so for panels with a width-to-depth ratio greater than 0.5.

(vii) Compressive strains correspond to the zone of concave curvature occupying the inner part of the subsidence profile of the subsidence curve between the points of maximum tilt (see Fig. 7.9.)

(viii) Maximum compressive strains vary according to panel width and depth. The maximum value occurs at a distance of c. 0.3-times the depth inside the ribside for critical and supercritical width panels. For subcritical width panels the position of the maximum value moves gradually towards the ribside as the width-to-depth ratio decreases. At a width-to-depth ratio of 0.4, it is 0.2 times the depth from the ribside, that is, it is coincident with the panel centre.

These components of mining subsidence are of varying degrees of importance depending on the structure concerned. Thus, for land drainage and water-courses in general, where possible ponding and flooding may occur, the degree and pattern of ground lowering is of the greatest importance. Tilts affect tall buildings, chimneys, pylons, retaining walls and bridge piers. They also may impair the functioning of roads and railways by altering cambers, super-elevations and gradients. Many items of industrial machinery, such as heavy presses and overhead cranes, are also sensitive to minor changes of level. Buildings and bridges suffer crushing or cracking from the effects of compressive or tensile ground strains. Bearing surfaces of bridge decks and roof joists may be lost if excessive lengthening occurs. In general, the cost of precautionary measures is the responsibility of the developer; the cost of repairing damage may influence the development and profitability of underground mines, although there is usually no direct requirement to mine so as to avoid damage. In Britain the National Coal Board has substantial experience in the assessment of the scale of damage in relation to the magnitude of subsidence strains (compressive or tensile) which can occur to different-sized structures; see Fig. 8.6 and Table 8.6 (p. 253).

Since total extraction methods are usually employed at considerable depth (now rarely less than 100–150 m in Britain) the subsidence troughs differ from those associated with pillar failure. The subsidence troughs associated with longwall workings tend to have gently inclined limits, cracking is restricted to certain specific conditions and settings, and the floor of the trough usually shows a smooth profile. Whilst the form of longwall mining shown in Fig. 7.1(b) is a relatively recent development, the method has in places been employed in a less mechanized manner for many years; in the English Midlands, for instance, since the mid- to late eighteenth century. Similar surface effects therefore obtain in those areas of early mining in which pillars were largely or wholly removed on retreat.

Geotechnical interest is directed both to the prediction of future effects and the assessment of the effect of ground movements due to previous total extraction workings. It is possible to predict the location and magnitude of subsidence effects with far greater precision than is possible with room and pillar workings, although in essence the basic surface effects are similar. The subsidence (ground lowering) movements are generally predictable to within ± 10 per cent, although a slightly lower accuracy may be found when movements are very small (as near the limits of influence). In their English East Midlands study, the N.C.B. (Shadbolt *et al.*, 1973) found that 25 per cent of all cases of subsidence undergo a measure of abnormal movement (see also section 7.3.2).

7.3.1 *Estimating mining subsidence*

There are two common methods of estimating mining subsidence, namely the zone method which is favoured in continental Europe, and the dimensionless empirical method which is widely used in Britain.

7.3.1.1 *Zone method.* The zone (or circle) method estimates surface subsidence using a series of zones or circles around a surface point, the outer radius circumscribing the area of influence. Subsidence is proportional to the sum of the percentage of each zone or circle that has been extracted at depth. The method readily allows for the estimation of subsidence due to the working of irregular plan shapes. Empirical zone or subsidence factors are required for the estimation; the zone method is dependent on proper selection of these factors since predictions made in one coalfield need not conform to those in another. These factors have to be applied to the proportion of coal extracted in each zone or circle before values can be summed to give total subsidence values. Marr

(1975) uses seven zones, each zone having a width equal to 10 per cent of the depth of working (see Fig. 7.10); the subsidence at a surface point is given by the expression:

$$S = A^n a + B^n b + C^n c + D^n d, \quad \text{etc.}$$

where

    A, B, C, D etc. are proportions of zone area extracted,
    a, b, c, d etc. are the zone factors, and
    n is a constant found to be c. 2.3 for British coal-bearing strata.

Other adjustments include an allowance for pillars between worked areas so that realistic estimates are made of the reduced subsidence that is experienced, compared with that derived from purely additive effects, together with an allowance for seam dip. It is not possible to use the zone method directly to calculate surface strains.

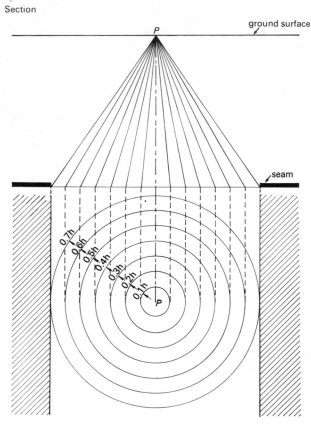

Section

ground surface

Plan

Figure 7.10 Zones of working at depth that may cause surface movement.

7.3.1.2 *British empirical method*. The dimensionless empirical method has been developed in Britain principally by Orchard (1954, 1957) of the National Coal Board. It is detailed in the N.C.B. *Subsidence Engineers' Handbook* (2nd ed., 1975) and is well summarized by Shadbolt (1977). This method relies on established dimensionless relationships which are shown in graphical form in Fig. 7.11 and were noted briefly in the introduction to section 7.3. These relationships are based on precisely surveyed surface subsidence data for rectangular panels. When, as often happens, panels are not uniformly rectangular, modifications are necessary to assess the effects of changes in shape. Although somewhat subjective in complex areas, satisfactory estimates may be made by experienced persons. These relationships can be used not only to estimate subsidence but also to estimate surface ground strains. The present authors prefer the empirical method, since it accommodates strains, but would recommend that where detailed estimates of surface subsidence are essential and the areas of working are highly irregular, the *zone method* should be used.

It is possible to program micro-computers to calculate ground lowering and strains by employing the subsidence development relationships shown in Fig. 7.11. The basic data include:

(i) The prediction of potential subsidence (ground lowering) expressed in terms of maximum subsidence for various panel widths and depths irrespective of panel length (as shown in Fig. 7.11(a)). (The values given in the *Subsidence Engineer's Handbook* relate specifically to ground that has already been undermined in other seams; they require factoring by 0.9 for previously unworked areas.)

(ii) The prediction of subsidence as a function of the rates of face advance (panel length) to depth as shown in Fig. 7.11(b).

(iii) The prediction of subsidence profiles by means of estimating the proportion of maximum subsidence in relation to panel width and distance from the panel axis, both factors being expressed with respect to panel depth (see Fig. 7.11(c)).

(iv) Adjustments for ribside pillars and for seam dip using known relationships. (Pillars may influence subsidence so that surface movements are not purely additive. Seam dip, apart from offsetting surface effects down-dip from the plan position, causes the strain pattern to become asymmetric with the up-dip tensile strains reduced and the down-dip tensile strains correspondingly increased.)

(v) The prediction of maximum strains using the relationship shown in Fig. 7.11(d).

(vi) The prediction of strain 'profiles' by means of estimating the proportion of maximum tensile or compressive strains in relation to panel width and distance from panel centre line, both factors being expressed with respect to panel depth (see Fig. 7.11(e)).

(vii) Adjustments for compression at panel centres and for ground slope.

From these relationships surface lowering (or subsidence), horizontal ground strain, and if necessary, tilts, may be calculated for any point for single workings, or, by addition, for multiple workings in several seams. Strains and tilts may be calculated in two orthogonal directions parallel and normal to the panel centre line. In practice, values for a grid of points are readily determined and these values may be subsequently contoured. It is feasible to consider horizontal strains in *any two orthogonal directions* (not necessarily parallel and perpendicular to the panel centre line) such as may be appropriate for bridges or pipelines. Figure 7.12 illustrates typical computer output based on the dimensionless empirical approach. Other methods of considering three-

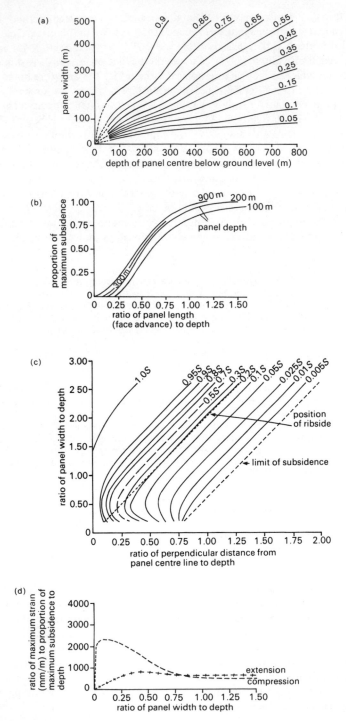

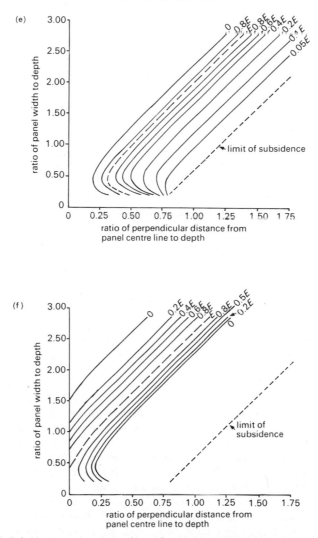

Figure 7.11 Subsidence-strain relationships After N.C.B. (1975). (*a*) Maximum subsidence as a proportion of extracted seam thickness in relation to panel depth and width. (*b*) Proportion of maximum subsidence related to the ratio of panel length (face advance) to depth, for various depths. (*c*) Proportion of subsidence at any point on cross section to subsidence over panel centre line. (*d*) Maximum strains in proportion to width and depth of panel. (*e*) Proportion of tensile strain at a point on cross section to maximum tensile strain on that section. (*f*) Proportion of compressive strain at a point on cross section to maximum compressive strain on that section.

dimensional movements and strains are available (see Burton, 1977) and are appropriate to areas with anomalous movements (Knipe, 1983). Finite element and boundary element approaches to subsidence simulation are also used (for example, Hocking, 1978).

7.3.2 *Other factors influencing subsidence*

It is generally held that subsidence movements at a point on the ground surface commence when mining extraction begins within the area of influence (Fig. 7.9) and cease when workings have left the area of influence. Most of the subsidence (up to 95%) occurs within this period, or within a few days if workings cease within the area of influence. Ground movements that continue after the workings have advanced beyond the area of influence are called residual subsidence and were formerly thought to be largely complete within a few weeks or months. It is now considered that in some settings this residual subsidence period may be substantially delayed and may comprise a larger proportion of total subsidence than was originally appreciated (Orchard and Allen, 1974). The critical settings appear to be those with more massive strata, notably the Pennant measures of South Wales and the Permo-Triassic strata of the English Midlands, Yorkshire and Co. Durham. Reasons for this delay probably relate to arching or 'holding-up' within the more competent strata; the working of a second coal seam can in places produce unexpectedly large ground movements probably due to disruption of the arching in the massive strata that was established during the working of an earlier seam. Collins (1977) has undertaken a number of long-term investigations which show that: (a) residual subsidence may take place over 2–4 years, (b) that the percentage residual subsidence is greater over deeper workings, and (c) that percentages occasionally can greatly exceed 5%, varying between the panel centre and ribsides, and being proportionally greater over the ribsides (although actual movements are greatest at the centre). Collins (*op. cit.*) reports one case in which residual subsidence amounted to 34–36% of total subsidence and occurred during a two-year period.

Geological factors may act to modify anticipated ground movements, although it is generally considered that these effects are rather limited. Lee (1966) noted the impact of faulting in the normal subsidence profile and showed that appreciable steps in this profile can arise especially when longwall workings lie beneath a fault plane that subcrops within the area of influence. These effects are greatly heightened with major faults which may act as boundaries to collieries, as in South Wales where fault stepping is in places associated with fissuring which may be up to 2 m wide and more than 30 m deep. Details of the magnitude of such fissures are given in the Technical Tribunal Reports (1969) on the Aberfan tip disaster. Whilst fault stepping movements are difficult to quantify, and occur in a minority of cases, multiple seam workings on the upthrow side of a fault, in which the panels are aligned parallel to the fault and are relatively wide, appear to favour anomalous movements.

As noted earlier, joints and superficial structures may act to concentrate surface strains, particularly where panel ribsides correspond to master joint directions. Whittaker and Breeds (1977) have reviewed the differences in subsidence behaviour between ground composed entirely of Coal Measures strata (plus superficials) and that found in the concealed British Coalfields in which Permian limestones and Triassic sandstones overlie Coal Measures strata. Whilst no major differences have been noted, the Permo-Triassic strata are subject to a slightly higher than normal number of movements which depart from the average, showing somewhat reduced subsidence compared with Coal Measures strata. This is possibly due to block behaviour in the younger rocks (the differences being most marked with the Triassic sandstones which show noticeable irregular tensile strains) and where narrow panels are worked. It is widely recognized that normally-consolidated superficial deposits and unconsolidated fill material can be subject to anomalously high surface subsidence, far exceeding that predicted for normal strata.

In summary, reasonable estimates of surface subsidence, ground strains and tilts can be made using available computer programs which model empirically-established relationships. The essential data for any analysis includes the plan layout of the mine workings, the extracted thickness of mineral, the seam dip and working levels and ground surface levels within the area of influence. It is always prudent to establish the full extent of earlier mining and in particular the surface positions of known faults, landslips and areas of fill; additional precautions may be necessary in any area which could include features likely to give disproportionately large subsidence effects.

7.4   Mine entrances

These are in the form of either shafts or adits. Shafts are vertical or near-vertical mine entrances which may be single, in pairs or groups, and in plan are generally round, oval or rectangular. Adits or drifts are inclined mine entrances (also referred to as inclines or slants) often sloping downwards at gradients between 1 in 2.5 and 1 in 4. Both shafts and adits may be lined or partly lined, although shafts are more commonly cased in iron, steel, concrete, brick or stonework; adits may rely on wood or steel props or arches for support.

Mine entrances in operating mines rarely cause surface subsidence hazards. Problems chiefly arise from abandoned entrances, many of which are unknown or incorrectly located on existing records. The nature of the problems associated with shafts and adits is well covered in a recent N.C.B. (1982) publication. The geotechnical basis of the problem of backfilled or partly-backfilled shafts and the risks of settlement or liquefaction/mobility of the shaft filling is adequately covered. It includes the effects of groundwater pressures and a review of factors likely to disturb the equilibrium in infilled shafts. Collapse of shaft infilling can cause substantial surface cratering especially in areas with weak or waterlogged superficials when depressions, having the form of very large crown-holes, may develop if the shaft lining fails.

It is generally recognized that shaft cappings, and particularly infillings, are susceptible to changes in groundwater conditions and to static and dynamic loading. Collapses may be related to:

(a) Marked changes in groundwater level, particularly in unusually wet and dry weather, but also changes in soakaways, burst pipes, and so on.
(b) Dumping of materials near or over shafts.
(c) Construction near or over shafts.
(d) Vibration from plant or traffic.
(e) Blasting and seismic shocks.
(f) Mining subsidence.

Combustion of coaly infill in old shafts with the evolution of noxious gases can ultimately give rise to surface movements and may even have resulted in deaths in adjacent properties (see Newcastle upon Tyne *Evening Chronicle*, 18 January 1978).

As with shallow partial extraction workings, it is not possible to predict the timing of collapse. The extent of collapse is a function of the nature and depth of the superficial deposits and the depth of highly-weathered bedrock materials; movements rarely exceed twice the thickness of superficial deposits. Problems with adits are generally more akin to those found with shallow room and pillar workings in more steeply inclined strata.

7.5 Investigation and treatment

There is an extensive literature both on investigations of old, active and proposed mine workings and on the necessary ground treatments: Wardell and Wood (1965); Taylor (1968); Price *et al.* (1969); Malkin and Wood (1972); Littlejohn (1979); N.C.B. (1982). Consequently this section can only briefly indicate the broad approaches which are usually adopted.

The careful collation of old mining records is of paramount importance in assessing the likely presence and location of mine workings. In the opinion of the present authors a detailed knowledge of the local geological sequence and sedimentological variations is equally important. Due regard must be given to the names given to mineral seams; it is not uncommon to find old mine plans using different names for the same seam and the same name for several seams (Symons, 1980). Seam thickness and the character of the immediate roof and floor are frequently important guides to seam identification since they are commonly the only geological observation other than seam dip or faulting on an abandoned mine plan. Most British coal plans are now held by the N.C.B. but it is not unusual to find that abandonment plans of contiguous clay or ironstone workings are held elsewhere (for example by the Library of the U.K. Department of Industry). Much use should also be made in researching topographic plans and estate plans through local libraries, museums and copyright libraries. These several sources of information are adequately reviewed by Dumbleton and West (1976).

Subsequent investigation should always include rotary drilled boreholes, the various methods being outlined in Bell (1975). The present authors consider rotary air-flush drilling to be most suitable with some core recovery, although open-hole drilling is generally adequate after the sequence has been fully established. Whilst geophysical techniques are widely available it is generally found that none of the surface methods is wholly reliable with respect to the identification and location of old workings and shafts. It may safely be assumed that when such techniques are reliable they will then be rapidly adopted by the surface coal-mining industry, since in Britain it is probable that annual investment in surface coal exploration greatly exceeds that by the construction industry in site investigations of mining areas. Down-hole geophysical methods, which are more reliable with respect to the characterization of materials, are currently unable to identify critical voids which are more than 1 to 2 m from a borehole.

Borehole investigation depths in shallow mining areas should always be sufficient to identify shallow workings that in the most extreme conditions could influence the ground surface. Deep-level foundations will require correspondingly deeper investigations. Particular attention should be given to loss of returns, core brokenness (for example, RQD or fracture frequency) and the identification of the floor of former workings. Taylor (1968) and Price *et al.* (1969) give a number of examples of the spacing of boreholes appropriate to specific projects; it is not unusual to employ spacings of 20 m and even 1 to 2 m in critical areas. A grid layout should always be avoided since it could correspond with the pillars in partly extracted workings. If workings are not encountered it can be unwise to assume that they are not present unless one is well-

(a)

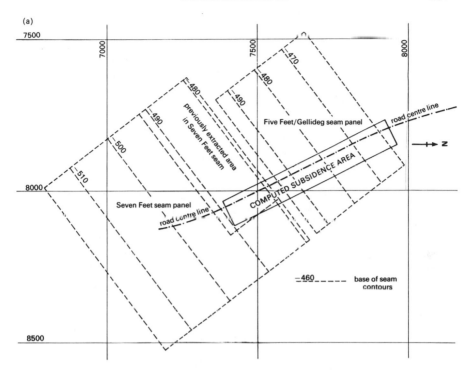

(b)

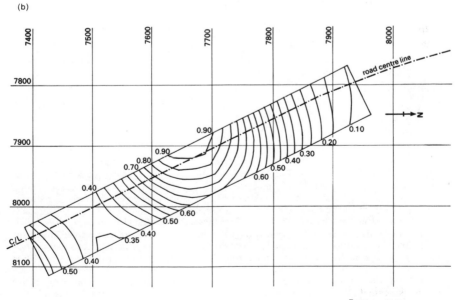

[Figure 7.12

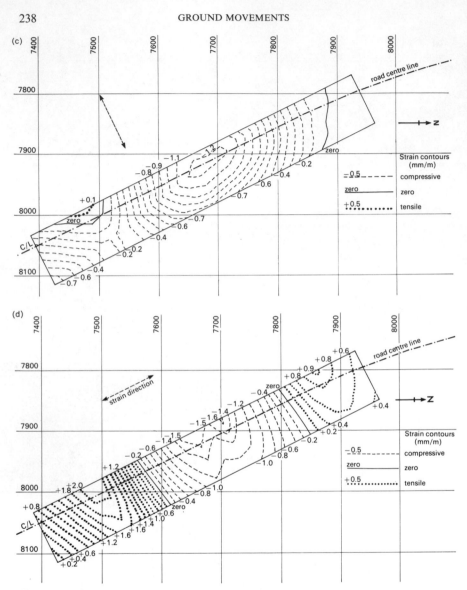

Figure 7.12 Ground lowering and strain prediction (by computer) showing one of 18 contiguous mining areas straddling the route of a proposed road. The requirement was to assess the possible amount of subsidence and associated strain during the construction period, to enable potential extra quantities of fill and possible structural movements to be included in the budget. (*a*) shows the general spatial arrangement of the road and intended workings. Two panels were to be worked, one in the Seven Feet seam and the other in the underlying Five Feet/Gellideg seam. Workings had already commenced in the Seven Feet seam, so that the subsidence and strain from this part of the mine had to be discounted from the total panel effects. Seam dips were ESE at gradients of up to 1 in 11. Due to the depth of working (600–650 m) both panels were of sub-critical width and length. (*b*) shows ridge and trough subsidence pattern. The centre of the trough due to the Five Feet/Gellideg panel is displaced south-eastwards 60 m because of seam dip. (*c*) and (*d*) plot the anticipated strains perpendicular and parallel to the road centre line. Note the peak tensile strains produced by the close proximity of NW ribside of Seven Feet panel and SE ribside of the Five Feet/Gellideg panel.

acquainted with the mining history of the area. Other techniques that are often useful include engineering geomorphological mapping and air-photograph interpretation. The *in-situ* examination of old workings is at times necessary, but strict safety precautions must be observed at all times. The risks of roof or floor collapse, explosion, fire and suffocation are real and the appropriate authorities must always be contacted before excavating deep inspection shafts or trenches or attempting to enter old workings.

Investigation of old mine workings should not be restricted to shallow workings if it is possible for discontinuities in the bedrock to have an adverse effect on the proposed engineering works. Such discontinuities are commonly associated with zones of high tensile strain around superimposed panels, especially at the limits of mine takes. High porewater pressures may develop beneath superficial deposits if bedrock aquifers are interconnected. Deep excavations, as in surface mining operations, may experience slope stability problems as a result of tensile strains above ribsides, the development of elevated cleft-water pressures and the development of contemporaneous shears along clay bands (*see* Stimpson and Walton, 1970). An example of slope instability promoted by longwall mining is given by Walton and Taylor (1977) and is shown in Fig. 7.13.

In areas with extensive deep mining, the patterns of residual extension and compression should be determined. Locations in which considerable extensions may have arisen should be investigated with the installation of sub-rockhead piezometers, especially if major earthworks such as cuttings or embankments are proposed.

Deep excavations in coalfields can promote dewatering and eventual re-flooding of contiguous room and pillar workings, and areas of former total extraction. It is sometimes suggested, albeit incorrectly, that hydrostatic support can be withdrawn if groundwater is removed (Instn. Civil Engineers, 1977). The present authors are of the opinion that ground movements in shallow room and pillar workings may perhaps be accelerated because of slaking of roof and pillar materials on dewatering. With areas of total extraction it is reasonable to assume that the broken ground and goaf will behave as a poorly-compacted fill.

Model studies of the caved zone behind the working face and the ground surface (for example, Farmer and Altounyan (1980) are of importance in attempting to understand changes in mass permeability and resulting groundwater conditions. Pertinent to this question is the recent flooding of the first production face in the new Selby coalfield in Britain.

In proposed areas of built development the aforementioned methods and approach can be used to ascertain the likelihood and magnitude of ground movements. In active mining areas combative subsidence precautions are customarily taken with structures, whilst with shallow workings various treatment and remedial measures are appropriate. These precautions are only briefly considered here since they are adequately covered elsewhere (in the references cited and in Chapter 8). Structural solutions are chiefly applied in active or proposed longwall mining areas. They include the use of flexible structures and special foundations which on one hand may include heavy reinforcement and on the other may reduce the deformation by using a sandwich of low frictional materials. These matters are adequately covered in the N.C.B. *Subsidence Engineer's Handbook* (1975); by Dept. Transport (1982) and in chapter 8. Whilst structural precautions, ranging from reinforced strip or raft foundations, to meshes and grids at the formation level of roads, are appropriate in certain settings where ground movements might arise from shallow room and pillar workings (Dept. Transport,

J

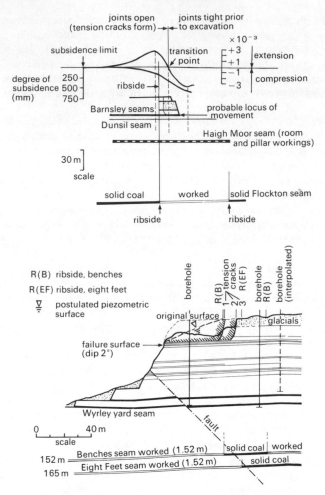

Figure 7.13 Slope failures induced by longwall mine workings. (a) Showing disposition of surface strains and subsidence profile due to Flockton seam workings (Yorkshire). Subsidence gradient over slip area − 1 in 90 (0.6°). (b) Section through slide in Staffordshire. The initial failure was followed by second movement. Note extrapolated ribsides in relation to tension cracks. After Walton and Taylor (1977).

1982), the more usual precautions include the grouting, stowing or excavation and removal of old workings voids. Methods and techniques are adequately outlined in Price *et al.* (1969) and Littlejohn (1979).

It will be appreciated that the appraisal of both deep and shallow mine openings in respect of existing or proposed development is a detailed subject, calling for substantial experience in the range of ground conditions that could exist as a result of former or proposed workings.

References

Arrowsmith, E.J. and Rankilor, P.R. (1981) Dalton by-pass; site investigation in an area of abandoned haematite mine workings. *Q.J. Engineering Geology* **14** (3) 207–218.

Beer, G. and Meek, J.L. (1982) Design curves for roofs and hanging-walls in bedded rock based on 'Voussoir' beam and plate solutions. *Trans. Inst. Min. Metall. (Sect.A:Min.Industry)* **91**, A18–22.

Bell, F.G. (Ed.) (1975) *Site Investigations in Areas of Mining Subsidence*. Newnes-Butterworths, London.

Bruhn, R.W., Magnuson, M.O. and Gray R.E. (1980) Subsidence over abandoned mines in the Pittsburgh Coalbed. *Proc. Conf. on Ground Movements and Structures, UWIST, Cardiff*, 142–156.

Burton, D. (1977) A three dimensional system for the prediction of surface movements due to mining. *Proc. Conf. on Large Ground Movements and Structures, UWIST, Cardiff*, Pentech Press 209–228.

Cameron, D.W.G. (1956) Menace of present day subsidence due to ancient mineral operations. *J. Royal Inst. Chartered Surveyors (Scottish Supplement)*, **19** (3) 159–171.

Carter, P., Jarman, D. and Sneddon, M. (1980). Mining subsidence in Bathgate, a town study. *Proc. Conf. on Ground Movements and Structures, UWIST, Cardiff*, 101–124.

Challinor, P.G. (1976) The mode of collapse of room and pillar workings. Unpublished MSc. Advanced Course in Engineering Geology Dissertation/Thesis, University of Durham.

Collins, B.J. (1977) Measurement and analysis of residual mining subsidence movements. *Proc. Conf. on Large Ground Movements and Structures, UWIST, Cardiff*, 3–29.

Department of Transport (1982) *Design of highway structures in areas of mining subsidence*, Dept. Standard BD 10/82, Roads and Local Transport Directorate London, 21 pp.

Duval, W.I.I. (1977) General principles of underground openings in competent rock. *Proc. 17th Rock Mechanics Symp.*, AIME, New York, 101–111.

Dumbleton, M.J. and West, G. (1976) *Preliminary Sources of Information for Site Investigations in Britain*. Departament of the Environment RRL Report LR403.

Evening Chronicle (Newcastle Upon Tyne) 18 Jan. 1978. 'Experts tackle gas menace'.

Evans, W.H. (1941) The strength of undermined strata. *Trans. Inst. Min. Metallurgy, London*, **50**, 475–532.

Farmer, I.W. and Altounyan, P.F.R. (1980) The mechanics of ground deformation above a caving longwall face. *Proc. Conf. on Ground Movements and Structures, UWIST, Cardiff*, 75–91.

Galloway, R.C. (1969) *History of Coal Mining in Great Britain*. David and Charles Reprints, London.

Gray, R.E. and Meyers, J.F. (1970) Mine subsidence and support methods in the Pittsburgh area. *J. Soil Mechanics and Foundation Division, ASCE* **96** (SM4) 1267–1287.

Hocking, G. (1978) Stress analysis of underground excavations incorporating slip and separation along discontinuities. In *Recent Advances in Boundary Element Methods* (ed. C.A. Brebbia), Pentech Press, London and Polymouth, 195–214.

Hoek, E. and Brown, E.T. (1980) *Underground excavations in rock*. Inst. Min. Metall., London.

Institution of Civil Engineers, (1977) Ground subsidence. *The Institution*, London.

Jones, O.T. and Davies, E.L. (1919) Pillar and stall working under a sandstone roof. *Trans. Instn. Min. Engrs.* **76**, 313–329.

Knipe, C. (1983) Personal communication.

Lee, A.J. (1966) The effect of faulting on mining subsidence. *Mining Engineer* **125** (71) 417–127.

Littlejohn, G.S. (1979) Consolidation of old coal workings. *Ground Engineering* **12**, (May), 15–21.

Malkin. A.B. and Wood, J.C. (1972) Subsidence problems in route design and construction. *Q. J. Engineering Geology* **5**, 179–192.

Marr, J. (1975) The application of the zone area system to the prediction of mining subsidence. *Mining Engineer* **133**, 53–62.

National Coal Board (1975) *Subsidence Engineer's Handbook*. Mining Department, National Coal Board, London.

National Coal Board (1982) *The Treatment of Disused Mine Shafts and Adits*. Mining Department, National Coal Board, London.

Orchard, R.J. (1954) Recent developments in predicting the amplitude of mining subsidence. *J. Royal Inst. Chartered Surveyors* **33**, 864.

Orchard, R.J. (1957) Prediction of the magnitude of surface movement. *Colliery Engineering* **34**, 455.

Orchard, R.J. (1964) Partial extraction and subsidence. *Mining Engineer* **123**, 417–427.

Orchard, R.J. and Allen, W.S. (1974) Time dependence in mining subsidence. In *International Symposium on Minerals and the Environment*, Inst. Min. Metall. London, 643–649.

Pearson, G.M. and Wade, E. (1967) The physical behaviour of seat-earths. *Proc. Geol. Soc. London* **1637**, 24–33.

Piggot, R.J. and Eynon, P. (1977) Ground movements arising from the presence of shallow abandoned mine workings. *Proc. Conf. on Large Ground Movements and Structures, UWIST, Cardiff*, Pentech Press, 749–780.

Price, D.G. Malkin, A.B. and Knill, J.L. (1969) Foundations of multi-storey blocks with special reference to old mine workings. *J. Engineering Geology* **1** (4) 271–322.

Shadbolt, C.H. (1977) Mining subsidence. *Proc. Conf. on Large Ground Movements and Structures, UWIST, Cardiff*, Pentech Press, 705–748.

Shadbolt, C.H., Whittaker, B.N. and Forrester, D.J. (1973) Recent developments in mining subsidence engineering. *Royal Inst. Chartered Surveyors, East Mids. Counties Minerals Div. 6th General Meeting*, 19 March 1973, Nottingham.

Stimpson, B. and Walton, G. (1970) Clay mylonites in English Coal Measures. Their significance in open-cast slope stability. *Proc. 1st Int. Cong. Int. Assoc. Eng. Geol. Paris*. **2**, 1388–1393.

Symons, M.V. (1980) Preliminary site investigations in old coal mining areas—problems of correlating coal seam names. *Proc. Conf. on Ground Movements and Structures, UWIST, Cardiff*, 211–240.

Szechy, K. (1967) *The Art of Tunnelling*. Akademiai Kiado, Budapest.

Tincelin, E. (1958) *Pressions et deformations de terrain dans les mines de fer de Lorraine*. Jouve Editeurs, Paris.

Taylor, R.K. (1968) Site investigations in coalfields: the problem of shallow mine workings, *Q.J. Engineering Geology* **1** (2) 115–133.

Taylor, R.K. (1975) Characteristics of shallow coal-mine workings and their implications in urban redevelopment areas. In *Site Investigations in Areas of Mining Subsidence* (ed. F.G. Bell), Newnes-Butterworths, London.

Terznghi, K. (1946) Rock defects and loads of tunnel supports. In *Rock Tunnelling with Steel Supports* (eds. R.V. Proctor and T.L. White) Commercial Shearing and Stamping Co., Youngstown, Ohio, U.S.A.

Walton, G. and Taylor, R.K. (1977) Likely constraints on the stability of excavated slopes due to underground coal workings. *Conf. Rock Eng. Newcastle* (ed. P.B. Attewell). British Geotech. Soc., 329–349.

Wardell, K. and Wood, J.C. (1965) Ground instability problems arising from the presence of old, shallow mine workings. *Proc. Midl. Soil Mech. Found. Engng. Soc.* **7**, 5–30.

Welsh Office (1969) *A Selection of Technical Reports submitted to the Aberfan Tribunal*. HMSO, London.

Wittaker, B.N. and Breeds, C.D. (1977) The influence of surface geology on the character of mining subsidence. *Conf. on Geotechnics of Structurally Complex Formations, Capri*. Associazone Geotechnica Italiana, vol. 1, pp. 459–468.

8 Structural design and ground movements

J.D. GEDDES

8.1 Introduction

When designing buildings, bridges, and other structures the engineer should attempt to ensure that they continue to fulfil their functional requirements during the whole of their design life. The requirements may range from the provision of load-carrying capacity to the maintenance of watertightness and general weatherproofing, and from aspects of safety to aesthetic matters. In meeting many of these needs, a consideration of structural displacements and the deformations and distortions to which they give rise is of fundamental importance. Structural Codes of Practice put due emphasis on the limitation of deflections, as well as ultimate load-carrying capacity, in the design of individual structural elements and complete frameworks. The introduction of limit state design has also focused attention on the need to consider such issues. Movements, particularly those in buildings, occur as a result of many factors and influences, and attitudes to their severity are likewise conditioned by a large number of factors— technical and human, aesthetic and financial. Alexander and Lawson (1981) in their study of design for movement in buildings have reviewed the current position and Table 8.1 lists limiting values for deflections in a variety of building elements. Most of these are expressed, not in absolute values, but as proportions of spans or heights, that is, deflection ratios.

Because of the technical and financial difficulties inherent in carrying out a full interactive analysis involving foundation soil or rock, substructure and superstructure, such calculations are seldom undertaken and limits of the type illustrated tend to be regarded as superstructure- and structural element-orientated. The customary role of the foundation engineer is to ensure that structures designed according to current non-interactive procedures will not be subjected to damaging foundation movements. When analysing such problems, in the vast majority of instances the gross movements and the differential movements of the foundations are caused by the loading imposed on the supporting soil or rock by the dead and live loads of the structure. This loading and the structural displacements and distortions are thus interrelated. However, in the majority of the site conditions with which this book is concerned such direct relationships do not exist, the movements to which structural foundations and, through them, their superstructures are subjected are for the most part caused by stress changes in the supporting strata brought about by natural processes and human activities such as mining, tunnelling and excavating. The presence of the structure may locally modify the values of the movements produced by these 'external' agencies by the

243

Table 8.1 Acceptable deflection limits for structural elements (after Alexander and Lawson, 1981)

Element	Criterion	Allowable deflection
Vertical deflections		
Beam	Steel beam total deflection	Span/200
	Reinforced concrete beam	Span/250 or 30 mm
	Cracking of brick or blockwork partition*	Span/500 or 15 mm
	Cracking of lightweight partition*	Span/350 to Span/360 or 20 mm
	Live load visible deflection +	Span/360
	Upward deflection because of precamber	Span/300
Floors or roofs	Differential settlement	Span/250 to Span/500 depending on cladding
	Timber flooring	Span/330
	Paved or asphalt covering	Span/250
	Flexible short span roof sheeting	Span/125
	Movement of sensitive equipment (e.g. generator)	1 in 750 slope (for example)
Cantilever	Visible deflection +	Span/180
	Cracking on cladding* (relative movement along edge)	Span/250 to Span/500, depending on cladding
Gantry girder	Inefficient travel of overhead crane +	Span/700
Lateral deflections		
Column	Side-sway of multi-storey building +	Height/1000 recommended
	Failure of frame with diagonals	1 in 600
	Racking of walls or infills of masonry structure	Height/500
	Single storey or low-rise flexible frame	Height/300
	Visible deflection of canopy roof	Height/250
Mullions	Bending of support to glazing	Span/175
Gantry girder	Crane rail separation +	Span/500

Note: All deflections are serviceability limits under worst total loading except as below:
* Installation after de-propping of floors
+ Imposed short-term loading only

processes of structure-soil interaction but this does not fundamentally change the essentially different nature of the problems posed. It thus frequently happens that the size and distribution of the movements imposed on the structure are completely different from those normally associated with 'conventional' construction.

8.2 Sensitivity of buildings to foundation movements

In the 'conventional' design of buildings it is now usual to think in terms of placing limits on settlements so that structural distress and damage to cladding and finishes will be avoided. Reliable criteria of this kind are very difficult to establish. Judgements often

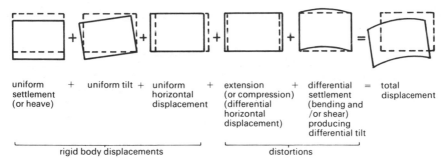

uniform   +   uniform tilt +   uniform   +   extension   +   differential   =   total
settlement   horizontal   (or compression)   settlement   displacement
(or heave)   displacement   (differential   (bending and
                                                                   horizontal   /or shear)
                                                                   displacement)   producing
                                                                                            differential tilt

|_____| |_____|
              rigid body displacements   distortions

Figure 8.1 Components of movement.

tend to be subjective, because what is visually acceptable to one person may be unacceptable to another. Criteria can only be determined through extensive performance surveys of existing buildings and the latter are infinitely variable in geometry, materials, finishes, structural arrangement, environmental setting and functional use. At the present time information is lacking in the quantity and quality needed to establish precise guide lines for acceptable movements of foundations.

However, a number of studies have gone some way towards establishing the forms which such criteria might take. The earliest study of significance was that undertaken by Skempton and MacDonald (1956) in which they attempted to draw conclusions from an analysis of existing documented histories of movement and related damage or non-damage. This work followed earlier suggestions by Terzaghi and Peck (1948) and Meyerhof (1953), and it has been followed in turn by the proposals of Polshin and Tokar (1957), Bjerrum (1963), Grant et al. (1974), and Burland and Wroth (1975 a, b). These efforts were directed at the determination of criteria for damage caused by vertical movements and their derivatives. Much less attention has been given to horizontal movements.

Figure 8.1 shows in simplified form how the total displacement of a structure may be considered to consist of five components. Three of these are rigid body displacements and two are distortional components. The latter are produced by divergences within the structure from the uniform movements exemplified by the former. They are the main potential sources of damage. For simplicity, only movements in a single vertical plane are illustrated.

8.2.1 Vertical movements

In the work which led to their recommendations, Skempton and MacDonald surveyed the performance of 98 buildings, and from a consideration of purely vertical movements after the elimination of any overall tilt effect they suggested the limiting criteria shown in Table 8.2. The buildings studied were of traditional steel and reinforced concrete frame types and structures with loadbearing walls. The damage indices they identified (Fig. 8.2) were (1) the maximum settlement under any stanchion, ρ_{max}, (2) the greatest differential settlement between any two stanchions, $(\delta\rho)_{max}$, and (3) the maximum 'angular distortion' occurring in any bay, $(\delta\rho/L)$, this being defined as the difference in vertical settlement between any two adjacent stanchions (after allowance for overall tilt) divided by the horizontal distance between them. This latter concept

Table 8.2   Settlement limitations (Skempton and MacDonald, 1956)

Criterion		Independent footings	Rafts
Angular distortion ($\delta\rho/L$)		1/300	1/300
Greatest differential settlement (($\delta\rho$)$_{max}$)	Sands	$1\frac{1}{4}$ in (30 mm)	$1\frac{1}{4}$ in (30 mm)
	Clays	$1\frac{3}{4}$ in (45 mm)	$1\frac{3}{4}$ in (45 mm)
Maximum settlement (ρ_{max})	Sands	2 in (50 mm)	2–3 in (50–75 mm)
	Clays	3 in (75 mm)	3–5 in (75–125 mm)

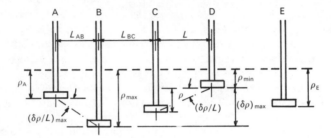

Figure 8.2   Movement indices (after Skempton and MacDonald, 1956).

was related to damage caused by shear distortion rather than bending deformation. As in the classic text by Terzaghi and Peck (1948), they also differentiated between foundations on sands and clays and between independent footings and rafts, attributable to the different settlement patterns associated with these soils and types of foundation (that is, a tendency to a randomness for sands and independent footings and towards a bowl-shaped settlement pattern for clays and rafts). In almost all cases, damage was primarily functional and aesthetic rather than structural in character. Their limiting value of 1/300 for angular distortion was associated with cracking in walls and partitions. A ratio of 1/150 was deemed necessary to cause structural damage. For design they suggested a limiting ratio of 1/500 with a further reduction to 1/1000 for very sensitive structures.

Bjerrum (1963) added to this data and presented the tentative schedule shown in Table 8.3. He adopted the idea of a limit of 1/150 for structural damage and 1/500 for the safe limit against cracking of walls. Further data were provided by Grant *et al.* (1974) who contributed case histories of a further 95 buildings. Their analysis appeared to confirm the value of 1/300 as a reasonable damage limit.

The parameters adopted by the foregoing workers have been criticized by a number of authors as an over-simplification. In particular the effect of curvature of ground and structure is not clearly described and this is of great importance in relation to the behaviour of loadbearing brick walls subjected to sagging and hogging profiles. Figure 8.3 illustrates the need for discrimination. The two lower diagrams show how a deflected profile can be obtained by simple bending or by shear displacements. In practice

Table 8.3 Tentative damage criteria (Bjerrum, 1963)

Angular distortion $(\delta\rho/L)$	
1/750	Limit where difficulties with machinery sensitive to settlements are to be feared
1/600	Limit of danger for frames with diagonals
1/500	Safe limit for buildings where cracking is not permissible
1/300	Limit where first cracking in panel walls is to be expected Limit where difficulties with overhead cranes are to be expected
1/250	Limit where tilting of high, rigid buildings might becomes visible
1/150	Considerable cracking in panel walls and brick walls Safe limit for flexible brick walls height/length $< \frac{1}{4}$ Limit where structural damage of general buildings is to be feared

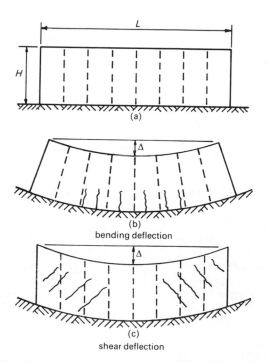

bending deflection

shear deflection

Figure 8.3 Subsided profile obtained by (b) bending, (c) shear.

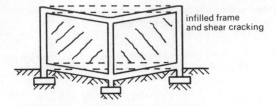

Figure 8.4 Framed structure with shear distortion.

one or other mode may predominate depending on the structural geometry, the construction materials and other factors. In other instances both mechanisms may be involved. Figure 8.4 shows, in a simple way, the manner in which an infilled frame may conform to a settlement profile by shear distortion.

To cater for the range of ground and structural movements with which this book is concerned, as shown in Fig. 8.1, an extended range of parameters has to be used to give a more full description of the geometry of displacement and deformation. The upper part of Fig. 8.5 shows the displacement profile produced by the movement, with horizontal and vertical components, of a number of points. The latter may represent positions on a ground surface, positions on the soffit of a continuous foundation or the centre points of independent foundations. In the case of representation of ground movements, the profile may relate to the ground movements produced by structural loading, to those which would be produced by external factors such as mining or excavation in the absence of a supported structure, or to the latter after modification due to the presence of a structure. The lower part of the figure shows the quantities derived from the vertical components of movement. In the majority of

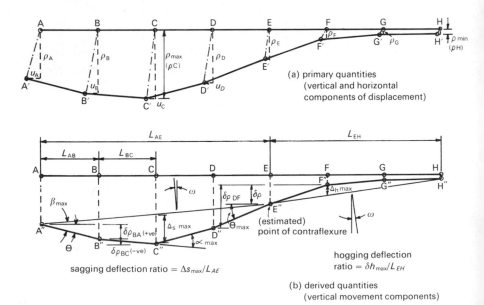

Figure 8.5 Patterns of movement.

practical circumstances the presence of the horizontal movements does not substantially change the geometry described in this way. The presence of horizontal movements must not be ignored in any consideration of structural effects.

The terms adopted are as follows (after Burland and Wroth, 1975a, b; Burland et al., 1977).

(a) Settlement ρ and differential settlement or relative settlement $\delta\rho$ are shown in Fig. 8.5a and 8.5b. For maximum settlement, ρ_{max} is used, and the difference between ρ_{max} and the minimum settlement ρ_{min} is denoted by $(\delta\rho)_{max}$. Where heave (negative settlement) is exhibited the symbol ρ_h may be used.

(b) Rotation θ is the change of gradient of a line joining two reference points.

(c) Angular strain is denoted by α. The value at **B** is given, for example, by

$$\alpha_B = \frac{\delta\rho_{BA}}{L_{AB}} + \frac{\delta\rho_{BC}}{L_{BC}} = \frac{\rho_B - \rho_A}{L_{AB}} + \frac{\rho_B - \rho_C}{L_{BC}}$$

It is positive when it describes 'sag' or upward concavity and negative when it describes 'hog' or downward concavity.

(d) Relative deflection Δ is the displacement relative to the line connecting two reference points a distance L apart. The same sign convention is used as in (c).

(e) Deflection ratio (sagging ratio or hogging ratio) is denoted by Λ_s/L and Λ_h/L. Determination of its maximum value and the location of the latter may not be easy when a curved profile is involved.

(f) Tilt is denoted by ω and it describes the rigid body rotation of the structure or a well-defined part of it. The term is most appropriately used in relation to the movements of raft type or connected foundations and has less relevance when applied to separate footings. In relation to ground movements, tilt usually refers to rotation of the line joining two surface points. It is thus synonymous with θ given in (b).

(g) Relative rotation or angular distortion β is the rotation of the line joining two reference points relative to the tilt. The term 'angular distortion' was used by Skempton and MacDonald, but it is now more commonly used when only shear distortion is involved, otherwise the term 'relative rotation' is more apt.

(h) Differences in horizontal displacements u give rise to length changes δu. Where these take place over a length L the average horizontal strain is denoted by $\varepsilon = \delta u/L$.

These definitions are of general validity but relate only to in-plane displacements. They may be used for framed structures with independent foundations, for framed structures on raft or continuous foundations and for continuous structures such as brick walls. The movements may be those predicted or measured at discrete points. For framed structures they will in general be those at the bases of stanchions, for continuous elements they will represent discrete points on a continuous curve and hence finite difference approximations to mathematical quantities such as gradients and curvature.

A major element of difference between this system and the earlier one of Skempton and MacDonald was the introduction of the concept of deflection ratio (Δ/L) as a measure of potential damage. Polshin and Tokar (1957) are typically credited with its first use. Thus, although they regarded maximum values of the relative rotation (angular distortion) to be 1/500 for steel and reinforced concrete infilled frames, with further limits of between 1/140 and 1/1000 for end bays of columns with brick cladding on sands and hard clays and on soft clays respectively, and a ratio of 1/200 where there is no infill or no danger of damage to cladding, they laid down much stricter criteria for load-bearing brick walls based on deflection ratio. Thus, for the latter with ratios of length

Table 8.4 Allowable total and differential settlements (angular strain) of foundations (Wilun and Starzewski, 1975)

Class of building and structure	Type of building or structure	Maximum allowable final settlement $(\rho_{max})(mm)$	Maximum allowable angular strain calculated for three, collinear, adjoining points or foundations of a structure (α_{max})
1	Massive structures of considerable rigidity about horizontal axes founded on rigid mass concrete foundations or cellular or rigid reinforced concrete rafts	150–200 (6 in–8 in)	Maximum differences of settlements at various points of the structure should not cause tilting of the foundation greater than $1/100 \div 1/200$ of the ratio of the smallest dimension of the foundation in plan to the height of the structure.
2	Statically determinate structures with actual pin joints (three pinned arches, single-span steel trusses, etc.) and timber structures	100–150 (4 in–6 in)	$\frac{1}{100} - \frac{1}{200}$
3	Statically indeterminate steel structures and loadbearing brickwork construction with reinforced concrete ring beams at every floor level, with longitudinally reinforced concrete strip foundations and with cross walls of at least 250 mm (9 in) thickness and spaced at not more than 6 m ($\sim$ 20 ft) centres and reinforced concrete frame-structures with columns at less than 6 m centres and founded on strip or raft foundations	80–100 ($3\frac{1}{2}$ in–4 in)	$\frac{1}{200} - \frac{1}{300}$
4	Structures of class 3 but not satisfying one of the stated conditions and reinforced concrete structures founded on isolated footings	60–80 ($2\frac{1}{2}$ in–$3\frac{1}{2}$ in)	$\frac{1}{300} - \frac{1}{500}$
5	Prefabricated structures consisting of large slab or block elements	50–60 (2 in–$2\frac{1}{2}$ in)	$\frac{1}{500} - \frac{1}{700}$

(1) Smaller values that are quoted relate to public buildings, dwellings, or buildings with structural members or finishes particularly sensitive to differential settlement; larger values relate to taller buildings of considerable rigidity about horizontal or to structures which can accept such movements.

(2) In special cases (such as gantry beams, high-pressure boilers, special storage tanks, silos under differential loading, etc.) allowable maximum or differential settlements or both should be taken as specified by service- or mechanical engineers or by manufacturers.

to height equal to or less than 3, the maximum deflection ratio for no cracking is given as 1/3000 and 1/2500 on sand and soft clay respectively, increasing to 1/2000 and 1/1400 for the same soils and for length to height ratios equal to or greater than 5.

Meyerhof (1953) had earlier suggested values of 1/300 for the differential settlement between adjacent columns of open frames of encased steel and reinforced concrete at cracking (angular distortion), 1/1000 for the design of frames with infilling and a deflection ratio of 1/2000 for the design of loadbearing brick walls.

In reaching these conclusions on limiting deflection ratio, Polshin and Tokar linked the onset of visible cracking to a specific tensile strain (ε_{lim}), dependent on the material involved. They used a value of 0.05 per cent for unreinforced brick walls. Burland and Wroth (1975a) extended this principle and showed how bending and shear modes could be encompassed in a simple model. In this they chose a value for ε_{lim} of 0.075 per cent. Their results produced values of 1/2500 for a length/height ratio of 1 and 1/1250 for a ratio of 5; of similar order to those of Polshin and Tokar. However, they made the important distinction that cracking in a hogging mode is likely to occur at ratios about half these latter values. This conclusion based on experimental evidence reported from a few field cases is explained by the fact that in a sagging mode the tensile effects at the lower part of a structure (foundations) are to some extent restrained by the normal and shear stresses at the foundation/soil interface whilst in a hogging mode such restraints are absent on the upper surface. Although further work is required to develop these ideas, this approach appears to be capable of generating advances in our knowledge of limiting deformations insofar as the performance of brick masonry type structures is concerned.

Table 8.4, based on Polish practice (Wilun and Starzewski, 1975) offers further suggestions related to limitations on maximum allowable settlements (ρ_{max}) and angular strain (α_{max}) for a range of building structures.

All the foregoing criteria related to visible cracking and damage must also be viewed in relation to aesthetic, serviceability and functional requirements. These will vary considerably between structures and the engineer must make appropriate judgements based on all known facts. Table 8.5 (BRE Digest 251, Dept. of the Environment, 1981) provides a proposed classification of degrees of damage to walls with particular reference to visual impact and the ease and nature of the repairs needed.

8.2.2 Horizontal movements

Although horizontal movements often occur there is relatively little published information of a quantitative kind on the damage associated with them. The primary source may be regarded as the National Coal Board (1975) which has presented the data in Table 8.6. The table proposes a correlation between damage and change of length of a structure. Figure 8.6, which accompanies the table, shows damage zones associated with this concept. The individual curves are rectangular hyperbolae, such that length of structure × ground strain = (constant) change of length. Such a relationship is only valid when average ground strain = average structural strain. In general this will not be true.

The preceding sections show that although the position is changing there are too few case histories available to permit a firm indication of tolerable vertical and horizontal movements and differential movements. With the exception of Table 8.6 the information provided in the tables and text has been gathered from a study of the performance of structures under load-related displacements and deformations. These can nevertheless

Table 8.5 Classification of visible damage to walls with particular reference to ease of repair of plaster and brickwork or masonry (Dept. of the Environment, 1981)

Degree of damage	Description of typical damage* (ease of repair in italic type)	Approximate crack width (mm)
0 Negligible	Hairline cracks of less than about 0.1 mm width are classed as negligible.	≯ 0.1**
1 Very slight	*Fine cracks which can easily be treated during normal decoration.* Perhaps isolated slight fracturing in building. Cracks rarely visible in external brickwork.	≯ 1.0**
2 Slight	*Cracks easily filled. Redecoration probably required. Recurrent cracks can be masked by suitable linings.* Several slight fractures showing inside building. Cracks not necessarily visible externally; *some external re-pointing may be required to ensure weathertightness.* Doors and windows may stick slightly.	≯ 5.0**
3 Moderate	*The cracks require some opening up and can be patched by a mason. Repointing of external brickwork and possibly a small amount of brickwork to be replaced.* Doors and windows sticking. Service pipes may fracture. Weathertightness often impaired.	5 to 15** or a number of cracks ≤ 3.0
4 Severe	*Extensive repair work involving breaking-out and replacing sections of walls, especially over doors and windows.* Window and door frames distorted, floor sloping noticeably+. Walls leaning+ or bulging noticeably, some loss of bearing in beams. Service pipes disrupted.	15 to 25** but also depends on number of cracks
5 Very severe	*This requires a major repair job involving partial or complete rebuilding.* Beams lose bearing, walls lean badly and require shoring. Windows broken with distortion. Danger of instability.	usually > 25** but depends on number of cracks

*It must be emphasized that, in assessing the degree of damage, account must be taken of the location in the building or structure at which it occurs and also the function of the building or structure.
**Crack width is one factor in assessing degree of damage and should not be used on its own as a direct measure of it.
+ Local deviation of slope, from the horizontal or vertical of more than 1/100 will normally be clearly visible. Overall deviations in excess of 1/150 are undesirable.

furnish guidelines which enable the engineer to obtain an appreciation of the possible damage potential of non-load induced ground movements in relation to existing structures and planned structures.

8.3   Approaches to construction

The first requirement for a successful design, or for an analysis of the safety of or likely damage to an existing structure, is a knowledge of the nature, timing, extent and size of the ground movements to be experienced. Other chapters show that current knowledge is not always capable of providing all or some of the latter information and in some instances it is improbable that it ever will be. The range of potential problems is vast and because of space limitations the present chapter can only deal with the issues posed in terms of general principles. The engineer faced with a specific problem must therefore be prepared to exercise his full range of knowledge and experience in seeking a solution.

Table 8.6 National Coal Board (1975) classification of subsidence damage

Change of length of structure	Class of damage	Description of typical damage
Up to 0.03 m	1 Very slight or negligible	Hair cracks in plaster. Perhaps isolated slight fracture in the building, not visible on outside.
0.03 m–0.06 m	2 Slight	Several slight fractures showing inside the building. Doors and windows may stick slightly. Repairs to decoration probably necessary.
0.06 m–0.12 m	3 Appreciable	Slight fracture showing on outside of building (or one main fracture). Doors and windows sticking; service pipes may fracture.
0.12 m–0.18 m	4 Severe	Service pipes disrupted. Open fractures requiring rebonding and allowing weather into the structure. Window and door frames distorted; floors sloping noticeably; walls leaning or bulging noticeably. Some loss of bearing in beams. If compressive damage, overlapping of roof joints and lifting of brickwork with open horizontal fractures.
More than 0.18 m	5 Very severe	As above, but worse, and requiring partial or complete rebuilding. Roof and floor beams lose bearing and need shoring up. Windows broken with distortion. Severe slopes on floors. If compressive damage, severe buckling and bulging of the roof and walls.

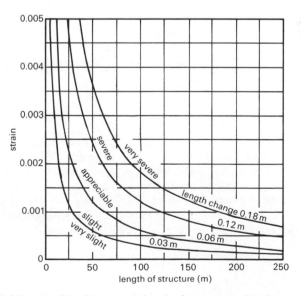

Figure 8.6 Relationship of damage to length of structure and horizontal strain.

There are four primary ways in which successful solutions may be found to construction problems created by ground movements, potential, past, present or future. These are:

   (i) avoidance
  (ii) ground treatment
 (iii) modifications to methods of working
 (iv) the adoption of structural measures.

Sometimes a combination of these approaches may be used.

The first three are dealt with briefly in this section and the fourth is treated in greater detail later.

8.3.1 *Avoidance*

A process of avoidance may be adopted when the engineer is otherwise unable to control the situation with which he is or will be confronted, either on the grounds of technical impracticability or of cost. Avoidance can be practised in two ways.

(a) By taking action which removes the cause of the ground movements. This form of avoidance is frequently, but not exclusively, adopted when human activity is responsible for creating the problem. Examples are furnished by the sterilization of part of a coal seam to protect a surface area, the elimination of a deep basement to a new building which might hazard the safety or serviceability of an adjacent existing building, the requirement that backfilled areas may not be built over for a specified period after completion of the backfilling operations, and the physical removal of unstable or highly compressible soils.

(b) By a decision to avoid construction in the affected area. This category frequently involves natural, geological phenomena as well as man-made hazards. Examples are provided by the hazard zoning of landslide areas, and the sterilization for building purposes of areas underlain by old shallow mineworkings or limestone cavities.

8.3.2 *Ground treatment*

With ground treatment, attempts are made to reduce or eliminate the production of movements before a structure is built or to protect an existing structure. Techniques employed include dynamic consolidation, vibro-flotation, chemical consolidation, pre-loading of highly compressible fills, perhaps in association with filter drains, and the use of electro-osmosis and drainage provision to stabilize slopes and excavations. Where underground mining or quarrying cavities exist with a risk of possible collapse, ground treatment may involve the filling of the old workings using a variety of materials and techniques.

8.3.3 *Modifications to methods of working*

This approach has as its objective the reduction or elimination of movements by a suitable selection or change of technique in relation to the human activity responsible for the creation of movements. Examples are furnished by the careful selection of a retaining technique to be adopted in the construction of a deep excavation and by the use of procedures such as the longwall partial extraction method in which the pattern of underground coal extraction is altered. It can be very successful in practice and has been used many times for the protection of important structures. When tunnelling in soil beneath built-up areas it is generally necessary to reduce surface movements by the selection of a method of driving appropriate to the given ground conditions.

8.3.4 *The adoption of structural measures*

If the foregoing procedures are not adopted or are used as a part solution, the design of the structure must be carried out to cater for the expected total or remaining movements. In section 8.2 the sensitivity of structures and structural elements to movements is considered. The data presented therein provide a guide to the need for special design measures. Existing construction may require structural measures for its protection, ranging from substantial permanent reinforcing to the incorporation of temporary supports.

8.4 The effects of ground movements on structures and related facilities

The conditions described in other chapters typically generate ground movements which in general can be described in terms of vertical and horizontal components (Fig. 8.5). Moreover, these movements may vary from one plan position to another, and with time. The overall effect is to produce a differential vertical settlement of the ground surface, giving rise to a subsidence profile which displays surface tilting associated with concave and/or convex curvature. Differential horizontal movements produce lengthening and/or shortening of elements of the ground and these are often referred to as tensile and compressive ground strains. The movement patterns may be described using the terminology set out in section 8.2.1. In addition, it is sometimes useful to refer to the radius of curvature of the ground at specific positions.

When the effects of these movements on structures are considered, two features must be borne in mind. In the first instance it must not be assumed that the ground movements produced in the presence of a structure will be the same as those with a structure absent. Local modifications to the latter take place, to meet the requirement that at the interfaces between the foundations and the ground the forces developed on the ground by the structure are equal and opposite to those developed by the ground on the structure. In addition, overall equilibrium requirements must be satisfied. The criterion for a solution is equilibrium of forces and not necessarily equality throughout of foundation movements and ground movements. Although in many instances contact between ground and structure will be maintained during the period of ground movement, there will be other situations in which gaps open between ground and structure at an interface and slip type displacements may also occur. Determination of the extent of these modifications is a complex matter and one which has not been fully resolved. It involves considerations of the relative stiffness of ground and structure and their stress/strain behaviours. Hence, when reference is made to movements it is essential to indicate which of these three categories is being discussed, that is, structural or ground, the latter being subdivided into those without the structure and those with the structure in place.

Secondly, it is important to realise that it is the variation of the final modified vertical and horizontal ground displacements over the areas of contact with the structure or other facility that determines their potential to cause damage. When the horizontal and vertical movements are uniform over the area occupied by the structure no changes are produced in the forces which act between ground and structure prior to the imposed ground movements. In such a circumstance the structure would merely be translated to a new horizontal and/or vertical position.

It is therefore to be expected that the imposed ground movements involving tilting, curvature in a vertical plane, differential vertical movements and horizontal strains will produce similar effects on the structures or services supported (see Fig. 8.1), although the scale of the effects will be modified. Vertical strains may also be present in the ground and interact with deeply-embedded structures such as piles.

8.4.1 *Vertical movements*

Tilts or changes of gradient are of particular importance in relation to the design of services such as sewers, water-supply pipelines, roads and railways because gradients form part of the design criteria. It is clearly necessary that desired minimum gradients are maintained after movements have occurred and that reversed gradients are not generated. In the case of tall narrow structures, tilt may affect their overall stability by lessening their factor of safety against overturning, and in extreme cases a stability failure could take place because of the creation of an eccentrically loaded foundation. The differential tilting of adjacent structures or unconnected structural elements may lead to a complete closing of gaps between them, with the transmission of perhaps damaging forces. Tilting may also affect the operation of industrial plant such as cranes and other machinery which might require level operating conditions.

Ground curvature is responsible for the production of bending effects and torsional twisting of three-dimensional structures. The extent to which a structure can conform to a subsidence profile caused by external agencies is a function of its flexibility in bending. With an infinite flexibility and hence an inability to redistribute support reactions, the latter remain as they were before the movements occurred, the subsidence profile is not altered locally and the structure takes up a profile which is identical to that of the ground. The imposed subsidence movements thus become synonymous with the structural movements. At the other extreme, with infinite flexural rigidity, the structure remains undeformed by the attempted imposition of the subsidence movements. The latter are modified at the ground/structure interface by local 'flattening' giving rise to a redistribution of the supporting reactions at the ground/structure interface. Physical contact between ground and foundation may be lost over part of the length. In this event the three movement profiles referred to earlier are all different. The redistribution of reactions in turn leads to changes in the distribution of bending and twisting moments and shearing forces on the structure which must therefore be catered for in design, usually with increased structural costs. The three-dimensional character of the movements should always be borne in mind and consideration of behaviour should not be restricted to a single vertical plane. In addition to causing bending moments in structural elements, ground curvature may also produce substantial distance changes between separate structural components, particularly when measured at some height above the ground/foundation interface level. Thus beams may lose contact with their designed bearing surfaces or may override them if inadequate provision is made for length changes of this kind. The particular danger of hogging curvature has also been referred to earlier in connection with brick-bearing continuous structures.

Buried structures such as pipelines, culverts and piles subjected to vertical movements may be exposed to substantial increases in the vertical loadings to which they are normally subjected. These are produced by the obstructions created by the presence of the structures to the free downward movements of the ground.

8.4.2 *Horizontal movements*

Horizontal ground movements increase or reduce the horizontal forces acting on vertical or inclined walls such as bridge abutments, culvert walls and the walls of buried tanks. They may produce failure as a result of compression or, in the case of liquid-retaining buried structures, failure as a result of the relaxation or complete withdrawal of lateral support if this has not been allowed for in design. The forces developed can be expected to take on values between those of the active and passive conditions, and depend on the compressibility of the structure and of the soil delivering the load, and upon the amount by which the ground is tending to contract or expand.

Horizontal movements also produce shearing stresses at the interface between the supporting soil and the soffit of ground-supported foundations and slabs. These movements and consequential stresses have the effect of creating tensional and compressive forces and related moments due to their eccentricity. They can also produce rotation and bending in plan when applied over an area of contact. The magnitude of such forces can be very great, and since they are not usually of consequence in the conventional design of structures and their foundations, their unexpected presence can be highly damaging. As described in section 8.6.1 their magnitude is a function of the soil and interface properties, the contact pressures, the adhesion and angle of skin friction and the relative slip or displacement between the ground and the slab being supported. The slip depends on the extensional/compressive properties of the structure and the ground strain being experienced. Because of the non-reversible nature of the mechanism relating slip to developed shear force, it is possible for tensile strains to be developed in a structure whilst the ground is experiencing compressive strains (Geddes, 1980).

These are the primary effects of imposed ground movements. Some of the methods of quantitative evaluation of these effects are now considered, dealing with the influence of vertical movements first and then the influence of horizontal movements. In practice, both components may be present and any design method must make allowance for this. In addition, different approaches are needed, depending upon whether the design of a new structure is being undertaken or whether the safety and functional stability of an existing structure is under consideration. In the former case, extreme limiting conditions might be adopted for design purposes because of the possible unpredictability of future imposed movement patterns, whereas in the latter condition there may be a greater possibility of evaluating the movements. However, the engineer is then sometimes faced with serious gaps in the general knowledge of soil/structure behaviour. Space limitations preclude anything other than a short presentation of the issues involved, but greater guidance can be obtained from the publications listed at the end of this chapter.

8.5 Design for vertical movements

Differential vertical movements give rise to shear distortion, bending distortion (due to ground curvature) and tilting. Tolerable limits have been cited in section 8.2. The potential effects of vertical movements can be substantially reduced by the simple expedient of subdividing the structure into a number of smaller sub-units with complete physical separation between them, or with connections which offer no restraint to relative movements between adjacent units but which at the same time permit the maintenance of functional requirements, such as watertightness and thermal insulation. Such subdivision also has the advantage that it reduces many of the problems which can occur due to normal causes such as shrinkage, thermal expansion, and creep. The presence of these normal movements must not be overlooked (Alexander and Lawson, 1981). Attention to detail is vitally important, otherwise the beneficial effects provided by such features as movement joints may be totally negated. With a reduced length and/or width, a smaller area of the subsidence zone is covered, thus reducing differentials over the structure. It is essential to provide for different tilting of adjacent units, differential vertical movements between units and differential horizontal movements to avoid the development of forces for which the designer has made no provision.

With respect to the redistribution of vertical forces acting on a structure, two types of foundation arrangement need to be considered: (i) a continuous foundation (strip or raft) over the length (or plan area) of the structure, and (ii) a series of independent footings, connected through the structural framework.

8.5.1 *Continuous foundation*

There are usually two limiting conditions which have to be considered: (i) when the foundation occupies a position straddling the minimum radius of ground curvature in a hogging mode, and (ii) when a similar condition exists in a sagging mode. Each may be associated with the concept of an infinitely rigid foundation or a flexible foundation.

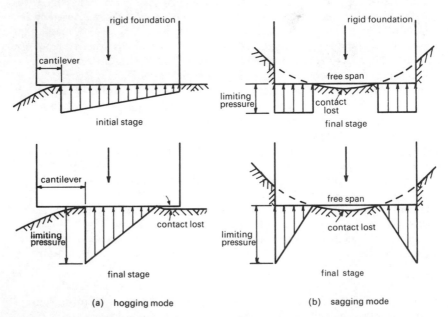

(a) hogging mode   (b)   sagging mode

Figure 8.7 Rigid foundation model (after Mautner, 1948).

Figure 8.7 (*a*) shows the model advanced by Mautner (1948) and the Institution of Structural Engineers (1949). A rigid foundation is positioned so that a hogging ground movement results in the partial withdrawal of support from one side of the foundation, so resulting in a cantilevering action with its resultant induced moments and shears. As the ground curvature is increased, the extent of the cantilevering effect also tends to increase with the ground pressure distribution changing from a trapezoid to a triangle. The foundation is assumed to act as an eccentrically loaded foundation and, apart from the settlement conditions, attention must be paid to the possibility of a bearing capacity type failure due to the changed loading conditions. Figure 8.7 (*b*) illustrates Mautner's model when a rigid foundation occupies a position of sagging curvature, with the creation of a void over a central portion of the foundation. He postulates that one may assume a triangular contact pressure distribution or a uniform one. Again it is assumed that increased curvature will result in loss of contact over a greater length. In making use of this method it is suggested that design should be based on a 'worst possible case' approach because of difficulties in the prediction of actual ground movements. This develops when the cantilever and free beam spans are at a maximum and Mautner suggests that this occurs when the maximum contact pressure reaches a limit, the yield limit, determined by substantially increased load-induced settlement for small load increments. At this yield contact pressure any attempt to alter the loading would produce penetration of the foundation into the soil, and hence reduce the free span. The design thus centres on the choice of a suitable yield-pressure value for different soil conditions.

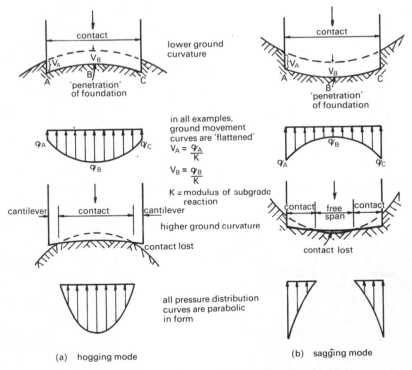

(a) hogging mode

(b) sagging mode

Figure 8.8 Flexible foundation model (after Wasilkowski, 1955).

Mautner also suggests that the permissible bearing pressure should not exceed the yield pressure or one-half of the ultimate bearing capacity of the soil, whichever is lower, and that advantages can be gained by the use of a design contact pressure which is relatively high in relation to the yield pressure.

Wasilkowski (1955) takes a different approach (Fig. 8.8). By introducing the concept of a modulus of subgrade reaction it is possible to calculate the amount by which the curved imposed soil profile is 'flattened' to allow bedding of a flexible foundation. Throughout the calculations it is assumed that contact pressure distributions follow parabolic curves. The Wasilkowski method can also be applied in the case of rigid foundations by modifying the algebraic expressions involved in the calculation. Some guidance is provided on the selection of the soil modulus value, based on Polish practice, but as yet there appears to be little published data which confirms the accuracy of the calculations. This approach, and that of Mautner, usually leads to heavy design moments and shears unless the structure is very light, and these require deep ground beams or grillages. The effects are reduced for smaller curvatures, when the soil modulus is relatively low, and when the structure is more flexible. With both these approaches, unless the orientation of the structure is known with some precision in relation to future ground movements, analyses should be performed with cantilevering and free beam action in different vertical planes across the foundation.

8.5.2 *Independent footings*

The size of individual foundations will frequently be such that they occupy only a small portion of a subsidence zone. For practical purposes they may be considered to settle uniformly, with a uniform contact pressure acting on them when axially loaded.

The basis of a calculation method may be illustrated by reference to the simple example of foundations supporting an indeterminate structural framework. If the ground beneath each of the foundations settles by a different amount this then leads to a redistribution of the support reactions. The structural foundations will attempt to follow the subsiding ground but will be resisted in this by the flexural stiffness of the superstructure. As a first approximation, with the individual foundation movements being set equal to the imposed ground movements at their respective positions, the modified support reactions can be determined. However, where the latter are increased in value it can be expected that further structural settlement will take place. Where foundation reactions are reduced, a slight, or zero, expansion will ensue. These movements can be calculated by the use of a modulus of subgrade reaction and the initial imposed ground movements altered accordingly. This will in turn entrain a further alteration in the distribution of support reactions and the process can be repeated until convergence occurs. The solution is best achieved by use of a matrix formulation, and elasto-plastic behaviour of the supported structure may be catered for. Checks need to be made to ensure that no foundation loses contact with the ground under it. It is also obvious that, for framed structures, less distress will be caused when the structure is determinate or has high flexibility, and when the superimposed ground curvature is small and the compressibility of the foundation soil is high.

The most well-known example of a flexible system is the CLASP form of construction, used extensively in the United Kingdom. Its satisfactory functioning has been confirmed on a large number of sites subjected to mining subsidence movements, and it is clearly suitable for use on other sites where different patterns of movement may be experienced.

It has been fully described in a number of publications (Gibson, 1957; Lacey and Swain, 1957; Bell, 1977; Geddes 1977b). The emphasis throughout is on articulation, with structural stiffness being provided by the controlled action of diagonal springs placed in a number of the rectangular portal bays. Great attention has been paid to the design of the structural and cladding details to ensure that resistance to the development of movements is minimized.

It is also possible to produce a design in which neither major distortion nor the imposition of large forces has to be catered for. This requires the provision of support systems which remain statically determinate during movements, for example by the use of three-point articulated supports.

8.6 Design for horizontal movements

8.6.1 *Subgrade restraint and friction forces*

If one considers a single concrete foundation slab resting freely on the surface of a soil, it is clear that if the soil below it is displaced uniformly and relatively slowly in a horizontal direction then the foundation will simply move with the soil and no horizontal shear forces will be developed at the interface. If, however, the foundation is restrained from moving and the process is repeated, a force will be produced by the build-up of frictional and/or adhesive stresses on the interface. The sum of such forces will equal the force of restraint, which could, for example, be provided through a structural framework connecting with the foundation slab. Experiments show that the development of the force is a function of the amount of slip that takes place at the interface, as well as other factors such as normal pressure on the plane of slipping. Without slip, or relative movement,

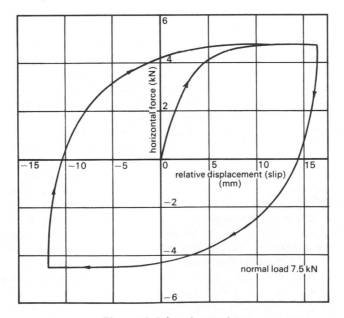

Figure 8.9 Subgrade restraint.

there is no force. Figure 8.9 shows a typical relationship of the kind that would be observed if the concrete slab rested on a relatively dense sand layer. It will be noted that a critical displacement of about 12 mm was needed in this instance to develop the maximum force. When the direction of slip (relative movement) was reversed a number of times, the hyteresis curve shown in the figure was obtained. This is a common form of behaviour. For frictional soils the ratio between the maximum friction force and the normal force is often found to be about 0.66, but there can be substantial variations from this value. The presence of adhesion forces at the interface also affects the ratio.

If two such foundation slabs are joined by a non-deforming structural frame or connection, so that the distance between them does not alter, the ground between them would be required to change in length by 24 mm to produce the maximum force of approximately 4.7 kN in the connection because a relative slip of 12 mm (in opposite directions) would be needed at each foundation to produce the equal and opposite forces needed for equilibrium. If the ground shortened by this amount a compressive force would be developed in the structure, and if the ground lengthened a tensile force would result. The minimum ground strain needed to produce the maximum force would depend, however, on the distance between the foundations. For a rigid connection the same maximum force would be produced by a ground strain of 2 mm/m on a structure 12 m long as by a ground strain of 1 mm/m on a structure 24 m long. Whether these strains would be damaging would depend on the ability of the structural frame (or connection) to sustain the load of 4.7 kN in tension or compression.

If the connection is deformable under compressive or tensile loads, the change in length of the ground between the foundations would then have to be increased by the amount of the structural shortening or lengthening under the load to produce the same maximum force as before. This means that deformable structures require correspondingly greater ground strains to develop the same loads.

Once the critical relative movements have been developed, further increases in ground strains do not produce increased loads. The extra ground movements simply produce greater slippage at the foundation/soil interface. Insofar as horizontal movements are concerned, if the structure is undamaged at the critical displacement stage it will remain undamaged for increased ground straining.

It follows that for a given foundation and vertical loading, the horizontal force produced can be reduced by selecting an interface material which has a low coefficient of friction and/or which has a high value of critical relative displacement. This gives rise to the concept of a sliding layer, which also has the effect of eliminating adhesion forces.

The same kind of mechanism can be assumed to operate when differential movements (strains) occur over the area covered by a single foundation slab or raft. In this instance the relative slips occurring over the foundation produce shearing stresses which aggregate to give internal forces which must be self-balancing in the absence of external horizontal forces. Hence, for a symmetrically-loaded foundation on a uniformly strained soil, with the critical relative displacement exceeded everywhere, symmetry and equilibrium require that the maximum force produced on the centre-line of the foundation slab is given by one-third of the total weight applied at foundation level (for a coefficient of friction of 2/3). This is an upper limit because, for the conditions described, it would not be physically possible simultaneously to exceed the critical displacement at every point on the foundation. It can easily be demonstrated, for example, that there will be a position for which the relative displacement must be zero. For design purposes, however, such an over-estimate may be fully acceptable and prudent.

It may also be shown (Geddes, 1977a, 1980) that the hysteresis effect described earlier can produce an apparent anomaly in the nature of the forces developed during mining. Where a tensile ground strain is followed by a compressive ground strain, a structure supported on a frictional soil undergoes tensile straining followed by compressive straining, followed by a further tensile straining. Thus, increasing structural tensile forces and strains can be developed in a condition where the ground is undergoing compressive straining. This feature and the phenomena described in the earlier paragraphs of this section clearly demonstrate that imposed ground strains are not necessarily identical in magnitude or sign with the structural strains they cause. For an undamaged structure it is only when the structure displays extensional or compressional flexibility, and when the ground straining does not reduce after reaching a maximum, that ground strains and structural strains move towards equality.

Design for horizontal friction forces thus involves a knowledge of the vertical loading imposed by the structure and the coefficient of friction and adhesion forces acting at the foundation/soil interface. It is customary to assume that maximum forces will be produced, either tensile or compressive or both, and to proportion and strengthen the structure accordingly. Account should also be taken of the fact that friction and adhesion forces will not necessarily be aligned along the major or minor axes of a structure, and design should thus allow for various orientations of the horizontal ground movements with respect to these axes. These forces, applied as a shear at the soffit of a foundation, act eccentrically with respect to the foundation slab and superstructure, thus producing bending moments additional to those created by vertical ground displacements. Design must cater for the variety of combinations of bending effects and horizontal force effects.

Insofar as existing structures are concerned, where it is not possible to isolate them from anticipated horizontal ground movements, an estimate of likely effects can only be obtained by performing a calculation of the kind outlined earlier taking into account the cycle of movements expected (Geddes, 1977a, 1980). Table 8.6, giving the results of field experiences by the National Coal Board, may be of predictive value subject to the reservations given in the text.

Horizontal forces can be minimized by the adoption of a lightweight superstructure and by the incorporation of a sliding layer to eliminate adhesion, and with the characteristics already described. As with vertical movements, subdivision of the structure into a number of sub-units with physical separation between them will reduce the horizontal forces generated. Projections of the ground slab below the soffit should be avoided.

8.6.2 Horizontal earth pressures

Measures should also be taken to reduce or eliminate the compressive forces which may otherwise be developed on vertical or near-vertical substructure faces. Compressible media, such as expanded polystyrene blocks inserted between the soil and substructure, have been used for this purpose. Temporary or permanent peripheral trenches have also proved effective in reducing the development of passive type earth pressures. Otherwise, allowance should be made in the limit for the creation of passive thrusts on substructures. These are clearly many times the customary active pressures. Thus, for a frictional medium with an angle of shearing resistance of 30°, the ratio of passive thrust to active thrust is 9 : 1 and for a value of 40° the ratio is about 21 : 1. Attempts to cater for such pressures will typically prove to be prohibitive in design terms and/or cost and it is preferable to reject the incorporation of substructures such as basements.

Where it is possible, without risk to the safety of a retaining structure, to allow 'forward' sliding (but not overturning) under increased horizontal earth pressures in order to reduce the compressive forces which would otherwise be applied to a structure, this may be adopted. The design of the structure is then determined by the limiting value of available sliding resistance. Due attention must be paid to the provision of suitable gaps to allow movements of this kind to take place.

8.7   Aspects of practice

Application of the principles and techniques referred to in the text may be illustrated with reference to some of the procedures used in relation to building structures and bridges. Dynamic movements are excluded from consideration.

8.7.1   *Building structures*

For proposed buildings, as with construction on any site, comprehensive data must be collected concerning the subsoil, the overall geological setting, the groundwater conditions, filled areas and other factors likely to be of relevance in the design of the structure. Careful consideration should be given to those aspects which are likely to produce ground movements on the site, making use of the techniques described in other chapters. This stage gives rise to varying degrees of accuracy and certainty in the estimation of possible ground movements. The greater the reliability of the predictions, the more easily can an appropriate economic solution be found. The less predictable the movements are, the more necessary it becomes to a design on a worst-condition basis or to adopt a mixed or non-structural solution. A decision must then be made about the general approach to be adopted. The alternatives are described in section 8.2, in the order in which typically they would be considered. If a structural solution is sought, it will usually be based on a fully-rigid or a highly-flexible design. In general, statically indeterminate structures, such as fixed arches and rigid frames which are sensitive to relative movements of supports, either horizontally or vertically, should not be selected. Freely-supported structures, with adequate bracing and ties at roof, floor and foundations are recommended.

Large buildings should be of restricted height and divided into small units not larger than about 20 m by 20 m in plan. The units should be independent, with construction gaps proportioned to cater for the expected movements without closure. The separation of the units should be carried through to include the foundations, and measures should be incorporated to maintain such independence. Adequate stiffening should be provided at roof, floor and foundation levels to maintain structural integrity. For light structures, the foundations may consist of concrete slabs, 15 cm to 40 cm in thickness, placed at shallow depth and laid on a granular material. Projections below the slab should be avoided. When the loading is more substantial, or multi-storey buildings are involved, cellular ribbed rafts or grillage type substructures may be used, again placed on granular material to allow slip to take place.

Small, lightly-loaded buildings are best supported on reinforced concrete slabs, although reinforced strip foundations may suffice in some instances. The building of rows or terraces of houses should be avoided, houses being provided only on the basis of detached or semi-detached units. Connections to outhouses should be avoided. Access roads and paved areas should not extend to the faces of buildings, or, if this is unavoidable, a flexible material should be used for these purposes.

Measures should be incorporated to cater for the development of tensile and compressive forces in foundation slabs, and to reduce the build-up of forces on buried components. Careful attention should be paid to the passage of services through the fabric of the building, drains and other rigid fittings being taken out through external walls and not through foundation slabs or rafts.

Door and window openings should be kept to a minimum to avoid undue weakening of walls, and too many openings should not appear in line, otherwise a weak plane will develop through a building. Brittle materials such as plaster should be avoided whenever possible.

When the direction of maximum curvature can be pre-determined, buildings should, if possible, be orientated so that their shorter axes coincide with that direction.

Structures covering small areas can sometimes be supported on three-point bearing systems, designed to allow relative settlements and rotations without damage.

Buildings may also be provided with jacking points so that re-levelling can be carried out in order to restore original alignments after the cessation of movements. It is always much cheaper to design and construct for such a need than to have to provide this kind of facility after movements have taken place.

When existing buildings are considered, the structural options open to the engineer are much more limited in number. Attention has frequently to be centred on non-structural solutions, structural measures being restricted to local temporary or permanent reinforcement or strutting.

8.7.2 *Bridges*

The supports of a highway bridge may have to cater for the effects of differential vertical displacements, differential tilts both longitudinally and transversely, differential horizontal movements longitudinally and transversely and differential plan rotation, depending on the agency producing the movements. Bridges which will be subjected to such combinations of movements are most easily and economically designed on the basis of simply-supported spans provided with bearings capable of sustaining the movements. For small amounts of transverse tilt, beam-slab construction without diaphragms, including prestressed concrete box beams, is likely to be satisfactory. When larger transverse tilts are involved, beams of lower torsional stiffness, such as steel beams or pre-stressed concrete I-beams, are usually more suitable. With extreme values of differential transverse tilt, a three-point support may be needed.

Statically-indeterminate structures should be avoided. Cantilever and suspended-span designs are suitable for longer span bridges but appropriate allowance must be made for horizontal movements to take place without danger and for transverse tilts to be accommodated. Considerable attention should be paid to the provision of bearings and expansion joints to ensure continued safe functioning under the various movements imposed.

The possible need for jacking pockets and local reinforcement, to allow for future re-levelling, should be borne in mind.

The procedures for the design of foundations and substructures follow those described earlier, with due attention being paid to the requirement to cater for lateral pressures and the development of friction forces due to horizontal ground strains.

More detailed treatments of some of the topics briefly referred to here are provided by Sims and Bridle (1966), National Coal Board (1975), Geddes and Cooper (1962), Department of Transport (1982), the Institution of Civil Engineers (1959 and 1977), the Institution of Structural Engineers (1949 and 1977) and the Ministry of Works (1951). Extensive bibliographies appear in these publications and others listed in the References and in the *Proceedings* of the 1st and 2nd Conferences on Ground Movements and Structures held in Cardiff in 1977 and 1980.

Notation

k	Modulus of subgrade reaction
L	Length
q	Foundation contact pressure
u	Horizontal displacement
v	Foundation penetration into subgrade
Δ	Relative deflection
α	Angular strain
α_{max}	Maximum angular strain
β	Angular distortion (relative rotation)
β_{max}	Maximum angular distortion
ε	Horizontal strain
ε_{lim}	Limiting tensile strain
ρ	Settlement
ρ_h	Heave (negative settlement)
ρ_{max}	Maximum settlement
ρ_{min}	Minimum settlement
$(\delta\rho)$	Differential settlement
$(\delta\rho)_{max}$	Greatest differential settlement
θ	Rotation (change in gradient of a line joining two reference points)
ω	Tilt

References

Alexander, S.J. and Lawson, R.M. (1981) *Design for Movement in Buildings*. CIRIA Technical Note 107, Construction Industry Research and Information Association, London.

Bell, S.E. (1977) Successful design for mining subsidence. *Proc. Conf. on Large Ground Movements and Structures, UWIST, Cardiff*, Pentech Press, pp. 562–578.

Bjerrum, L. (1963) Contribution to discussion. Session VI, *Proc. European Conf. on Soil Mechanics and Foundation Engineering, Wiesbaden*, II, pp. 135–137.

Burland, J.B. and Wroth, C.P. (1975a) *Settlement of Buildings and Associated Damage*. Building Research Establishment Current Paper CP 33/75.

Burland, J.B. and Wroth, C.P. (1975b) Settlement of buildings and associated damage. *Proc. Conf. on Settlement of Structures, Cambridge*, pp. 611–654.

Burland, J.B. Broms, B.B. and de Mello, V.F.B. (1977) Behaviour of foundations and structures. *Proc. 9th Int. Conf. on Soil Mechanics and Foundation Engineering, Tokyo*, pp. 495–546.

Department of the Environment (1981) *Assessment of damage in low-rise buildings*. Building Research Establishment Digest **251**, H.M.S.O.

Department of Transport (1982) *Design of highway structures in areas of mining subsidence*. Departmental Standard *BD* 10/82, Roads and Local Transport Directorate, London.

Geddes, J.D. and Cooper, D.W. (1962) Structures in areas of mining subsidence. *The Structural Engineer* **40**, 79–93.

Geddes, J.D. (1977a) The effect of horizontal ground movements on structures. *Proc. Conf. on Large Ground Movements and Structures, UWIST, Cardiff*, Pentech Press, pp. 623–646.

Geddes, J.D. (1977b) The behaviour of a CLASP-system school subjected to mining movements. *Proc. Conf. on Large Ground Movements and Structures, UWIST, Cardiff*, Pentech Press, pp. 579 596.

Geddes, J.D. (1980) Subgrade restraint and shearing force effects due to moving ground. *Proc. 2nd Conf. on Ground Movements and Structures, UWIST, Cardiff*, Pentech Press, pp. 288–306.

Gibson, D. (1957) Buildings without foundations. *J. Royal Inst. British Architects* **65** (2) 47–53.

Grant, R. Christian, J.T. and Vanmarcke, E.H. (1974) Differential settlement of buildings. *J. Geotechnical Engineering Division, ASCE*, **100** (GT9) 973–991.

Institution of Civil Engineers (1959) Report on mining subsidence. *The Institution*, London.

Institution of Civil Engineers (1977) Ground subsidence. *The Institution*, London.

Institution of Structural Engineers (1949) Report on mining subsidence and its effect on structures. *The Institution*, London.

Institution of Structural Engineers (1977) Structure-soil interaction: A state of the art report. *The Institution*, London.

Lacey, W.D. and Swain, H.T. (1957) Design for mining subsidence. *Architects' Journal* **126**, 557–570 and 631–636.

Mautner, K.W. (1948) The design of structures in areas subjected to mining subsidence. *The Structural Engineer* **26**, 35–69.

Meyerhof, G.G. (1953) Some recent foundation research and its application to design. *The Structural Engineer* **31**, 151–167.

Ministry of Works (1951) *Mining subsidence—effects on small houses*. National Building Studies Special Report No 12.

National Coal Board (1975) *Subsidence Engineers' Handbook*, (2nd edition).

Polshin, D.E. Tokar, R.A. (1957) Maximum allowable non-uniform settlement of structures. *Proc. 4th Int. Conf. on Soil Mechanics and Foundation Engineering, London*, **1**, 402–405.

Sims, F.A. and Bridle, R.J. (1966) Bridge design in areas of mining subsidence. *J. Institution of Highway Engineers* **13** (6) 19–32.

Skempton, A.W. and MacDonald, D.H. (1956) Allowable settlement of buildings. *Proc. Institution of Civil Engineers* **5** (III) 727–768.

Terzaghi, K. and Peck, R.B. (1948) *Soil Mechanics in Engineering Practice* (1st edn.). Wiley, New York.

Wasilkowski, F. (1955) Complete protection of structures against damage due to mining subsidence. (Translation No. 55 by Cement and Concrete Association, London).

Wilun, Z. and Starzewski, K. (1975) *Soil Mechanics in Foundation Engineering*, Vol. 2, (2nd edn.) Surrey University Press, London.

9 Mineralogical controls on volume change

R.K. TAYLOR and J.C. CRIPPS

9.1 Introduction

The minerals which can promote ground movements by virtue of volumetric changes arising from physicochemical reactions, including growth or solution, are most commonly found in finer-grained detrital and chemically precipitated sedimentary rocks. These types constitute more than 60 per cent of all sedimentary rocks (Potter *et al.*, 1980). In addition to clay mineral hydration, it is the instability of chemically or biochemically precipitated (soluble) minerals in present weathering environments which promotes deformation problems in the near-surface zone.

Detrital clay minerals are either stable or metastable in modern weathering environments, because they have achieved at least some degree of stability towards weathering on the parental land-mass from which they have been eroded. Ground movements peculiar to clays and mudrocks (which form *c.* 80 per cent of fine-grained sedimentary rocks) are a function of the small particle size of the contained clay minerals and concomitant hydration processes. The importance of *inter-* and *intra*particle shrinkage and swelling of clay minerals in relation to the engineering problems generated cannot be over-emphasized. Although the balance of this chapter tends to be directed towards soluble minerals, this is because clays and clay mineral behaviour forms a common thread running through many chapters of the book. Moreover, a review of soluble minerals is particularly relevant to the upsurge in construction activities in the Middle East during the last 10 years. On a world-wide scale however, the problems posed by clays and mudrocks far exceed those of soluble rocks and minerals.

Soluble precipitate minerals include sulphates such as gypsum/anhydrite, which tend to be stratigraphically restricted, and carbonates which are formed widely in many depositional and diagenetic environments. In the geological succession the afore-mentioned minerals are associated in evaporite deposits together with salts such as halite (rock salt). Although in temperate climates dissolution of these minerals is the norm, higher evaporation rates elsewhere can lead to elevated ionic strengths in the pore-waters and precipitation is possible. Coastal and inland sabkhas of the Middle East, the Arabian Gulf and parts of North Africa are cited in the text as modern depositional environments where 'evaporite minerals' have caused construction difficulties.

Pyrite is the most widespread of the sulphide minerals, and is found in hydrothermal ore veins, contact metamorphic rocks, in igneous rocks as an accessory, as well as in dark-coloured (organic) mudrocks. It oxidizes rapidly, the exothermic reaction being comparable with hydrocarbon combustion (Curtis, 1976). Spontaneous ignition of metal mine tailings in warm climates, acid mine drainage and the exacerbation of heating in the old generation of (loose) colliery tips in the U.K., are evidence of its instability. The biogenic oxidation process of pyrite is complex, and subsequent ionic reactions producing secondary sulphates are denoted by numerous instances of heave in black shales and compacted mudrocks, both in Europe and North America.

Crystal growth, hydration and solution have already been alluded to in the context of minerals identified as promoting some degree of deformation on their own account. These processes are considered quantitatively in the chapter in terms of the likely uplift pressures which volumetric increase may contribute. Solution rates are considered and mention is also made of ground movements induced by chemical (mineral) reactions in some man-made slags used as foundation fills.

Although frost heave (Jones, 1980) is conveniently explained largely in terms of capillary theory, both this phenomenon, and the larger-scale permafrost development of northern climes, are not considered in this chapter. They are more an attribute of contained (freezing) pore-water and groundwater, rather than mineral constituents *per se*.

9.2 Clay mineral expansion

Volume changes in response to water-content variations are characteristic of clays and many mudrocks, with *clay minerals* playing the significant part in this behaviour. Because of their large specific surface areas, these minerals are always hydrated in nature with *adsorbed water*. The properties of a clay are dependent on the thickness of this hydration shell which will vary according to the clay mineral group and species. It is pertinent first to consider briefly the main clay types, using the stratigraphical variations in the United Kingdom and the U.S.A. (Fig. 9.1) as convenient vehicles for this purpose.

9.2.1 Clay minerals

The basic building blocks of clay minerals are silicon-oxygen tetrahedra, aluminium-oxygen octahedra and magnesium-oxygen octahedra. Combining these molecules to form sheet structures of silica, gibbsite and brucite, respectively (as in Fig. 9.2) indicates how the structure of clay minerals is developed. Ionic substitutions complicate this simple model, so it is more correct to refer to gibbsite-like and brucite-like sheets. By stacking gibbsite-like and silica sheets together the common (1:1) clay mineral, kaolinite, is produced. The linkages between the 2-layer combinations are provided by hydrogen bonds and secondary valence forces. The simplified structure illustrated in Figure 9.2 implies that the electrical charges are neutral, but this is not the case in nature. Isomorphous substitution of one atom for another (for example, aluminium for silicon) will lead to a net negative unit charge. Isomorphous substitutions, surface crystal defects and broken bonds determine the size of the negative charge which is all-important in clay mineral hydration.

A tubular form of kaolinite (halloysite) is also illustrated since this type imbibes water between the sheets. Also shown diagrammatically in Fig. 9.2 are the 2:1 (3-layer) clay minerals, muscovite/illite (mica), smectite group (for example, montmorillonite) and chlorite. The first two groups are composed of a gibbsite-like sheet sandwiched between two silica sheets. It will be noted that potassium (K) forms an important bond between the sheets in micas. Chlorite is rather different in that the 2:1 layers are regularly intergrown

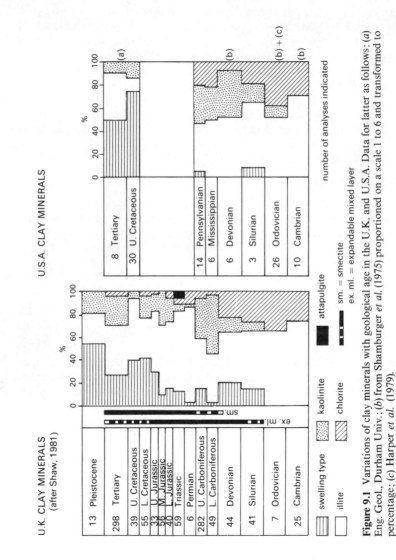

Figure 9.1 Variations of clay minerals with geological age in the U.K. and U.S.A. Data for latter as follows: (*a*) Eng. Geol., Durham Univ.; (*b*) from Shamburger *et al.* (1975) proportioned on a scale 1 to 6 and transformed to percentage; (*c*) Harper *et al.* (1979).

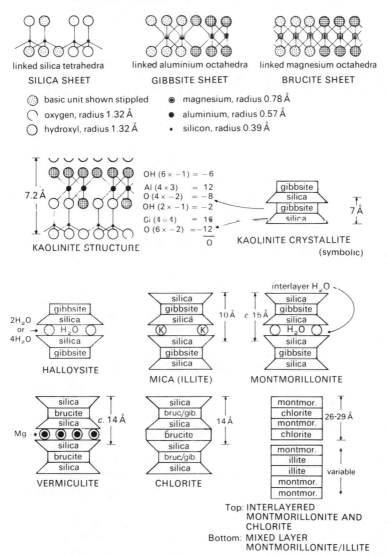

Figure 9.2 Showing basic building blocks, sheet structures and symbolic composition of common clay minerals.

with a brucite-type sheet of magnesium ions which are coordinated octahedrally by hydroxyls.

It is important to distinguish between those clay minerals which swell by *intra*crystalline as opposed to *inter*crystalline means. In the first case water penetrates between the unit layers of the crystal, whereas in the second case water is restricted to the external surfaces of the minerals. Of those minerals depicted in Fig. 9.2, smectite shows the greatest affinity for water. Sodium-saturated montmorillonite, which is not common in nature, will expand some 800 times by volume and dissociate into platelets of the same

K

order as the unit cell (10Å). Vermiculite (Fig. 9.2), which is a weathering product of certain micas and chlorite, and is found in some clay soils, also exhibits intracrystalline swelling. Swelling chlorites have been identified (Stephen and MacEwan, 1950; Honeybourne 1951), notably in the Keuper Marl of the United Kingdom. Common to both successions in Fig. 9.1 are expandable, mixed-layer (interstratified) clay minerals. These are commonly regular or random interstratifications of mica and smectite or chlorite-smectite as illustrated in Fig 9.2. Research data imply that on balance randomly interstratified illite-smectite is probably the more usual type.

The dimensions normal to the crystal sheets (for example, $d = 10$Å for mica in Fig. 9.2) change in swelling clay minerals according to the type of interlayer cation. Table 9.1 shows that the maximum intracrystalline swelling and the temperature required to remove inter-sheet water are a function of the surface charge density and its distribution in the structure of the specific clay mineral type. From smectite through vermiculite to mica the maximum swelling decreases and the dehydration temperature increases with increasing density of surface charge.

Swelling characteristics with dipolar organic liquids and thermal behaviour are properties used in the X-ray diffraction (XRD) identification of clay minerals (for example, Brindley and Brown, 1980).

Only the principal mineral types have been referred to here, since they are represented in the geological successions of Fig. 9.1. It should also be pointed out that *allophane*, which contains about equal parts of silica (SiO_2) and alumina (Al_2O_3) and has a high cation exchange capacity (Table 9.1), is insufficiently crystalline to be identified by XRD analysis.

Burial diagenesis models of Segonzac (1970) and Weaver and Beck (1971) show that illite (mica) and chlorite may be expected to increase at the expense of kaolinite and expandable clays over geological time. This is borne out by the data illustrated in Fig. 9.1 for both the U.K. and the U.S.A. Ground movements exacerbated by *expandable* clay minerals *per se* would not be expected in rocks older than Silurian age. However, the higher incidence of expandable minerals in Tertiary and Upper Cretaceous clays and mudrocks of North America is adjudged to be reflected by more numerous 'problem types' in the U.S.A. (Attewell and Taylor, 1973). At the extreme there are beds and bands of smectite-rich clay with volcanic affinities in both countries—for example, Wyoming bentonites (U.S.A), and Fuller's earth of Upper Jurassic and Lower Cretaceous age (U.K.).

9.2.2 *Clay mineral hydration*

All clays show some swelling on wetting. The degree is dependent on the mineral species, their orientation or fabric, the valency of exchangeable cations, the salt concentration of the pore-water, the degree of consolidation and the cementation and bonding by interstitial carbonates, oxides and postively charged organic matter.

The water molecules on the surface of a clay mineral are strongly held. Although H^+ in the water molecule does not exist in solution it does so in the context of the hydronium ion, $H^+ \cdot H_2O$. Hydrogen bonding to negatively charged oxygen and hydroxyl ions on the mineral surface accounts for the first two or three water monolayers.

Initial swelling of dehydrated clays is characterized by sorption of these first few monolayers and by exchangeable cations adsorbing water molecules. The *exchangeable cations* attracted to internal and external mineral surfaces originate in the pore solutions

Table 9.1 Effect of surface charge density on swelling and dehydration of the exchangeable cations (largely from Davey, 1972)

Interlayer cation	Montmorillonite (surface charge density = 3×10^4 e.s.u/cm²)		Vermiculite (surface charge density = $c.6 \times 10^4$ e.s.u/cm²)		Mica (surface charge density = 10^5 e.s.u/cm²)	
	Swelling, $d(001)$Å	dehydration temperature °C	Swelling $d(001)$Å	dehydration temperature, °C	Swelling, $d(001)$Å	dehydration temperature °C
Li⁺	≫ 40.0	—	≫ 40.0	—	—	—
Na⁺	≫ 40.0	150	14.8	150	10.0	—
Mg²⁺	19.0	—	14.6	—	—	—
Ca²⁺	19.0	250	15.4	250	10.0	—
K⁺	15.0 and ≫ 40.0	—	11.0	—	10.0	350
Cation Exchange Capacity, me/100 g	80–100		150		10–40	

Note: The C.E.C. of kaolinite ranges from 3–15 me/100 g, that of halloysite 5–50, chlorite 20 and allophane up to 70 me/100 g.

Table 9.2 Expansion and shrinkage characteristics of some clays/mudrocks (see Figure 9.4)

Lithology	Plasticity index %	Clay size content %	BRE (1980) shrinkage	Expansive clay class Williams & Donaldson (1980)	CEC me/100 g
Yazoo clay	64	85	High/v. high	V. high	99
Kincaid shale	41–53	68	High	V. high	72
Nacimiento shale	13	21	Low	Med.	26
" "	21	54	Med.	Low	16
" "	34	34	Med/high	V. high	47
Dawson shale	51	63	High/v. high	V. high	65
Fox Hills shale	21	61	Med.	Low	20
London Clay (brown)	53	65	High	V. high	32
London Clay (blue)	40	53	Med/high	V. high	26

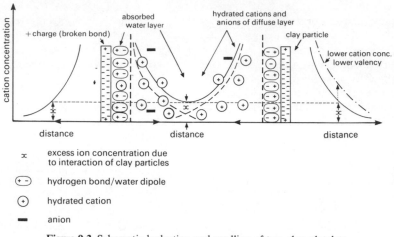

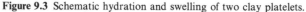

Figure 9.3 Schematic hydration and swelling of two clay platelets.

(electrolytes). The fact that they are exchangeable can in turn influence the degree of intraparticle (see Table 9.1) and interparticle (Fig. 9.3) swelling. The main cations in order of dominance (in nature) are calcium (Ca^{2+}), magnesium (Mg^{2+}), potassium (K^+) and sodium (Na^+). Cations of similar valency may replace one another, and although some cations may become fixed, or replaceable only with difficulty in some clay minerals (e.g. K^+ in degraded illites), they can be arranged in terms of replacing ability:

$$Li^+ < Na^+ < H^+ < K^+ < NH_4^+ \ll Mg^{2+} < Ca^{2+} \ll Al^{3+}$$

For example K^+ may replace Na^+ although the latter is more easily replaced by Mg^{2+} or Ca^{2+}. The cation exchange capacity (CEC) is readily determined (*see* Chapman (1965) for the ammonium acetate method), and the value is equivalent to the amount of negative charge per unit weight. Typical ranges are given in Table 9.1.

A balance between coulombic electrical forces and thermal diffusion largely excludes exchangeable cations from the inner adsorbed water layer. In this way the *diffuse double layer* of Fig. 9.3 becomes relevant to *inter*particle swelling. Importantly, it is *monovalent* cations of *low* concentration which generate the most *extended* diffuse layers and enhanced swelling. An *increase* in valency and concentration *reduces* the diffuse layer. If, therefore, the electrolytes in pore solutions are concentrated, swelling will be reduced (see section 9.2.4).

Cation exchange has been adopted in a number of situations in an attempt to reduce swelling. The addition of lime (Ca) to reactive Na-clays is a case in point.

Exceedingly high swelling pressures are commonly associated with an elevated exchangeable Na^+ value (for example Taylor (1978), Fig. 6.7). This is best expressed as *exchangeable sodium percentage*, where ESP = [(Exchangeable Na)/CEC] 100%. ESP values greater than about 15 serve to identify these highly susceptible clays.

9.2.3 *Diffuse double layer swelling*

Yong and Warkentin (1966) conclude that at spacings of less than 15Å there is a net attraction between clay mineral particles due to the dominance of van der Waals' force which acts between small particles. The cations in this case do not form separate double layers but are uniformly distributed between any two clay particles. For spacings greater than 15Å or so, (beyond any van der Waals influence) diffuse ion-layers form and particles will exhibit a net repulsion. Diagrammatically in Fig. 9.3 it is illustrated that the cation concentration is higher between the particles, so water is attracted by their osmotic activity. Consequently, swelling is induced by the concentration gradient and not by mineral surface properties.

Swelling pressures can be calculated from double layer theory by adopting a number of simplifying assumptions. An example which could be applicable to a micaceous clay is illustrated as an Appendix to the chapter. According to Verwey and Overbeek (1948) reasonable agreement with measured laboratory values has been obtained for clays. More surprisingly, the same has been demonstrated by Madsen (1979) for rocks of Miocene and Jurassic age.

The dependence of swelling pressure on volume change necessitates the use of special laboratory procedures and oedometers for accurate measurement (see Obermeier, 1974). Not all of the collected data shown in Fig. 9.4 conform with the ground rules. Nevertheless, the results provide a view of measured swell pressures (normal to bedding) for a wide variety of clays and mudrocks of different age. For some samples information is available to predict volume change susceptibility using Atterberg limits and clay size data (i.e. following Driscoll chapter 10, this volume). The younger samples, in particular the Yazoo Clay and Kincaid shale (of high CEC), are known to swell but the pressures (Table 9.2) are (relatively) much lower than in other clays and mudrocks. One probable reason is the underestimation of swell pressure by conventional oedometer measurement. Probably of more significance, however, is the increase in swelling pressure with present depth of burial, which Fleming et al. (1970) observed for the Palaeocene and Upper Cretaceous mudrocks from the U.S.A. shown in Fig. 9.4. Caution should be exercised in comparing near-surface overconsolidated clay and mudrock samples

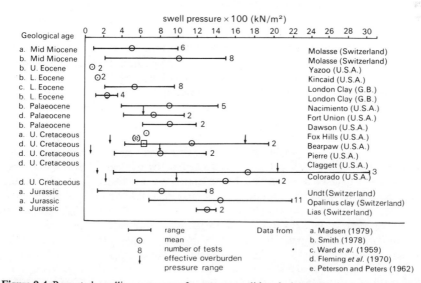

Figure 9.4 Reported swelling pressures for overconsolidated clays and mudrocks of varying geological age.

from the weathered zone with more deeply buried materials. Reduction in confining pressure (and of any stress difference, K_0) are complicating factors. Specimens should first be loaded to overburden pressure before measurement of swelling pressure. Although repeated wetting and drying (simulating weathering processes) does not affect the almost reversible volume change of high-swelling montmorillonitic soils, the same would not describe those with non-expandable minerals. Warkentin and Bozozuk (1961) observed that volume changes in low-swelling clays decreased to an equilibrium value.

9.2.4 *Moisture retention (soil suction)*

In the near-surface, *partly saturated* zone, percolating meteoric water is retained within the voids of rocks and soils. The energy with which this water is held has been described by a number of themodynamic models (see Gillott, 1968).

For practical purposes the total potential of soil moisture, ψ, may be written (after Richards, 1967) as:

$$\psi = z - h_t = z - h_m - h_s$$

If a soil is salt-free, then

$$\psi = z - h_m = z + u$$

where: z = gravitational component (cm water)

   h_t = total suction

   h_m = matrix suction (cm water = $-u$ cm water)

   h_s = solute suction

   u = pore-water pressure

Total suction therefore comprises two components:
(a) *matrix suction*, made up of mineral surface attractive forces for water and cations, together with surface tension (capillary) effects, and (b) *solute* or osmotic suction which arises from the soluble salts in the pore-water. The latter component is identical in concept to the osmotic approach considered in the previous section. Indeed, Richards (*op. cit*) in summarizing Australian experience makes the point that considerable heave can be promoted by reducing solute suctions (i.e. increasing the diffuse double-layer swelling) by repeated watering of gardens with solute-free water. This may involve only relatively small changes in matrix suction. The relationship between matrix and solute suction with swelling pressure of some reactive Australian clays has recently been investigated by Pile (1980).

Suction/water content relationships for U.K. clays are considered in chapter 10. It should be mentioned that on the pF suction scale (where pF $= 0 (= 1$cm water) at saturation, and pF7 is equivalent to oven drying) the division of soil water into 'adsorbed', 'capillary' or 'gravitational' becomes purely arbitrary (see Kohnuke, 1946).

9.3 Soluble minerals

Ground heave and settlement, respectively, are liable to occur due to the precipitation or removal of rock or soil components by groundwater. Heave is also possible if the addition of water during the hydration of certain minerals (for example, anhydrite) causes expansion. Conversely, a reduction in volume can lead to settlement during dehydration, although this process appears to be less important in practice. It is useful to consider the circumstances in which soluble evaporite minerals occur before relating this directly to the controls on ground movement. The active deposition of soluble minerals occurs in present-day arid environments and is denoted by ancient evaporites, such as, for example, those of the Permo-Trias of the United Kingdom, which were formed under similar conditions.

Certain weathering processes, particularly the oxidation of the non-detrital minerals of the pyrite group, will customarily give rise to a wide variety of sulphate minerals.

9.3.1 *Arid environments*

For present purposes it is convenient to describe arid environments as those in which net evapotranspiration exceeds total precipitation. Generally these areas constitute either arid or extremely arid deserts (McGinnies *et al.*, 1968). Under such circumstances evaporation of groundwater causes increasing salinity and dissolved salts are precipitated. A continuing supply of groundwater transports soluble components to the near-surface zone where evaporation causes precipitation. Construction difficulties under these conditions are reviewed by Evans (1977) with reference to the Middle East.

Ground movements are liable to occur due to crystal growth pressures exerted during precipitation, the removal in solution of evaporitic minerals and hydration changes in certain minerals. Active sub-surface diagenetic processes including dolomitization are reported by Bathurst (1975) to be the cause of heaves of up to 0.3 m to 1 m. Other problems which frequently result in expansion occur as a result of chemical reactions between groundwater containing high percentages of dissolved salts and construction materials. Attention has been given to these aspects by many authors including French and Poole (1976), Fookes and Collis(1976 *a,b*), Fookes *et al.* (1981), Fookes and Higginbottom (1980*a,b*), and Netterburg (1979*a*). Data summarizing acceptable levels of total salt contamination in concrete have been collected together by Frearson (1983) who suggests equivalent Cl^- limits of 0.02 per cent in fine aggregate to 0.06 per cent for coarse aggregate, or 0.06–0.9 per cent of the amount of cement used. In the case of SO_3^{2-}, 0.4–1.0 per cent is suggested for aggregates where the latter limit refers to manufactured aggregates. The SO_3^{2-} limit expressed in terms of the cement content is 4–5 per cent. Care should be taken in applying these figures since much depends on the form of the sulphate present (see *BRS Digest* 90, 1968; *BRE Digest* 174, 1975; *BRE Digest* 250, 1981).

Particular problems caused by the removal in solution of minerals are probable in conditions of decreasing groundwater salinity. A common situation in which this might occur is in the excavation for harbour and bridge works where tidal drawdown leads to exchange of concentrated groundwater with comparatively less saline sea-water. Rice (1983) reports a settlement of 90 mm in 2 years for a dry dock in Dubai. Pertinent to the Dubai setting are veinlets of gypsum shown in Fig. 9.5 which occur in unconsolidated deposits at Mina Jebel Ali Harbour. Partial removal of gypsum during excavation dewatering has formed 'gap-toothed' veins in a period of less than two months (Bush, 1983). Clearly, the amount of potential settlement depends on the amount of mineral available for solution. The rate of removal may be assessed in particular situations, as in section 9.3.4. Preventive measures usually involve the construction of imperme-able barriers, or grouting, to inhibit groundwater movements.

In some arid regions interstitial precipitation of soluble minerals produces metastable soils (Kleine, 1962) which are liable to large and rapid settlements if wetted (similar to fills,

Figure 9.5 Gypsum veinlets in unconsolidated deposits at Mina Jebel Ali Harbour, Dubai (by courtesy of P.R. Bush).

see chapter 2). Elsewhere evaporitic salts may serve to cement surface layers and produce a variety of duricrusts (Goudie, 1972, 1973). Common types are calcrete (or caliche), gypcrete, gypcrust, silcrete, and so on, where the prefix refers to the type of soluble mineral. Fookes (1978) distinguishes -cretes as surface hardened bedrock and -crusts as softer accumulations on unconsolidated deposits.

In the absence of surface infiltration, evaporation losses from above the water table are replenished by capillary moisture. The height to which water may rise in a soil due to capillary forces depends primarily on pore size and character. The greatest capillary rise occurs in soils with narrow, straight, well-connected pores. Lane and Washburn (1946) have conducted laboratory experiments to establish the magnitude of capillary rise in materials of different gradings. These data are compared with field experience of Fookes *et al.* (1981) in Fig. 9.6. Thus it can be seen that in fine-grained materials capillary rises of up to about 3m can occur.

The precipitation of soluble minerals within the capillary fringe is likely to cause heave due to crystal growth pressures. Many cases of affected roads and other surfaced areas, particularly lightly trafficked parts, are well documented. General engineering advice is provided by a number of authors including Tomlinson (1978), Tobin (1980) and Fookes (1978). The effects of related aspects are described by Blight (1976) and Netterberg (1979*b*). In many of these cases the origin of soluble minerals is the construction material (including calcrete), whilst in others salts have been carried upward by capillary groundwater and precipitated. Fookes and French (1977) note that the most serious effects occur in the lower half of the capillary fringe.

The enhancement of solute concentration may be prevented by inhibiting surface evaporation by the application of impermeable coatings. Netterberg and Loudon

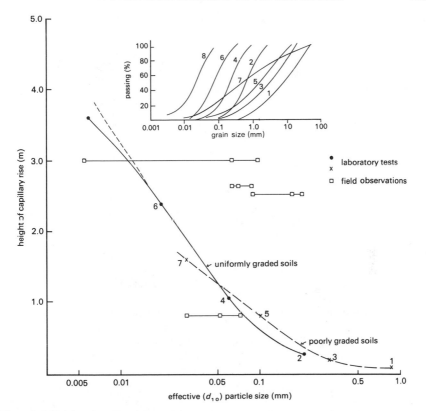

Figure 9.6 Height of capillary rise in laboratory tests and field observations. (After Lane and Washburn, 1946; Fookes *et al.*, 1981).

(1980) describe a series of laboratory tests in which saline capillary water was evaporated from sealed and unsealed simulated road profiles. Blight (1976) determines the susceptibility of surfacings to excessive evaporation in terms of the mean flow velocity transmitted through the surfacing:

$$v = \frac{k}{D} \times h$$

where v is the mean flow velocity (m/s), k is the coefficient of permeability (m/s), D, is the thickness of the layer (m) and h is the head of water (m). According to South African experience, damage does not occur if $k/D < 30$ per μsec. Various types of road surfacing preparations are considered by Blight (*op. cit*) who concludes that the best protection is afforded by at least 30 mm of hot rolled asphaltic concrete. In fact this may not be entirely satisfactory, since Cole and Lewis (1960) draw attention to the process of slow evaporation through even a 'perfect' bitumen seal.

In practice the actual mineral type precipitated depends on the composition of the groundwater, the temperature and the relative humidity (Kinsman, 1976; Hardie, 1967). In arid areas, low-lying coastal areas, known as sabkhas, are important sites of evaporitic mineral formation (Shearman, 1963; Bush, 1973). In these, evaporative

Table 9.3   Principal salts produced by the evaporation of sea-water (after Riley and Chester, 1971)

Principal salt precipitated	Percentage of total of dissolved salts, %	Weight of liquid remaining %	Specific gravity of brine*
Calcium carbonate and dolomite	1	50	1.14
Gypsum	3	10	1.214
Halite	70	3.9	1.236
Sulphates and chlorides of sodium, magnesium and potassium, e.g. Carnallite	26	—	—

(* S.G. of sea-water before evaporation = 1.026)

groundwater losses are replenished mainly by infiltration of sea-water blown inland during storms, or recharged directly from the sea. Thus the minerals precipitated occur within the sequence shown in Table 9.3. *Inland sabkhas* form in areas in which saline continental groundwaters are evaporated (Glennie, 1970). Appropriate conditions occur in the landward parts of some coastal sabkhas and also in inland drainage basins in arid regions. Fookes and Collis (1975a) distinguish two types of inland drainage basin in which soluble minerals form: *salt playas* which are salt lakes subject to periodic desiccation and *salinas* which are depressed land areas where capillary groundwater rises to the surface. Unfortunately, some uncertainty in the literature arises with the universal use of these terms. The mineralogy of inland sabkhas is variable since it is dictated by the catchment and source of groundwater (see Greensmith, 1971).

The nomenclature and occurrence in north Africa, the Middle East, Asia and Australia of saline soils and sabkhas are discussed by Ellis and Russel (1974) (see also Ellis, 1973; Cooke, 1981; Doornkamp *et al.*, 1980). Important examples also exist in north America (Kinsman, 1969). Sabkhas are effectively flat areas of low relief so the extreme vertical exaggeration should be borne in mind when considering the disposition of carbonate minerals, gypsum and anhydrite shown in Fig. 9.7. Halite occurs as ephemeral crusts up to 20 mm thick, together with other minerals including polyhalite, magnesite and sylvite. The groundwater table lies at depths of between 1 m and 1.5 m. It appears that the surface is in deflation equilibrium at the top of the capillary fringe so that increases in elevation are subject to wind erosion and depressions become filled with aeolian sediment or evaporitic minerals. In the case of the Trucial coast, the chlorinity of groundwater generally increases in a landward direction from about 51‰ to 146‰ in the intertidal zone. A further increase to average values of about 160‰ occurs across the area subject to periodic flooding until it is diluted to about 107‰ by terrestrial waters (Butler, 1969). The stable forms of calcium sulphate with respect to groundwater salinity are discussed briefly in section 9.3.3.2. Conditions conducive to the conversion of gypsum to anhydrite and vice versa are found in the sabkha environment (Hardie, 1967). The transition involves a solid-state change of up to about 63 per cent of the volume of anhydrite which may be manifested

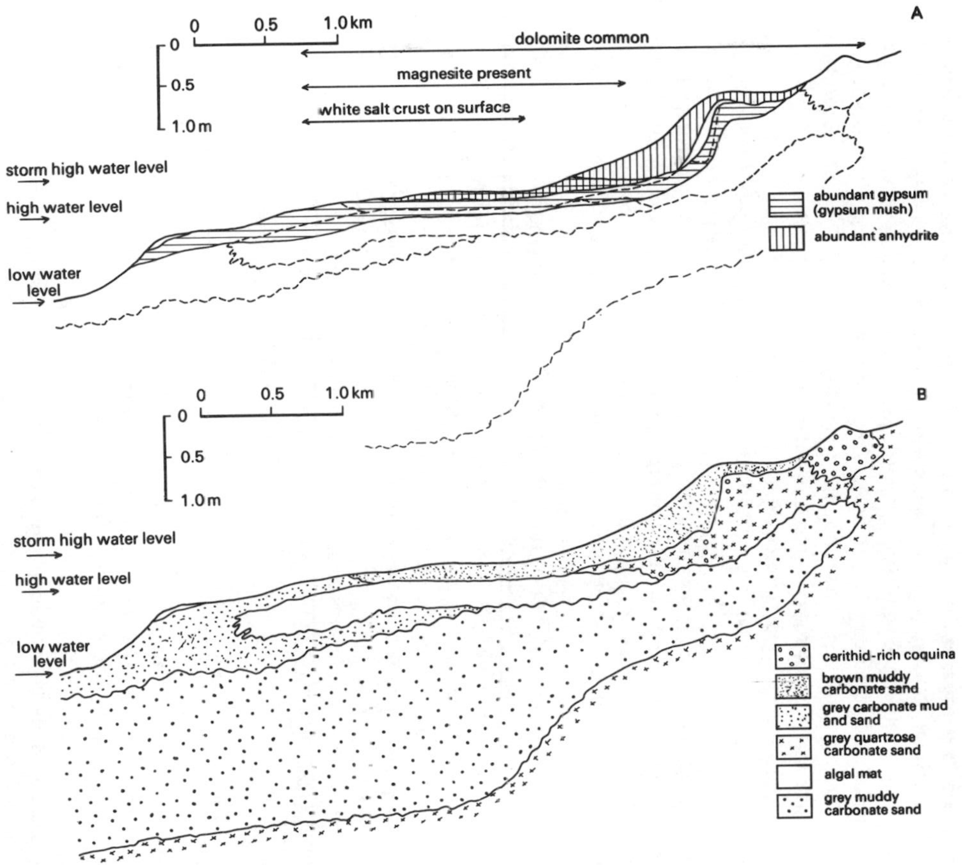
A
0
0.5
1.0km
dolomite common
0
0.5
1.0m
magnesite present
white salt crust on surface
storm high water level
high water level
low water level
abundant gypsum (gypsum mush)
abundant anhydrite
B
0
0.5
1.0km
0
0.5
1.0m
storm high water level
high water level
low water level
cerithid-rich coquina
brown muddy carbonate sand
grey carbonate mud and sand
grey quartzose carbonate sand
algal mat
grey muddy carbonate sand

Figure 9.7 Representative section across the coastal sabka, Abu Dhabi. *A*, showing evaporite distribution; *B*, lithological details. (After Evans *et al.*, 1969.)

as pressure ridges, anhydrite polygons and other forms of distinctive surface morphology.

Groundwater levels and composition exert important controls on ground movements in arid regions. Proposals which change either of these should be considered carefully in view of the possibility of precipitation, solution and changes of hydration state. Particularly important are increases in groundwater level or lowering of the land surface which may result in the evaporation of capillary moisture.

Methods of estimating the pressures exerted by growing crystals and during crystal hydration are dealt with in section 9.3.3, whilst solution rates are considered in section 9.3.4.

9.3.2. Oxidation of pyrite

Although the oxidation of pyrite itself involves expansion (Winmill, 1915) ground movements (in particular expansion) usually result from subsequent reactions between the oxidation products and construction materials, or other rock, soil or groundwater components. In fact, heaves as great as 0.3 m have been reported in dark-coloured mudrocks. Marine black shales are the most common source of sedimentary pyrite. Problems attributed to the oxidation of pyrite have been encountered in many parts of the world including the U.S.A. (Dougherty and Barsotti, 1972); Canada (Quigley and Vogan, 1970; Penner et al., 1973); Australia (Haldane et al., 1970; United Kingdom (Nixon, 1978); France (Millot, 1970) and Norway (Moum and Rosenqvist, 1959).

With the exception of the non-detrital minerals (including some carbonates) discussed in the previous section, most rock- and soil-forming minerals are relatively stable under normal surface conditions. However, as mentioned in section 9.1, certain minerals, in particular sulphides of iron, including pyrite and marcasite (FeS_2) and pyrrhotite (FeS), are especially reactive.

Factors contributing to pyrite oxidation problems may be considered with reference to Canadian experience with Ordovician black shales. Experience with Jurassic and Carboniferous shale fill in the United Kingdom is also pertinent. Quigley and Vogan (1970) describe heave of a lightly-loaded floor slab constructed on black shale in Ottawa, Canada, where bedrock formed a plateau surrounded by a 2.1 m-deep service duct. In another district of Ottawa, floor heave of up to 113 mm caused difficulties with electricity generators (Penner et al. 1973; see also Quigley et al., 1973). Both in these cases and in similar situations reviewed by Gillot et al. (1974), zones of degraded shale containing numerous veins of gypsum and jarosite were found. The underlying unweathered shale was generally reported to comprise calcite, pyrite, clay minerals, quartz, and carbonaceous matter, but no gypsum or jarosite. In some cases pyrite was found in both the unaltered and the altered shale, but at lower concentrations in the latter material. Quigley and Vogan (1970) concluded that expansion due to clay mineral hydration was insufficient to explain the large and continuing heave which they observed.

Floor heaves of several inches are reported by Nixon (1978) to have affected Upper Lias pyrite-bearing shales used as fill beneath houses in northern England. Again, altered shale with gypsum and jarosite crystallization was apparent. This particular problem shale is included with others in Table 9.4 which lists a number of United Kingdom pyritiferous mudrocks. In fact pyrite quantities as small as 0.1 per cent by weight are reported by Penner et al. (1973) to have caused heave. Nixon (1978) attributes the

Table 9.4 Occurrence of pyrite in United Kingdom mudrocks

Formation	Geological age	Location	Pyrite %	Reference
Bembridge Beds	Oligocene	Isle of Wight	NK	Nixon, 1978
Barton Clay		Fawley, Hants.	NK	Marsland and Butler, 1967
London Clay	Palaeocene	London Basin & Hants.	3.3	Burnett, 1974
		Herne Bay	0–4	Burnett and Fookes, 1974
Gault Clay	Cretaceous	Bucks.	1.0	Smith, 1978
Speeton Clay		Yorks.	NK	Univ. Durham
Fuller's Earth		Redhill, Surrey	0.5	Nixon, 1978
Weald Clay		Horsham, Sussex	NK	Nixon, 1978
Kimmeridge Clay	Jurassic	Kimmeridge, Dorset	4.0	Smith, 1978
Oxford Clay		Oxfords.	3–5	Smith, 1978
Oxford Clay		E & S England	5–15	Russel and Parker, 1979
Lr. Oxford Clay		E & S England	0–17	Jackson and Fookes, 1974
Fuller's Earth		Somerset	3.0	Smith, 1978
Whitby Shale		N. Yorks.	3–9	Nixon, 1978
Upper Lias Clay		Northants.	3–5	Smith, 1978
Lower & Middle Lias Clay		Lincs. & Glos.	NK	Nixon, 1978
Stonesfield Slate		Hampden, Glos.	NK	Nixon, 1978
Marl Slate	Permian	County Durham	0–4	Smith and Francis, 1967
Coal Measures Shale	Carboniferous	England	0.7–1.4	Collins, 1976
Edale Shale		Derbys.	NK	Vear and Curtis, 1981
Calciferous Sandstone		Philpstoun, Midlothian	1.1	Nixon, 1978
Carboniferous Limestone Series		Yorks., Derbys.	5–10	Nixon, 1978
Colliery Spoil (fine)		U.K.	0–12	Taylor, 1984
Colliery Spoil (coarse)		U.K.	0–1	Taylor, 1984
Tansley Shale		Derbys.	0.1–4.9	Spears and Amin, 1981

NK = amount not known

reactiveness of the Upper Lias shale to the small grain size (5–15 μm) of the pyrite which he notes is similar to that measured by Gillot and Eden (1975) for some heaving shales in Canada.

The oxidation of pyrite during weathering in the presence of water vapour, calcium carbonate and illite can give rise to a series of chemical reactions, as follows:

$$2FeS_2 + 7O_2 + 2H_2O \rightarrow 2FeSO_4 + 2H_2SO_4 \tag{9.1}$$

$$FeSO_4 + 2H_2O \rightarrow Fe(OH)_2 + H_2SO_4 \tag{9.2}$$

$$4FeSO_4 + O_2 + 2H_2SO_4 \rightarrow 2Fe_2(SO_4)_3 + 2H_2O \tag{9.3}$$

$$7Fe_2(SO_4)_3 + FeS_2 + 8H_2O \rightarrow 15FeSO_4 + 8H_2SO_4 \tag{9.4}$$

$$H_2SO_4 + CaCO_3 + 2H_2O \rightarrow CaSO_4 \cdot 2H_2O + H_2O + CO_2 \tag{9.5}$$

$$4(KAl_2Si_3O_8(OH)_2) + 12FeSO_4 + 54H_2O + 3O_2 \rightarrow$$

$$4(KFe_3(OH)_6(SO_4)_2) + 12Al(OH)_3 + 12Si(OH)_4 + 4H_2SO_4 \tag{9.6}$$

Reaction 9.3 depends on the activities of autotropic bacteria (see Kuznetsov *et al.*, 1963). In fact, the action of sulphide-oxidizing bacteria is utilized to enhance sulphide extraction from certain low-grade ores. The acidic (pH = 2.4) conditions required by the bacteria for rapid growth are generated by reactions 9.2 and 9.4 (see Penner *et al.*, 1973). Also of relevance is an optimum ambient temperature.

Experiments reported by Kuznetsov *et al.* (1963) indicate that pyrite oxidation is more rapid for fine-grained, amorphous and impure forms. Besides the oxidation of pyrite, the main effects of these chemical and biochemical activities are the breakdown of calcite (reaction 9.5) and clay minerals (represented here by illite—reaction 9.6) leading to the production of melanterite, gypsum, jarosite and limonite. Various other iron minerals including conquimbite, hematite and amorphous gels may also be generated and other reactions with base exchange may give rise to sulphate minerals such as sodium jarosite, alunite and a variety of intermediate forms. Under some circumstances other sulphates, or even natural sulphur, may be generated, especially at higher temperatures. Fasiska *et al.* (1974) report sulphates of titanium, aluminium, iron and chromium resulting from reactions with cations available from fly ash (see also Taylor and Hardy (1974) for colliery spoils).

Table 9.5 Expansion in terms of crystalline solid due to mineral alteration

Original mineral	Product mineral	Expansion %	Reference
Pyrite FeS$_2$	Jarosite	115	Penner *et al.*, 1973
	Melanterite	536	Shamburger *et al.*, 1975
	Anhydrous ferrous sulphate	350	Fasiska *et al.*, 1974
Calcite	Gypsum	103	Penner *et al.*, 1973
	Gypsum	60	Shamburger *et al.*, 1975
Illite	Alunite	8	Shamburger *et al.*, 1975
	Jarosite	10	Shamburger *et al.*, 1975

Oxidation of pyrite and the subsequent reactions generally involve volume increases, some of which are given in Table 9.5. These figures should be regarded only as approximate guides to the maximum possible expansion since precipitation rarely occurs at the site of mineral reactions. Pyrite alteration occurs in the presence of moist air so consequently the continuation of the reaction depends on a dynamic and open chemical environment in which products are removed in solution. Not all of these will be precipitated in the same place. Hence, the amount of heave depends on the supply of neoformed minerals. Also of pertinence are ground pressures tending to restrict crystal growth (see section 9.3.3.1).

Measures designed to prevent heaving due to pyrite oxidation have been considered by Penner *et al.* (1972) who point out that provision of underfloor drainage frequently affords an adverse supply of oxygenated water to bedrock. However, lowering the water table may have adverse effects if a capillary zone is introduced into areas previously fully saturated. Various measures to maintain low rockmass permeability and prevent the ingress of air have been suggested. Permanent saturation of the bedrock by raising groundwater levels may achieve the latter but may have other deleterious effects. Minimization of rockmass disturbance during excavation and immediate surface sealing are likely to be beneficial, particularly in tunnelling. Prevention of problems may also be achieved by designing foundations so that raised platforms subject to capillary wetting and differential loadings are eliminated.

In the case of existing buildings, removal of the affected shale is often impractical. In some cases space for heave has been created by the construction of a structurally-supported floor. Elsewhere grouting to reduce rockmass permeability and measures to maintain permanent saturation have been used. Penner *et al.* (1973) describe remedial measures designed to neutralize the acid conditions responsible for the rapid growth of pyrite-oxidizing bacteria by continuously flooding the area with a 0.1 N solution of potassium hydroxide. In fact this could result in settlement if soluble components are removed.

9.3.3 *Crystal pressures*

Following a thorough review of salt-weathering processes, Cooke (1981) concludes that in arid climates rock breakdown occurs mainly by three processes:

(a) Growth of crystals in confined spaces within the rockmass due to precipitation resulting from either evaporation or cooling.
(b) Increases in volume of minerals due to hydration.
(c) Differentials in the coefficients of thermal expansion of different rock components.

Various tests (ASTM C88-76, BS 1438: 1971, and Australian Standard 1141, 1974) have been devised to test the susceptibility of rocks to breakdown by crystal pressures. Although Cooke (1981) criticizes these for not modelling environmental changes adequately, they do demonstrate that the pressures produced can exceed the tensile strength of rocks.

Cooke and Smalley (1968) conclude that although many evaporate minerals found in arid areas have markedly different coefficients of linear expansion compared with most rocks, the pressures are too small to cause rock breakage. One possible exception is the case of halite in cracks near the surface. This implies that pressures are less than about 4 MN/m². Sperling and Cooke (1980) discuss the relative efficiency of rock breakdown by various salts common in arid areas; minerals with high solubilities are the most effective. Also pertinent are the formation of hydrates and the variation of solubility with temperature.

9.3.3.1 *Crystal growth pressures.* Attention has already been paid to the situations in which crystallization of soluable minerals occurs. A supply of soluble minerals and solution concentration or cooling gives rise to precipitation. Evans (1970) considers the phase boundary relationships in the salt-solution-rock system. Since growth requires the maintenance of a thin film of solution at the rock interface, the sum of the salt-solution and solution-rock interfacial tensions must be less than the salt-rock interfacial tension. According to Winkler and Singer (1972) the theoretical pressure produced by growth depends mainly on the degree of supersaturation:

$$P_c = \frac{RT}{V} \ln \frac{c}{c_s}$$

where P_c is the crystal pressure (kN/m²), R is the gas constant (8.3143 J/K mol), T is the absolute temperature (K), V is the molecular volume of the salt (litre/mol), c is the concentration of the solution (g/litre), c_s is the saturation concentration for the soluble mineral (g/litre). The values of crystal growth pressure quoted in Table 9.6 have been *recalculated* using this equation.

The pressures given in Table 9.6 are the maximum possible for a particular mineral and will be generated only if crystallization occurs in a porous material in which all the pore space is already occupied by solid material. Also, crystal habit affects pressure with acicular forms exerting more pressure than prismatic or tabular types (see Sperling and Cooke, 1980). In fact, although the crystal growth pressures for gypsum and anhydrite are high in comparison with other salts common in deserts they are not particularly effective agents in rock breakdown. Low solubility, only slight variation in solubility with temperature and the absence of hydrates with radically different solubilities reduce their destructive power.

Wellman and Wilson (1965) point out that to minimize the chemical free energy of crystallization processes, large crystals will tend to grow in preference to small ones. Large crystals continue to grow even against a restriction, so long as growth entails less

Table 9.6 Crystal growth pressures (MN/m²) for gypsum, anhydrite and halite (after Winkler and Singer, 1972).

| Mineral | Molecular weight g/mol | Molecular volume cm³/mol | Specific gravity | Concentration, c/c_s | | | | Temperature, °C | |
| | | | | $c/c_s = 2$ | | $c/c_s = 10$ | | $c/c_s = 50$ | |
				0°C	50°C	0°C	50°C	0°C	50°C
Anhydrite	136.14	46.0	2.96	33.9	40.3	113.5	134.3	192.5	229.2
Gypsum	127.00	54.8	2.36	28.6	33.8	95.0	112.5	161.6	192.5
Halite	58.54	27.8	2.17	56.1	66.3	186.9	221.9	316.6	378.7

work than that required to increase the surface area by extending the crystal into a narrow pore. Thus, for growth to continue

$$(P_l - P_s)dV < \sigma dA$$

where P_l is the pressure in the liquid, P_s is the pressure in the solid ($\equiv$ crystallization pressure), dV is the increase in volume, σ is the salt-solution interfacial tension, dA is the increase in area. This equation reduces to

$$P_l - P_s = \sigma \frac{dA}{dV}$$

and may be evaluated in terms of pore size of the medium in which crystallization occurs and the size of the crystals of soluble mineral. In rocks with a range of pores of different sizes, crystallization occurs first in the large pores, but this will not cause expansion until growth occurs in the small pores, unless this is prevented by restraining pressure. Hence, soils with large pores connected by networks of narrow pores are (comparatively) more susceptible to expansion by crystal pressures.

The literature includes many instances of heave resulting from crystal pressures. This process has been discussed with reference to pyrite oxidation in section 9.3.2, but estimates of resultant heave pressures vary widely, which probably reflects the complexity of the chemical reactions and the possible range of end products. Values range from 14kN/m^2 (laboratory tests; Anon, 1960), 74kN/m^2 (Quigley and Vogan, 1970) to as high as 500kN/m^2 (Sherrell, 1979). A consequential stress level of possibly up to 500kN/m^2 and an $8 \times$ volume increase in original sulphide is also cited by Knill (1975) for gypsum and melanterite formation.

It is apparent that many gypsum veins are formed from gypsum liberated as a result of anhydrite hydration (Holliday, 1970). Shearman et al. (1973) attribute the separation of the walls of the veins studied to hydrofracture followed by passive crystallization of gypsum. Bundy (1956) observes that gypsum veins in south-western Indiana contain fibrous crystals with their long axes arranged across the vein. Furthermore, since the crystals are strained and the thinnest ones occur in rocks with the smallest pore sizes, it can be argued that the veins were widened by crystal growth.

It might be expected that the relatively high theoretical crystallization pressures of gypsum and anhydrite in Table 9.6 would be accompanied by crystallization in an acicular habit rather than in the more common tabular and prismatic form. However, Gillot et al. (1974) illustrate the opening of cracks by short prismatic crystals orientated with their c-axes perpendicular to the bedding planes.

9.3.3.2 *Crystal hydration pressures.* Of particular importance in terms of ground movements is the hydration of anhydrite to gypsum. Also of pertinence to heave is the hydration of oxides of calcium and magnesium which occur in some slags (see section 9.4).

The conversion of anhydrite plus water to gypsum involves a net theoretical volume *reduction* of about 7 per cent (see Wohnlich, 1976). In fact, as noted by Chatterji and Jeffrey (1963), a 0.5 per cent increase occurs. A maximum volume *increase* of about 63 per cent results if the volume of the water is not taken into account. Ground movements affecting Das Island (Persian Gulf) have been attributed to anhydrite conversion, and although in many other cases swelling behaviour has been linked to hydration, it is apparent that this process is not invariably accompanied by expansion.

Table 9.7 Hydration pressures (MN/m^2) for bassanite ($CaSO_4\frac{1}{2}H_2O$) to gypsum convertion (after Winkler and Wilhelm, 1970).

Relative humidity %	Temperature, °C						
	0	10	20	30	40	50	60
100	224.9	199.1	177.8	161.5	136.8	117.5	94.1
90	202.7	181.4	159.2	143.6	117.3	97.1	73.4
80	184.4	162.4	139.0	122.3	95.3	74.8	51.8
70	162.1	139.8	116.0	98.8	71.1	49.5	25.7
60	139.3	114.3	89.6	70.5	42.8	20.2	0
50	108.6	83.6	58.3	40.1	8.9		

The lack of disturbance of bedding and textural studies cited by Holliday (1970) are indicative of volume for volume replacement in some situations (see also Fabre and Dayre, 1982).

The theoretical magnitude of pressures developed during hydration may be derived from a consideration of vapour pressures. Winkler and Wilhelm (1970) give the following relationship:

$$P_h = \frac{\bar{n}RT}{V_h - V_a} 2.3 \log \frac{P_w}{P'_w}$$

where P_h is the hydration pressure (kN/m^2), $\bar{n}$ is the number of moles of water gained during hydration, R is the gas constant (8.3143 J/K mol), T is the absolute temperature (K), V_a and V_h are respectively the molar volumes of the salt before and after hydration (litre/mol), and P_w and P'_w are respectively the vapour pressure of atmospheric water and of the hydrated salt (N/m^2). Winkler and Wilhelm (1970) calculated the hydration pressure for a number of salts; the values for bassanite-gypsum conversion are given in Table 9.7.

Difficulties have frustrated many attempts to measure hydration pressure experimentally and the results are not easily extrapolated to the process *in situ*. In early work by Bonnell and Nottage (1939) pressures of up to 6.9 MN/m^2 were measured during the hydration of sodium and magnesium sulphates in sand-salt mixtures. According to Müller-Salzburg and Götz (1976), Huder and Amberg (1970) determined a laboratory gypsum to anhydrite hydration pressure of 380 MN/m^2 which contrasts sharply with the counter-pressure of 196 kN/m^2 which prevented expansion in experiments conducted by Einfalt and Götz (1976). The theoretical maximum pressure of 20 MN/m^2 quoted by James (1982) should also be borne in mind. Care must be taken over the use of experimental hydration pressures since the results are affected by the duration of the hydration, temperature, relative humidity and other experimental details.

An indication of possible *in situ* hydration pressures is provided by Brune (1965) who documents some dramatic instances of anhydrite hydration in Texas, Oklahoma and New Mexico. On occasions, the associated heave is described as explosive in character, especially where water had been introduced. Hydration has been observed to occur at depths as great as 152 m with an effective overburden pressure as great as about 2 MN/m^2. Brune (1965) considers that pressures of up to about 14 MN/m^2 would be necessary to uplift and deform the overlying rock beam.

Measurements of the average thickness of certain gypsum beds indicate an increase of about 35 per cent (cf. = 63% theoretically possible) compared with anhydrite beds in the same formation (Brune, 1965). In this case conversion of anhydrite to gypsum occurred at a depth of about 152 m. Above depths of about 122 m the gypsum beds were successively thinned by solution until at 30 m they were practically removed. Shallower hydration of anhydrite in the Muschelkalk and middle Keuper of mid-Europe is described by Kaiser (1976) who quotes a depth of approximately 40–100 m. Above this lies a 5 to 15 m-thick zone of massive gypsum. Leaching of gypsum at shallower levels removes the mineral to depths of between 35 and 80 m (see also Holliday, 1970; Redfield, 1968).

Many Miocene anhydrite formations in the Alps are associated with swelling clay minerals and have been subject to large heaves. Fecker (1981) suggests that the affinity for water displayed by anhydrite can inhibit clay mineral swelling and hence reduce overall expansion. In many cases the precise cause of swelling is difficult to establish (Krause, 1976). Serrano *et al.* (1981) consider this problem with reference to a gypsum marl formation in northern Spain and attributed rapid (days) swelling to clay mineral effects compared with longer-term (weeks) changes for anhydrite.

Much research effort has been invested in estimating the environmental temperatures and pressure conditions for the hydration and de-hydration of calcium sulphate and other minerals. Experiments quoted by McDonald (1953) suggest that anhydrite would be the stable form of $CaSO_4$ in water at temperatures of over 42°C compared with thermodynamic calculations which yield a value of 40°C. This is reduced to about 34°C in sea-water sufficiently concentrated to precipitate gypsum (see also Zen, 1965). However, determinations based solely on solubility have been criticized: the transition point of anhydrite is approached from the undersaturated side of equilibrium since nucleation does not occur at temperatures lower than 80–100°C. From experiments in which precipitation and solution of synthetic gypsum and anhydrite occurred in solutions of known activity of water, Hardie (1967) concludes that in pure water the transition temperature would be 58 ± 2°C. Again this temperature is reduced for saline water. In concentrated sea-water precipitating gypsum it would be 52°C (chlorinity 65‰), 26°C for anhydrite (chlorinity 145‰) and 18°C for halite (chlorinity 160‰). These data suggest that conversion from gypsum to anhydrite and vice versa may occur in arid environments (section 9.3.1) although they practically rule out primary precipitation of anhydrite (see Kinsman, 1966).

9.3.4 *Solution rates of soluble minerals*

Soluble minerals are liable to be removed from rocks and soils by percolating ground water where solution generally increases flow rates. Ground movements occur due to the loss of support; unconsolidated deposits in particular may collapse. Much of the work on the solution of minerals has been carried out with reference to dams and reservoirs built on indurated carbonate rocks. In most of these cases seepage control is the main concern (see Erguvanli, 1979; Altug and Oymangrout, 1982; James, 1981; Wahlstrom, 1974; and Walters, 1962). Consideration of gypsum and anhydrite in these and other types of structure has been given by many authors including Eck and Redfield (1965), Nonveiller (1982), and Redfield (1968). In some cases excessive settlements have been attributed to the solution of gypsum and anhydrite (see Calcano and Alzura, 1967; Olive, 1957). The causes and effects of very large ground movements resulting from solution mining of halite in Cheshire, United Kingdom, are described by Bell (1975). In addition, the removal of soluble minerals by natural groundwater flow is discussed by

Table 9.8 Solution parameters of limestone, anhydrite, gypsum, halite and quartz in pure water at 10°C (after James and Kirkpatrick, 1980). c = concentration, t = time, A = area, $\bar{V}$ = volume of solution.

Mineral	Saturation concentration kg/m^3	Solution rate constant, K (flow velocity = 0.05 m/s)
Gypsum	2.5	0.2×10^{-5} m/s
Halite	360.0	0.3×10^{-5} m/s
Limestone	0.015 (0.4)*	0.4×10^{-5} m/s
Anhydrite	2.0	0.8×10^{-6} m⁴/kg sec
Quartz	0.01	$\dfrac{dc}{dt} = \dfrac{A}{\bar{V}} \times 10^{-7}$ kg/m²/s

* water containing dissolved CO_2

Hawkins, 1979 (Triassic deposits near Bristol); Dearman and Coffey, 1981 (Permian deposits, Teeside); and Poupelloz and Toulemont, 1981 (Paris).

Although solubility is an important factor in solution behaviour, James and Kirkpatrick (1980) point out that solution rate is a more significant parameter. These authors indicate that with respect to saturation concentration in pure water, limestone (calcite) and quartz have similar solubilities (Table 9.8) but contrasting solution rates. In

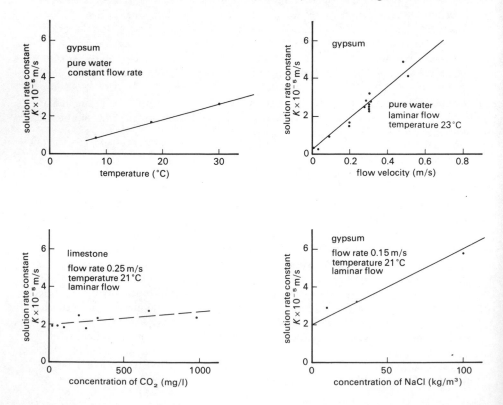

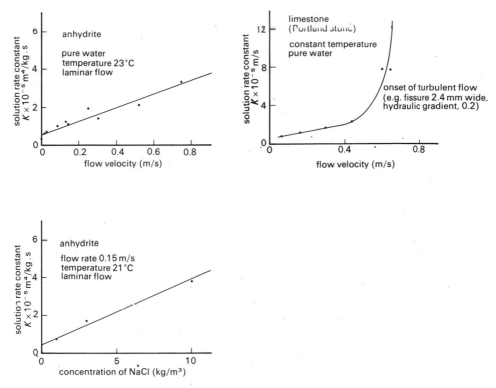

Figure 9.8 Variation of solution rate constant (K) of gypsum, anhydrite, and limestone with temperature, flow velocity, solution composition, and style of fluid flow. (After James and Lupton, 1978; James, 1981).

the case of quartz, solution is so slow that in practical terms the mineral is regarded as insoluble.

Experiments carried out by James and Lupton (1978) confirm that solution can be expressed by the equation

$$\frac{dM}{dt} = KA(c_s - c)^n$$

where M is the mass of mineral dissolved (g), t is time (seconds), A is the area of mineral exposed to solution (m^2), c_s and c respectively are the saturation concentration of mineral and concentration in solution (g/litre), n is a power ($n = 1$ for gypsum, halite and calcite, and $n = 2$ for anhydrite), and K is the solution rate constant (m/s for $n = 1$, m^4/gs for $n = 2$). Thus anhydrite displays second-order solution kinetics and in practice dissolves at a slower rate than gypsum or halite since before solution ions must be detached from a tightly-packed crystal lattice into a less well-ordered boundary layer.

The value of the solution rate constant, K, depends on a number of factors including temperature, flow velocity and solution constitution, as demonstrated in Fig. 9.8. The mechanisms of the dissolution process are discussed by James and Lupton (1978) who point out that high fluid flow rates and high ionic strength solutions reduce the thickness of the boundary layer. In practice, the solution rate constant may also depend on the form of a particular mineral and vary according to particle size and the presence of impurities. Eroskay (1982) notes that benthonic limestone is generally more susceptible to dissolution than the micritic or pelagic types he studied.

In practice the dissolution of a mineral depends on its sedimentary mode of occurrence and its relationship to groundwater. The possibilities may be summarized as follows:

(a) confined flow of water through either a fissure in, or permeable beds adjacent to, massive beds of soluble mineral.
(b) confined flow of water through a particulate permeable mixture of soluble minerals and other non-soluble materials.
(c) unconfined flow of water over the surface of either a massive bed of, or bed containing particles of, soluble minerals.

James and Lupton (1978) derive theoretical solution rates and cavity enlargement profiles for cases (a) and (b). Case (c) is more difficult to analyse due to the development of heterogeneous solute concentrations.

Of particular importance is the development of a 'runaway' situation in which an under-saturated solution reaches the exit of a flow system. Removal of soluble mineral then results in an enhanced flow rate and hence a rapid increase in the amount of material dissolved. Tests in which the widening of a hole drilled through solid mineral was monitored indicate that in the case of gypsum a short, steeply-tapered solution cavity is produced. This contrasts with the narrower but longer one formed in anhydrite. Thus anhydrite is much more susceptible to the onset of 'runaway' solution although precipitation of gypsum near the end of the flow channel may restrict the flow and inhibit solution (see Fabre and Dayre, 1982).

James and Kirkpatrick (1980) consider the dissolution of soluble minerals with respect to hydraulic structures with a life expectancy of 100 years. Table 9.9 shows the largest size of fissure allowable under assumed hydrological conditions. Also included is the limiting size of fissures which bring about migration of a cavity along the flow path. Grouting to reduce fissure width and hence flow velocity is generally adopted as a preventive measure. If cement grout is capable of penetrating fissures as small as 0.2 mm, then satisfactory control of solution can be achieved for gypsum and limestone, but may not be successful

Table 9.9 Maximum allowable size of fissures in massive deposits with fissures spaced at 1 m intervals due to pure water with an hydraulic gradient of 0.2 (after James and Kirkpatrick, 1980).

Mineral	Fissure width (mm)		
	No extensive enlargement of fissures in 100 years	Retreat of inlet of 0.1 m/year	Preventive measures
Gypsum	0.2	0.3	Grouting
Anhydrite	0.1	0.2 ⎫	Cut-off
Halite	0.05	0.05 ⎬	e.g. plastic concrete
Limestone	0.5 (0.4)*	1.5	Grouting

* Impure water capable of dissolving 300 mg/1 CaCO$_3$

in the case of anhydrite without additional preventive measures. The exacting standard required for halite is unlikely to be achieved in practice. The aforementioned authors provide further details of treatment and also discuss investigative methods. An interesting case, in which solution rate predictions are applied to the dissolution of massive gypsum forming a riverside cliff at Ripon, is considered by James et al. (1981). Gypsum solution problems in that area of Yorkshire are also described by Kendal and Wroot (1924).

James and Lupton (1978) present data useful for dealing with soluble particulate minerals in a permeable deposit. Clearly, the maximum potential settlement is related directly to the amount of soluble mineral present, but the authors derive equations for determining the width and rate of advance of the solution front, viz.:

frontal velocity
$$\bar{u} = \frac{Q/A^*}{1 + (N\rho\beta l_0^3 10^3)/c_s} \text{ m/s}$$

width of solution front
$$= \frac{\bar{u}\rho l_0 \times 10^3}{\alpha K c_s^n} \text{ m}$$

where Q is the flow rate (m^3/s) and A^* the area (m^2), N is the number of soluble particles per m^3, β is a volume coefficient for the soluble particles (volume $= \beta l_0^3$), l_0 is the initial length of the soluble particles (m), ρ is the density of the soluble mineral (Mg/m^3), c_s is the saturation concentration of the soluble mineral (g/litre), n is a power ($n = 1$ for gypsum, halite and calcite, $n = 2$ for anhydrite), K is the solution rate constant (m/s for $n = 1$; m^4/g sec for $n = 2$) and α is an area coefficient for the soluble particles (area $= \alpha l_0^2$). The dominant controlling factors are permeability, solution rate constant and solubility of the mineral. Clearly these concepts have direct application in the analysis of solution problems arising from the dilution of high-salinity groundwater or changes in groundwater conditions.

9.4 Ground movements due to industrial materials

Experience shows that various industrial by-products make excellent fills or aggregates for concrete and bitumen. Among the more important of these materials are slags from the production of iron and steel, although other slags have also been utilized on a smaller scale. Clearly, the composition of slags is variable and depends on the ore reduction process and composition of the constituents. Blastfurnace slag (whether air-cooled or otherwise), prepared for engineering applications, may be acceptable, subject to satisfactory performance in tests (see BS 1047, part 2, 1974). However, other types of slag, in particular those produced in open-hearth or electric-arc steel furnaces using basic fluxes are identified by Crawford and Burn (1969) as materials liable to exhibit instability (see also Spanovich and Fewell, 1970). Heaves as great as 33 per cent of the thickness of fill are reported by Bailey and Reitz (1970). Spontaneous heating (probably arising from hydration temperatures) in some iron and steel slag fills may also cause problems, as has been the case with many *loose* colliery discards. Although in colliery discards pyrite oxidation is an exothermic reaction, coal oxidation has been identified as the major heat source in practice (see also Building Failure Sheet 14).

The chemistry of iron and steel slags is complex and variable. Their swelling behaviour is generally attributed to the presence of oxides of calcium, magnesium and iron which expand due to hydration. Unfortunately, the literature includes little information regarding the reactions which accompany this process. Ramachandran *et al.* (1964) have calculated that the hydration of CaO to $Ca(OH)_2$ can result in an expansion of up to 198 per cent. In practice swelling is much less this. Observations indicate that the growth of $Ca(OH)_2$ occurs in pores and spaces previously occupied by CaO, giving a linear expansion of 8 per cent. Chemical analysis of swelling slag led Crawford and Burn (1969) to conclude that initial rapid expansion (4.5% per year) was due to the hydration of CaO. This was followed by slower expansion (2% per year) as MgO became hydrated.

Various tests intended to identify susceptible materials have been devised. Crawford and Burn (1969) and Emery (1980) suggest tests in which the reactions are accelerated by high temperature treatment (see ASTM C151–77). Analytical chemical determination of the mass percentages of CaO, MgO, SiO_2, Al_2O_3 and S are advocated by BS 1047, part 2, 1974. Acceptable material should pass one or both of the following:

(a) $CaO + 0.8 MgO \leq 1.2 SiO_2 + 0.4 Al_2O_3 + 1.75 S$

(b) $CaO \leq 0.9 SiO_2 + 0.6 Al_2O_3 + 1.75 S$

Preferably, polished specimens treated with 10 per cent $MgSO_4$ solution at 50°C are examined for evidence of etching (see also *BRE Digest* 222, 1979).

Crawford and Burn's (1969) paper generated considerable discussion which is itself reviewed by the same authors (1971). Of particular interest are comments by Spanovich and Fewell (1970) who refer to existing slags in western Pennsylvania, U.S.A. These authors suggest that stabilization may be achieved by restricting the penetration of moisture by bitumen grout injection. It is also suggested that an overburden of about 3 m is sufficient to prevent swelling. It is noteworthy that structural damage caused by expansive fill beneath non-loadbearing floor slabs has, on occasions, been attributed to the settlement of loadbearing parts of buildings. Weathering or acid treatment do not appear to be viable solutions to the problem since even slags as old as 30 years have proved troublesome.

Appendix: Swelling pressure due to interparticle repulsion in a micaceous clay

(see Yong and Warkentin (1966) for further details)

Let water content $= 30\%$
　surface area $= 100 \, m^2/g$ (mica)
　pore-water salt concentration $= 0.0005$ mol/litre KCl
　exchangeable ion—potassium (valency, unity)

Now　　$d = (100W)/S$

where　d = half-distance between parallel mica plates in Å　　　　　　**A9.1**
　　　W = water content %
　　　S = surface area in m^2/g,
whence $d = \dfrac{100 \times 30}{100} = 30\text{Å}$

The concentration of cations midway between 2 clay plates is

$$C = \pi^2/[z^2 B(d + x_0)^2 10^{-16}]$$　　　　　　**A9.2**

where C is in mol/litre

z = valency of exchangeable cations

d = half-distance in Å

x_0 = correction factor of between 1
and 4Å, depending on valency of
ions and charge density of clay

B is a constant ($= 10^{15}$ cm/mmol),
depending on temperature and
dielectric constant

whence
$$C = \frac{(3.1416)^2}{1 \times 10^{15} \times 32^2 \times 10^{-16}}$$

(taking x_0 as 2)

$C = 0.0964$ mol/litre

The swelling pressure,

$$P = RT(C - \gamma C_0)$$

At room temperature (20°C)

$$P = 8.3143 \times 293(0.0964 - 2 \times 0.0005)$$

$$P = 232\,kN/m^2$$

where R = gas constant (8.3143 J/K mol)

T = absolute temperature

γ = number of ions in pore water salt

C_0 = concentration of salt in
pore water in mol/litre

A9.3

Acknowledgements

Gratitude is expressed to many people for assistance with this chapter. Particular contributions were received from Dr P.R. Bush, Professor P.G. Fookes, Mr J.P.H. Frearson, Mr I.E. Higginbottom, Mr S.M.M. Rice and Dr D.A. Spears.

Notation

A area of mineral (m^2)

Å   angstrom unit ($= 10^{-10}$ m)

A^* cross-sectional area normal to direction of flow (m^2)

B a constant (cm/mmol; eqn. A9.2)

C concentration of cations (mol/litre)

C_0 concentration of pore-water salt (mol/litre)

c concentration of solution (g/litre)

c_s saturation concentration for a soluble mineral (g/litre)

D thickness of impermeable barrier (m)

d half-distance between clay platelets (Å)

dA increment of area of crystal due to growth (m^2)

dc increment of concentration (kg/litre)

dM increment of mass of mineral dissolved (g)

dt increment of time for dissolution (s)

dV increment of volume of crystal due to growth (m^3)

h head of water (m)

h_t total suction ($-$ cm water)

h_m matrix suction ($-$ cm water)

h_s solute suction ($-$ cm water)

K solution rate constant (m/s or m^4/gs)

K_0 lateral stress ratio for one-dimensional strain

k coefficient of permeability (m/s)

l_0 initial length of soluble particles (m)

N number of soluble particles per m^3

n a power expressing kinetic order of solution process

$\bar{n}$ number of moles water gained during hydration.

P_c crystal growth pressure (kN/m^2)

p_h crystal hydration pressure (kN/m^2)

P_l pressure in liquid (kN/m^2)

P_s pressure in solid (kN/m^2)

P_w vapour pressure of atmospheric water (N/m^2)

P'_w vapour pressure of hydrated salt (N/m^2)

pF   soil suction scale (pF0 $= 1$ cm water, pF6 $= 10^5$ cm)

Q volumetric flow rate (m^3/s)

R gas constant (8.3143 J/K mol)

S surface area (m^2/g)

T absolute temperature (K)

u pore-water pressure (cm water)

$\bar{u}$ rate of advance of solution front (m/s)

V molecular volume of precipitating salt (litre/mol)

$\bar{V}$ volume of solution (m^3)

V_a molar volume of salt prior to hydration (litre/mol)

V_h molar volume of hydrated salt (litre/mol)

v mean flow velocity (m/s)

W water content (%)

x_0 correction factor in Å (eqn. A9.2)

z valency of exchangeable cations

$\bar{z}$ gravitational component of head (cm water)

α area coefficient for soluble particles (area $= \alpha l_0^2$)

β volume coefficient for soluble particles (volume $- \beta l_0^3$)

γ number of ions in pore-water salt

ρ density of soluble particles (Mg/m^3)

σ salt-solution interfacial tension (kN/m)

ψ soil moisture potential (cm)

Chemical elements expressed in conventional notation.

References

Altug, S. and Oymangrout, J.V. (1982) Study of leakage on the east side of the Opyapinar reservoir, Turkey. *Bull. Int. Ass. Eng. Geol.* **25**, 117–124.

American Society for Testing and Materials (1976) Test for soundness of aggregates by use of sodium sulfate or magnesium sulfate. ASTM Standard C88–76.

American Society for Testing and Materials (1977) Test for autoclave expansion of portland cement. ASTM Standard C151–77.

Anon. (1960) Structures do not settle in this shale, but watch out for heave. *Engineering News— Record.* **164**, 46–48.

Attewell, P.B. and Taylor R.K. (1973) *Clay shale and discontinuous rock mass studies.* Final Report to European Research Office, U.S. Army, Contract No. DA-ERO-591-72-G005.

Australian Standard 1141, (1974) *Methods for sampling and testing aggregates.* Standards Association of Australia.

Bailey, J.J. and Reitz, H.M. (1970) Building damage from expansive steel backfill. *Soil Mech. & Found. Div. Am. Soc. Civ. Engrs.* **96**, 1810–1813.

Bathurst, R.G.C. (1975) *Carbonate Sediments and their Diagenesis.* Elsevier, Amsterdam.

Bell, F.G. (1975) Salt and subsidence in Cheshire, England. *Eng. Geol.* **9**, 237–247.

Blight, G.E. (1976) Migration of subgrade salt damages thin pavements. *Transp. Eng. J. Proc. ASCE.* **102**, 779–791.

Bonnell, D.G.R. and Nottage, M.E. (1939) The crystallization of salts in porous materials. *J. Soc. Chem. Ind.* **58A**, 16–21.

Brindley, G.W. and Brown, G. (1980) Crystal structures of clay minerals and their X-ray identification. *Mineral. Soc. Lond. Monogr.* **5**.

British Standard 1438 (1971) Specification for media for biological percolating filters. British Standards Institution, London.

British Standard 1047, part 2 (1974) Specification for air cooled blast furnace slag coarse aggregate for concrete. British Standards Institution, London.

British Standard 1377 (1975) Methods of test for soils for civil engineering purposes. British Standards Institution, London.

Brune, G. (1965) Anhydrite and gypsum problems in engineering geology. *Bull. Ass. Eng. Geol.* **3**, 26–38.

Building Failure Sheet 14 (1972) Solid floor on shale fill. *Building* **102**, 14 April.

Building Research Establishment (1975) Concrete in sulphate-bearing soils and groundwaters. *BRE Digest* **174**, H.M.S.O., London.

Building Research Establishment (1979) Fill and hardcore. *BRE Digest* **222**, H.M.S.O., London.

Building Research Establishment (1980) Low-rise buildings on shrinkable clay soils: Part 1. *BRE Digest* **240**, H.M.S.O., London.

Building Research Establishment (1981) Concrete in sulphate-bearing soils and groundwaters. *BRE Digest* **250**, H.M.S.O., London.

Building Research Station (1968) *Concrete in sulphate bearing soils and groundwaters. BRE Digest* **90**, H.M.S.O., London.

Bundy, W.M. (1956) Petrology of gypsum-anhydrite deposits in southwestern Indiana. *J. Sed. Petrol* **26**, 240–252.

Burnett, A.D. (1974) The modification and application of the quantitative X-ray diffraction method of Schultz (1964) to the mineralogical study of London Clay samples. *J. Soil Sci.* **25**, 179–188.

Burnett, A.D. and Fookes, P.G. (1974) A regional engineering geological study of the London Clay in the London and Hampshire basins. *Q. J. Engineering Geology* **7**, 257–295.

Bush, P.R. (1973) Some aspects of the diagenetic history of the sabkha in Abu Dhabi, Persian Gulf. *In* Purser, B.H. (ed.) *The Persian Gulf, Holocene Carbonate Sedimentation and Diagenesis in a Shallow Epicontinental Sea*, Springer–Verlag, Berlin, 395–407.

Bush, P.R. (1983) Formation and Destruction of Coastal Evaporites in the Middle East. In *Proc. Meeting Engineering Group of Geological Society of London* 11th January.

Butler, G.P. (1969) Modern evaporite deposition and geochemistry of coexisting brines, the sabkha, Trucial Coast, Arabian Gulf. *J. Sed. Petrol.* **39**, 70–89.

Calcano, C.E. and Alzura, P.R. (1967) Problems of dissolution of gypsum in some dam sites. *Bull. Venezuelan Soc. Soil Mech. Foundation Eng.*, July–Sept.

Chapman, H.D. (1965) Cation exchange capacity. *In* Black, C.A. (ed.) *Methods of soil analysis*: Part 2, American Society of Agronomy, 1571–1572.

Chatterji, S. and Jeffery, J.W. (1963) Crystal growth during the hydration of $CaSO_4\frac{1}{2}H_2O$. *Nature* **200**, 463–464.

Cole, D.C.H. and Lewis, J.G. (1960) Progress report on the effect of soluble salt on stability of compacted soils. *Proc. 3rd Aust.–New Zealand Conf. Soil Mech. Found. Eng.*, 29–31.

Collins, R.J. (1976) A method for measuring the mineralogical variation of spoils from British Collieries. *Clay Minerals* **11**, 31.

Cooke, R.V. (1981) Salt weathering in deserts. *Proc. Geol. Ass.* **92**, 1–16.

Cooke, R.V. and Smalley, I.J. (1968) Salt weathering in deserts. *Nature* **220**, 1226–1227.

Crawford, C.B. and Burn, K.N. (1969) Building damage from expansive steel slag backfill. *J. Soil Mechanics and Foundations Div. ASCE*, **95**, 1325–1334.

Crawford, C.B. and Burn, K.N. (1971) Building damage from expansive steel slag backfill. *J. Soil Mechanics and Foundations Div. ASCE*, **97**, 1026–1029.

Curtis, C.D. (1976) Stability of minerals in surface weathering reactions: a general thermochemical approach. *Earth Surface Processes* **1**, 63–70.

Davey, B.G. (1972) Some physico-chemical aspects of swelling clay soils. *In* James, B.J.F. (compiler) *Symposium on Physical Aspects of Swelling Clay Soils*, The University of New England, Armidale, N.S.W., 13–22.

Dearman, W.R. and Coffey, J.R. (1981) Effects of evaporite removal on the mass properties of limestone. *Bull. Int. Ass. Eng. Geol.* **24**, 91–96.

Doornkamp, J.C., Brunsden, D. and Jones, D.K.C. (eds.) (1980) *Geomorphology and superficial materials* IV. Bahrain Surface Materials Resources Survey, State of Bahrain.

Dougherty, M.T. and Barsotti, N.J. (1972) Structural damage and potentially expansive sulfide minerals. *Bull. Ass. Eng. Geol.* **IX** (2) 105–125.

Eck, W. and Redfield, R.C. (1965) Engineering geology problems at Sandford Dam, Texas. *Bull. Ass. Eng. Geol.* **3**, 15–25.

Einfalt, H.C. and Götz, H.P. (1976) Chemical transformation from anhydrite to gypsum. *Bull. Int. Ass. Eng. Geol.* **13**, 69–70.

Ellis, C.I. (1973) *Arabian Salt Bearing Soil (Sabkha) as an Engineering Material*. Laboratory report 523, Transport and Road Research Laboratory.

Ellis, C.I. and Russel, R.B.C. (1974) *The Use of Salt Laden Soils (Sabkha) for Low Cost Roads*. Paper 78/74, Transport and Road Research Laboratory.

Emery, J.J. (1980) Assessment of ferro-slag for fill applications. *Proc. Sym. Soc. Chem. Ind: Reclamation of Contaminated Land*, Paper F1/1.

Erguvanli, K. (1979) Problems on damsites in karstic areas with some consideration as to their solution. *Bull. Int. Ass. Eng. Geol.* **20**, 173–178.

Eroskay, S.O. (1982) Engineering properties of carbonate rocks and regions in Turkey. *Bull. Int. Ass. Eng. Geol.* **25**, 61–65.

Evans, G., Schmidt, V., Bush, P. and Nelson, H. (1969) Stratigraphy and geologic history of the sabkha, Abu Dhabi, Persian Gulf. *Sedimentology* **12**, 145–149.

Evans, I.S. (1970) Salt crystallization and rock weathering: a review. *Revue Géomorph. Dyn.* **19**, 153–177.

Evans, P.L. (1977) *The Middle East—an Outline of the Geology and Soil Conditions in Relation to Construction Problems*. Current Paper 13/78, Building Research Establishment.

Fabre, D. and Dayre, M. (1982) Propriétés géotechniques de Gypses et Anhydrites du Trias des Alpes de Savoie (France). *Bull. Int. Ass. Eng. Geol.* **25**, 91–98.

Fasiska, E., Wagenblast, N. and Dougherty, M.T. (1974) The oxidation mechanisms of sulphide minerals. *Bull. Ass. Eng., Geol.* **11**, 75–82.

Fecker, E. (1981) Influence of swelling rock on tunnelling. *Bull. Int. Ass. Eng. Geol.* **24**, 27–32.

Fleming, R.W. Spencer, G.S. and Banks, D.C. (1970) *Empirical Study of Behavior of Clay Shale Slopes*. N.C.G. Technical Report No. 15, U.S. Army Engineer Nuclear Cratering Group, Livermore, California (N.T.I.S. Springfield, Virginia). Vol. 1.

Fookes, P.G. (1978) Engineering problems associated with ground conditions in the Middle East: inherent ground problems. *Q. J. Engineering Geology.* **11**, 33–50.

Fookes, P.G. and Collis, L. (1975a) Problems in the Middle East. *Concrete* **9** (7), 12–17.

Fookes, P.G. and Collis, L. (1975b) Aggregates and the Middle East. *Concrete* **9** (11), 14–19.

Fookes, P.G. and French, W. (1977) Soluble salt damage to surfaced roads in the Middle East. *Highway Engineer* **24** (12), 10–26.

Fookes, P.G. and Higginbottom, I.E. (1980a) Some problems of construction aggregates in desert areas with particular reference to the Arabian Peninsula: 1. Occurrence and special characteristics. *Proc. Inst. Civ. Eng.* **68** (1) 39–67.

Fookes, P.G. and Higginbottom, I.E. (1980b) Some problems of construction aggregates in desert areas, with particular reference to the Arabian Peninsula: 2. Investigation, production and quality control. *Proc. Inst. Civ. Eng.* **68** (1), 69–90.

Fookes, P.G., Pollock, D.J. and Kay, E.A. (1981) Middle East concrete—rates of deterioration. *Concrete* **15** (9), 12–19.

Frearson, J.P.H. (1983) The contamination of concrete aggregates by evaporitic minerals in the Middle East. In *Proc. meeting, Engineering Group of Geological Society London*, 11th January.

French, W.J. and Poole, A.B. (1976) Alkali-aggregate reactions and the Middle East. *Concrete* **10** (1), 18–20.

Gillott, J.E. (1968) *Clay in Engineering Geology*. Elsevier, Amsterdam.

Gillott, J.E. and Eden, W.J. (1975) Concrete deterioration and floor heave due to biogeochemical weathering of underlying shale. *Can Géotech. J.* **12**, 372–378.

Gillott, J.E., Penner, E. and Eden, W.J. (1974) Microstructure of Billings Shale and biochemical alteration products, Ottawa, Canada. *Can. Géotech. J.* **11**, 482–489.

Glennie, K.W. (1970) *Desert Sedimentary Environments (Developments in Sedimentology, 14)*. Elsevier, Amsterdam.

Goudie, A.S. (1972) On the definition of calcrete deposits. *Z. Geomorphol.* **16**, 464–468.

Goudie, A.S. (1973) *Duricrusts in Tropical and Sub-tropical Landscapes*. Oxford Research Studies in Geography, Clarendon Press. Oxford.

Greensmith, J.T. (1971) Revision of Hatch, F.H. and Restall, R.H., *Petrology of the Sedimentary Rocks*. Thomas Murby, London.

Haldane, A.D. Carter, E.K. and Barton, G.M. (1970) The relationship of pyrite oxidation in rockfill to highly acid water at Carin Dam, A.C.T., Australia. *Proc. Int. Conf. Eng. Geol.* **2**, pp. 1113–1124.

Hardie, L.A. (1967) The gypsum-anhydrite equilibrium at one atmosphere pressure. *The American Mineralogist* **52**, 171–200.

Harper, T.R., Appel, G., Pendleton, M.M., Szymansk, J.S. and Taylor, R.K. (1979) Swelling strain development in sedimentary rocks in northern New York. *Int. J. Rock Mech. Min. Sci. & Geomech. Abstr.* **16**, 271–292.

Hawkins, A.B. (1979) Case histories of some effects of solution/dissolution in the Keuper rocks of the Severn Estuary. *Q. J. Engineering Geology* **12**, 31–40.

Holliday, D.W. (1970) The petrology of secondary gypsum rocks—a review *J. Sed. Petrol.* **40**, 734–744.

Honeybourne, D.B.(1951) Clay minerals in the Keuper Marl. *Clay Mins. Bull.* **1**, 150–155.

Huder, J. and Amberg, G. (1970) Quelling in Mergel, Opalmiston und Anhydrite Schweiz. *Bauzeitung* **83**, 975.

Jackson, J.O. and Fookes, P.G. (1974). The relationships of the estimated former burial depth of the lower Oxford Clay to some soil properties. *Q. J. Engineering Geology* **7**, 137–179

James, A.N. (1981) Solution parameters of carbonate rocks. *Bull. Int. Ass. Eng. Geol.* **24**, 19–25.

James, A.N. (1982) Engineering properties of evaporitic rock. *Bull. Int. Ass. Eng. Geol.* **25**, 125–126.

James, A.N., Cooper, A.H. and Holliday, D.W. (1981) Solution of the gypsum cliff (Permian, Middle Marl) by the River Ure at Ripon Parks, North Yorkshire. *Proc. Yorks. Geol. Soc.* **43**, 433–450.

James, A.N. and Kirkpatrick, I.M. (1980) Design of foundations of dams containing soluble rocks and soils. *Q. J. Engineering Geology* **13**, 189–198.

James, A.N. and Lupton, A.RR. (1978) Gypsum and anhydrite in foundations of hydraulic structures. *Géotechnique* **28**, 249–273.

Jones, R.H. (1980) Frost heave of roads. *Q. J. engineering geology* **13**, 77–86.

Kaiser, W. (1976) Behaviour of anhydrite after the addition of water. *Bull. Int. Ass. Eng. Geol.* **13**, 68–69.

Kendall, P.F. and Wroot, H.E. (1924) *The Geology of Yorkshire*. Printed privately, Vienna.

Kinsman, D.J.J. (1966) Gypsum and anhydrite of Recent Age, Trucial Coast, Persian Gulf. *2nd Symp. on Salt, Cleveland, Ohio*, Northern Ohio Geol. Soc., vol. 1, pp. 302–326.

Kinsman, D.J.J. (1969) Modes of formation, sedimental associations and diagnostic features of shallow water and supratidal evaporites. *Am. Ass. Pet. Geol. Bull.* **53**, 830–840.

Kinsman, D.J.J. (1976) Evaporites: relative humidity control of primary mineral facies. *J. Sed. Petrol.* **46**, 273–279.

Klein, I.E. (1962) Foundation and groundwater problems related to the occurrence of gypsum on hydraulic engineering works of the U.S. Bureau of Reclamation in the San Luis Unit of the Central Valley project in California. *1st Int. Symp. on Public Works in Gypsiferous Terrain, Madrid*.

Knill, J.L. (1975) Foundations on the Coal Measures. In Bell, F.G. (ed.) *Site Investigations in Areas of Mining Subsidence*, Newnes-Butterworths, London.

Kohnuke, H. (1946) The practical use of the energy concept of soil moisture. *Soil Sci. Soc. Am. Proc.* **11**, 64–66.

Krause, H. (1976) Sulphate rocks in Baden–Württemberg and their importance in relation to civil engineering. *Bull. Int. Ass. Eng. Geol.* **13**, 45–49.

Kuznetzov, S.I. Ivanov, M.V. and Lyalikova, N.N. (1963) *Introduction to Geological Microbiology*. (Trans.: Broneer, P.T. ed. Oppenheimer, C.H.) McGraw-Hill, New York.

Lane, K.S. and Washburn, D.E. (1946) Capillary tests by capillarimeter and by soil filled tubes. *Proc. Highw. Res. Bd.* **26**, 460–473.

Madsen, F.T. (1979) Determination of the swelling pressure of claystones and marlstones using mineralogical data. *Proc. 4th Cong. Int. Soc. Rock Mech.* Vol. 1, pp. 237–243.

Marsland, A. and Butler, M.E. (1967) Strength measurements on stiff fissured Barton Clay from Fawley (Hampshire). *Proc. Geotechnical Conf., Oslo*, pp. 139–145.

McDonald, G.J.F. (1953) Anhydrite gypsum equilibrium relations. *Amer. J. Sci.* **251**, 884–898.

McGinnies, G., Goldman, B.J. and Payshore, P. (eds.) (1968) *Deserts of the World*. Univ. of Arizona Press.

Millot, G. (1970) *Geology of Clays*. (Trans.: Farrand, W.R. and Paquet, H.) Springer–Verlag, New York.

Moum, J. and Rosenqvist, I.Th. (1959) Sulfate attack on concrete in the Oslo region. *J. Amer Concrete Inst. Proc.* **56**, 257–264.

Müller–Salzburg, L. and Götz, H.P. (1976) Heaving of invert in tunnels. *Bull. Int. Ass. Eng. Geol.* **13**, 51–53.

Netterberg, F. (1979*a*) Repair of salt-damaged surfacings on the Witwatersrand. *Proc. 3rd Conf. Asphalt Pavements, Durban*, Vol. 1, pp. 137–141.

Netterburg, F. (1979*b*) Salt damage to roads—an interim guide to its diagnosis, prevention and repair. *Institution of Municipal Engineers of S. Africa* **4** (9), 13–17.

Netterberg, F. and Loudon, P.A. (1980) Simulation of salt damage to roads with laboratory model experiments. *7th Reg. Conf. for Africa on Soil Mech. & Found. Eng., Accra*, 355–361.

Nixon, P.J. (1978) Floor heave in buildings due to the use of pyritic shales as fill material. *Chemistry and Industry*, 4 March, 160–164.

Nonveiller, E. (1982) Treatment methods for soluble rocks. *Bull. Int. Ass. Eng. Geol.* **25**, 165–169.

Obermeier, S.F. (1974) Evaluation of laboratory techniques for measurement of swell potential of clays. *Bull. Ass. Eng. Geol.* **11**, 293–314.

Olive, W.W. (1957) Solution-subsidence trough Castile Formation of Gypsum Plain, Texas and New Mexico. *Geol. Soc. of Am. Bull.* **68**, 351–358.

Penner, E., Eden, W.J. and Gillott, J.E. (1973) Floor heave due to biochemical weathering of shale. *Proc. 8th Int. Conf. Soil Mech. Found. Engng., Moscow*, Vol. 2, pp. 151–158.

Penner, E., Eden, W.J. and Grattan–Bellew, P.E. (1972) *Expansion of pyritic shales.* Can. Bldg. Digest No. 152, National Research Council of Canada.

Peterson, R. and Peters, N. (1963) Heave of spillway structures on clay shales. *Can Géotech. J.* **1**, 5.

Pile, K.C. (1980) The relationship between matrix and solute suction, swelling, pressure and magnitude of swelling in reactive clays. *New Zealand Instn. Civ. Engrs. Proc. Tech. Gps.* pp. 1–197 to 1–201.

Potter, P.E., Maynard, J.B. and Pryor, W.A. (1980) *Sedimentology of Shale.* Springer–Verlag, New York.

Poupelloz, B. and Toulemont, M. (1981) Stabilisation des terrains karstiques par injection. Le cars du Lutetien gypseux de la région de Paris. *Bull. Int. Ass. Eng. Geol.* **24**, 111–123.

Quigley, R.M. and Vogan, R.W. (1970) Black shale heaving at Ottawa, Canada. *Can. Geotech. J.* **7**, 106–112.

Quigley, R.M., Zajic, J.E., McKeyes, E. and Yong, R.N. (1973) Biochemical alteration and heave of black shale; detailed observations and interpretations. *Can. J. Earth Sci.* **10**, 1005–1015.

Ramachandran, V.S., Sereda, P.J. and Feldman, R.F. (1964) Mechanisms of hydration of calcium oxide. *Nature* **201**, 288–289.

Redfield, R.C. (1968) Brantley reservoir site—an investigation of evaporite and carbonate facies. *Bull. Am. Eng. Geol.* **6**, 14–30.

Rice, S.M.M. (1983) The influence of evaporites on the planning and construction of civil engineering projects in the Middle East. In *Proc. meeting, Engineering Group of the Geological Society London 11th January.*

Richards, B.J. (1967) Moisture flow and equilibria in unsaturated soils for shallow foundations. In *Permeability and Capillarity of Soils*, American Society for Testing and Materials, Philadelphia, Pennsylvania, pp. 4–34.

Riley, J.P. and Chester, R. (1971) *Introduction to Marine Geochemistry.* Academic Press, London and New York.

Russel, D.J. and Parker, A. (1979) Geotechnical, mineralogical and chemical interrelationships in weathering profiles of an over-consolidated clay. *Q. J. Engineering Geology*, **12**, 107–116.

Segonzac, G.D. de (1970) The transformation of clay minerals during diagenesis and low grade metamorphism: a review. *Sedimentology* **15**, 281–346.

Serrano, A.A., Oteo, C., Dapena, E. and Viñas, J.M. (1981) Analysis of swelling phenomena in a gypsum-marl formation. *Proc. 10th Int. Conf. Soil Mech. & Found. Engng., Stockholm*, Vol. 1, pp. 779–784.

Shamburger, J.H., Patrick, D.M. and Lutton, R.J. (1975) Design and construction of compacted shale embankments. Vol. 1. Survey of problem areas and current practices. *Report No. FHWA-RD-75-61 Prepared for Federal Highway Administration, Washnigton, D.C.* (N.T.I.S. Springfield, Virginia).

Shaw, H.F. (1981) Mineralogy and petrology of the argillaceous sedimentary rocks of the U.K. *Q. J. Engineering Geology*, **14**, 277–290.

Shearman, D.J. (1963) Recent anhydrite, gypsum, dolomite and halite from the coastal flats of the Arabian shore of the Persian Gulf. *Proc. Geol. Soc. Lond.* **1607** (2nd July 1963).

Shearman, D.J., Mossop, G., Dunsmore, H. and Martin, H. (1973) Origin of gypsum veins by hydraulic fracture. *Trans. Inst. Mining. Met.* **82**, Section B, B66–B67.

Sherrell, F.W. (1979) Engineering properties and performance of clay fills. In *Clay Fills*, Institution of Civil Engineers, London, 241.

Smith, D.B. and Francis, E.A. (1967) *Geology of the Country between Durham and West Hartlepool*. Memoir of the Geological Survey of Great Britian, H.M.S.O., London.

Smith, T.J. (1978) Consolidation and other geotechnical properties of shales with respect to age and composition. Unpublished Ph.D. Thesis, Univ. Durham.

Spanovich,M. and Fewell, R. (1970) Building damage from expansive steel back-fill. *J. Soil Mechanics and Foundation. Div. ASCE*, **96**, 1808–1810.

Spears, D.A. and Amin, M.A. (1981) Geochemistry and mineralogy of marine and non-marine Namurian black shales from the Tansley borehole, Derbyshire, *Sedimentology* **28**, 407–417.

Sperling, C.H.B. and Cooke, R.V. (1980) Salt weathering in arid environments. Part I. Theoretical considerations. *Papers in Geography* No. 8, Bedford College, London.

Stephen, I. and MacEwan, D.M.C. (1950) Swelling chlorite. *Géotechnique* **2**, 82–83.

Taylor, R.K. (1978) Properties of mining wastes with respect to foundations. *In* Bell, F.G. (ed.) *Foundation Engineering in Difficult Ground*, Newnes–Butterworths, London, pp. 175–203.

Taylor, R.K. (1984) *Composition and Engineering Properties of British Colliery Discards*. National Coal Board (in press).

Taylor, R.K. and Hardy, R.G. (1974) Sulphate species in colliery spoil banks. *Trans. Inst. Min. Metallurgy*, Section A, **83**, A123–A126.

Tobin, M.P. (1980) Factors influencing road design, construction and maintenance in the United Arab Emirates. *Proc. Inst. Civ. Engrs.* **68** (1) 27–38.

Tomlinson, M.J. (1978) Middle East—Highway and airfield pavements. *Q. J. Engineering Geology London* **11**, 65–73.

Vear, A. and Curtis, C.D. (1981) A quantitative evaluation of pyrite weathering. *Earth Surface Processes and Landforms* **6**, 191–198.

Verwey, E.J.W. and Overbeek, J.Th.G. (1948) *Theory of the Stability of Lyophobic Colloids*. Elsevier, Amsterdam.

Wahlstrom, E.F. (1974). *Dams, Dam Formations and Reservoir Sites*. Elsevier, Amsterdam.

Walters, R.C.S. (1962) *Dam Geology*. Butterworths, London.

Ward, W.H., Samuels, S.C. and Butler, M.E. (1959) Further studies of the properties of London clay. *Géotechnique* **9**, 33.

Warkentin, B.P. and Bozozuk, M. (1961) Shrinking and swelling properties of two Canadian clays. *Proc. 5th Internat. Conf. Soil Mech. Found. Eng. Paris*, **3A**, 851.

Weaver, C.E. and Beck, K.S. (1971) Clay water diagenesis during burial: How mud becomes gneiss. *Sp. Pap. Geol. Soc. Am.* **134**.

Wellman, H.W. and Wilson, A.T. (1965) Salt weathering: a neglected erosive agent in coastal and arid environments. *Nature* **205**, 1097–1098.

Williams, A.A.B. and Donaldson, G.W. (1980) Building on expansive soils in South Africa: 1973–1980. *Proc. 4th Int. Conf. Expan. Soils, Denver*, ASCE, pp. 834–844.

Winkler, E.M. and Singer, P.C. (1972) Crystallization pressure of salts in stone and concrete. *Bull Geol. Soc. Am.* **83**, 3509–3513.

Winkler, E.M. and Wilhelm, E.J. (1970) Salt burst by hydration pressures in architectural stone in urban atmospheres. *Bull. Geol. Soc. Am.* **81**, 567–572.

Winmill, T.F. (1915) Atmospheric oxidation of iron pyrites. *Part IV. Trans. Instn. Mining Engrs.* **51**, 500.

Wohnlich, M. (1976) Sulphate rocks in Switzerland. *Bull. Int. Ass. Eng. Geol.* **13**, 39–43.

Yong, R.N. and Warkentin, B.P. (1966) *Introduction to Soil Behaviour*. Macmillan Co., New York.

Zen, E–An (1965) Solubility measurement in the system $CaSO_4$-NaCl-H_2O at 35°, 50° and 70°C and one atmosphere pressure. *J. Petrol.* **6**, 124–164.

10 The effects of clay soil volume changes on low-rise buildings

R. DRISCOLL

10.1 Introduction

Clay soils suffer volume changes as their water content varies. The most important result of this is the effect of ground movements on man-made structures, particularly low-rise buildings (and especially dwellings).

The magnitude of the damage done to these various structures can be seen with reference to the problem in the U.S.A. where it has been estimated that the volume change of clay soils produces an annual repair bill of $8 billion (Earl Jones, 1979); this figure makes 'expansive clay' the largest annual natural 'disaster' in North America. British experience of problems deriving from these soils is far less detrimental for reasons that will become evident though, periodically, events occur that have costly repercussions. The most recent of these was the drought of 1975/76, the severest ever recorded in Britain (Royal Society, 1978), which occurred only four years after the introduction by British insurance companies of indemnity of privately-owned homes against damage arising from 'subsidence' of the ground. Claims made on 15 major insurance companies rose from 164 in 1971 to 20 922 in 1976, falling to 7454 in 1979. It has been estimated (Reece, 1980) that the total cost of subsidence damage to dwellings for the period 1971–1980 was £250 million (1980 prices), the majority of which resulted from the drought.

Soil-structure interaction, from whatever cause, is a complex combination of ground movement causing structural response, which in turn affects ground movement. In the case of movements resulting from clay shrinkage and swelling the interaction is especially complicated by the fact that ground movements do not result from applied foundation loadings, so that the stress changes that occur in the soil (and hence volume changes) are difficult to predict with the confidence usually ascribed to load-settlement calculations. Furthermore, prediction of structural response to shrinkage or swelling is complicated by the wide variety of structural forms, ages and materials incorporated in the housing stock in Britain especially.

10.2 Properties of expansive clays and factors affecting volume change

It is the size and molecular composition of clay particles that impart to clay its volume change capabilities. The small (customarily $< 2 \mu m$), plate-like clay particles have very large specific surfaces (surface area/volume) and are, therefore, able to adsorb large quantities of water on these surfaces.

L

Table 10.1 Mineralogy of some British clay soils (after Avery and Bullock, 1977).

Origin	Geology	Sample size	No. of sites	Sample depth below ground level (mm)	Clay fraction (% < 2 μm)	CEC* (me/100 g)	Ratio of clay** minerals
Eocene	London Clay	13	10	200–950	60 (8.9)	56(4.7)	1.1 :2.4 :6
"	Reading Clay	6	3	250–680	58 (10.6)	54(6.7)	1.3 :2 :6.7
Cretaceous	Gault Clay	8	6	180–610	57 (4.8)	53(5.9)	1.25:2 :6.5
Jurassic	Kimmeridge Clay	6	5	100–480	63 (12)	48(10)	1.4 :2.3 :5.3
"	Oxford Clay	10	7	130–790	63 (8.2)	55(11)	1.4 :2.5 :6.3
"	Upper Lias Clay	6	3	130–610	61	46	4.5 :2 :2.5
"	Middle Lias Clay	6	4	180–710	51 (7.6)	46(4.1)	3 :2 :1+
"	Lower Lias Clay	17	11	100–780	60 (8.4)	50(5.3)	1.7 :2.8 :4+
Triassic	Red Keuper Marl	13	11	130–580	52 (10)	38(7.7)	0 :6 :0+
Permian	Permian Marl	6	6	130–830	54.5(12)	35(7)	1 :6 :0+
Carboniferous	Various shales	12	11	230–760	51 (7)	30(7)	2 :6 :0.6+
Devonian	Various sandstone marls and shales	11	9	180–760	43 (5)	31(8)	0 :5 :1.5+

Notes: * Total cation exchange capacity
** Ratios of kaolinite: illite: mica-smectite Avery and Bullock did not identify any pure smectite (montmorillonite)
+ Traces of other minerals
(Standard deviations given in brackets)

It is demonstrated in chapter 9 that certain clay minerals, especially montmorillonite, a member of the smectite group, are also, by virtue of weak intermolecular bonding, able to hold water within the sheets of molecules forming the particles. Electrochemical forces act to attract water molecules to the clay particle surfaces, and combine with capillary forces in the small voids within the soil structure to set up suctions in the soil water. The removal of this water increases these suctions, so increasing effective stresses and strength but reducing volume. Conversely, the supply of water at low suction causes it to be drawn into the soil as water moves from zones of low suction to zones of high suction. This results in reduced suction, lower effective stress and strength but increased volume. The attraction of water molecules to particle surfaces is influenced by the capacity of the clay mineral to exchange positively charged ions (cations) with cations in the water. This 'cation exchange capacity' (CEC) may be used as a measure of volume change potential. Avery and Bullock (1977) have analysed a range of soil samples taken from shallow depth in some typical British clays. Their results, shown in Table 10.1, include the mineralogy of the soils, which consist largely of a mixture of kaolinite, illite and mica-smectite. It can be seen from the table that those clays with the mica-smectite mineral predominating have the larger CEC values. The geologically more recent Eocene clays are shown to have the greatest expansive potential. Weaver and Pollard (1973) suggested that CEC values greater than 10–15 me/100 g indicate some expansive potential. It can, therefore, be seen that on this basis all the soils described in Table 10.1 possess some potential expansion.

A far more practical means of assessing expansive potential is to relate it to the clay soil Atterberg limits. These properties have been widely used to identify expansive soils, for example in the U.S.A. by Holtz and Gibbs (1957) and in South Africa by Williams and Donaldson (1980), as shown in Fig. 10.1. The Building Research Establishment (1980)

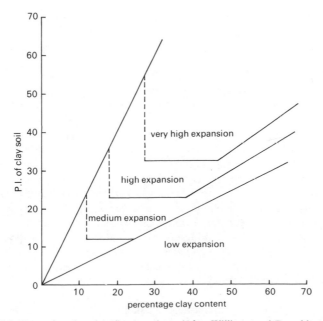

Figure 10.1 Expansive clay classification chart. (After Williams and Donaldson, 1980).

Table 10.2 Clay shrinkage potential.

Plasticity Index %	Clay fraction %	Shrinkage potential
> 35	> 95	Very high
22–48	60–95	High
12–32	30–60	Medium
> 18	> 30	Low

Table 10.3 Shrinkage potential of some common clays.

Clay type	Plasticity Index %	Clay fraction %	Shrinkage potential
London	28	65	Medium/high
London	52	60	Very high
Weald	43	62	High
Kimmeridge	53	67	High/very high
Boulder	32	—	Medium
Oxford	41	56	High
Reading	72	—	Very high
Gault	60	59	Very high
Gault	68	69	Very high
Lower Lias	31	—	Medium
Clayey silt	11	19	Low

has published a similar classification which has been reproduced in Table 10.2. The potential of some British clays, assessed using this Table, is shown in Table 10.3.

As has already been stated, the principal variable influencing actual soil volume change is the change in moisture suction. Large suctions can only be generated in the highly impermeable clay soils and then only by the large stresses induced by water evaporation from the soil surface. The cracked, dry appearance of clay ground in summer weather is well known but evaporation from bare clay surfaces is quickly curtailed by the decreased permeability of the surface of micro-cracks. This effect serves to limit the depth of influence of the climate, in the absence of vegetation, since the onset of winter rain terminates the process. It is only in climates comprising long, dry summers with limited winter rainfall that permanent drying of the ground can extend to considerable depths. In Britain, it is the roots of plants that conduct moisture from greater depths, to be transpired through leaf surfaces to the atmosphere. Some large, broad-leaved trees can transpire daily considerable volumes of water at the height of summer and it has been estimated that only when suctions of about 15 atmospheres (1.5 MN/m^2) are necessary to remove soil moisture will wilting and the cessation of transpiration commence. Unfortunately, as yet no readily available means exist for practical engineering use to measure *in-situ* soil moisture suction with good accuracy, especially at higher suctions. Many research devices exist to measure suctions in small laboratory samples; some, especially the thermocouple psychrometer may be used to measure *in-situ* suction. However, the psychrometer cannot measure suction below 1 atmosphere and cannot achieve a better resolution than 1 atmosphere. The various techniques for measuring

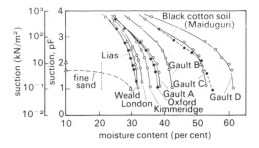

Figure 10.2 Relationship between suction and moisture content for undisturbed, expansive clay soils. (After Croney, 1977).

suction have been reviewed by Johnson (1977). Croney (1977) has published 'characteristic' curves of soil moisture content versus suction; these are shown in Fig. 10.2 for some undisturbed samples of the more expansive British clays. Also shown is a curve for fine sand that indicates an inability to sustain high suction and, hence, volume changes. The suction values are expressed in terms of 'pF', where pF is the logarithm to base 10 of the equivalent height of water in centimetres.

10.3   Prediction of ground movements and tree effects

It is important to identify three distinct circumstances in which volume change ground movements occur in the U.K. First, there is the 'open' clay site that is unaffected by large vegetation, usually trees. Secondly, there is the site where trees are growing close by, and, thirdly, the site where trees have been removed. Obviously the first is the simplest. Ward (1953) showed clearly that, under grass, ground movements extend to a depth of about 1.5 m. Movements at 1.0 m were felt to be sufficiently small to be tolerated by most structures. Ward developed the 'water shrinkage factor' ($\Delta w/\Delta z$); Δw is the summer soil moisture deficit (SMD) expressed, like rainfall, in units of length and derived, in this case, by integrating volumetric moisture contents in the moisture deficient profile, when compared to the profile at winter capacity; Δz is the surface movement. He measured $\Delta w/\Delta z$ values of about 4 for various sites on highly expansive clays. In the ideal case of a homogeneous, isotropic, saturated soil, a minimum value of $\Delta w/\Delta z = 3$ would apply. This value would give a pessimistic estimate of surface movement.

Values of SMD at various locations in Great Britain are published at two-weekly intervals throughout the summer by the Meteorological Office (1979). The estimated values of SMD are calculated from theoretical and empirical models of the evapotranspirational process developed by Penman (1963). The distribution of peak potential SMD in Britain is shown in Fig. 10.3. Using a water shrinkage factor of 3 would indicate an upper bound value of about 50 mm for ground surface movement in the S.E. of England, the area in which the most expansive soils occur. Ward (1953) measured surface movements at three different sites ranging from 41 to 58 mm; surface heave in excess of 35 mm was recorded at the Building Research Station (Driscoll, 1979) in the winter of 1976.

The discussion of the 'open' site has so far concerned only British ground and climatic conditions. The position in countries that experience greater extremes of climate is

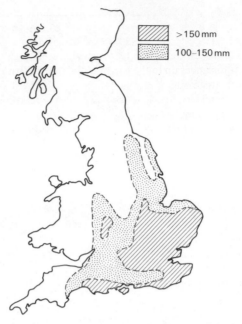

Figure 10.3 Distribution of peak summer potential soil moisture deficit (average for 1956–1975).

somewhat different. Much research on the behaviour of expansive clays has been carried out in the U.S.A., Australia, Israel, South Africa and many other countries. The main physical feature common to parts of these countries, but not to Britain, is that deep, desiccated ground profiles develop over many years in climates that comprise long dry summers followed by winters of insufficient rainfall to saturate the profile. The subsequent sealing of the ground surface by constructing roads, buildings and other structures allows moisture to accumulate beneath the covered area and results in a long-term heave of ground. Research in these countries has concentrated primarily on estimating heave from laboratory oedometer tests on 'undisturbed' samples (see for example O'Neill and Poormoayed, 1980). Tests of this type rely on assumptions of both a saturated final moisture content and one-dimensional heave that lead to overestimates of actual heave. More measurements of final *in-situ* suctions (and moisture contents) are required to improve predictions.

Since Britain does not experience sufficiently long, dry summers for deep profiles of permanently desiccated soil to develop except near large trees, the major uncertainty is the influence of such trees both in spatial extent and in the magnitude of desiccation. The case of existing trees influencing existing buildings probably presents the greatest uncertainty, whereas when trees have been removed before construction future tree influence is removed and positive building design measures can be taken.

The substantial effect that trees can have in increasing moisture deficits was first demonstrated in Britain by Ward (1947). Moisture content profiles in the soil both close to and distant from elm and poplar trees established that desiccation had occurred close to the trees. The differences between these profiles corresponded to an SMD of about 300 mm, suggesting a potential surface movement of 75–100 mm ($3 < \Delta w/\Delta z < 4$).

Further observations by Ward (1953) concerned the distance of trees (mostly elm and poplars) from buildings which had been damaged by movements of foundations undermined by tree roots. From his measurements, Ward concluded that, on the highly shrinkable clays, trees which transpired large quantities of moisture in summer could damage buildings at distances up to the height of the tree. In Britain doubts about the relationship between tree roots and foundation movements have generated much controversy involving builders, regulatory authorities, planners and architects. The distancing of trees from buildings on new developments, using the '1H' rule-of-thumb, is considered by many to be aesthetically unacceptable (Flora, 1978), while there has been a reluctance by builders to install deeper foundations. Additionally, allegations of damage to existing buildings by the root action of trees growing on neighbouring land have generated many cases of litigation. Recent attempts to produce more information on tree-root behaviour have produced only limited results. Reynolds (1979) conducted a review of world literature on tree roots but was unable to improve on Ward's conclusions. Reynolds recommended research to study moisture content changes and ground movements at various distances from selected trees on various clay soils. A limited study is in progress.

In another study, Cutler and Richardson (1981) analysed data collected at the Royal Botanical Gardens, Kew, over the period 1971–1979. The data, which were obtained from questionnaires sent to customers who had submitted tree roots for identification, yielded information on the distance from the tree to the damaged building, tree height, soil description, root size and other factors. Unfortunately, no information on depth of rooting or degrees of damage sustained was obtained. The results of the analysis are given for each tree species as the percentage of damage cases recorded for various distances between tree and damaged building. Table 10.4 has been taken from these results and shows a ranking of species in descending order of the incidence of the type of tree in damage cases. The table also shows the likely maximum height of each species for urban

Table 10.4 Tree species damage ranking.

Ranking*	Species	Max. tree height (m)	Max. distance recorded (m)	Max. distance for 75% of cases (m)	Max. distance for 90% of cases (m)
1	Oak	16–23	30	13	18
2	Poplar	25	30	15	20
3	Lime	16–24	20	8	11
4	Common ash	23	21	10	13
5	Plane	25–30	15	7.5	10
6	Willow	15	40	11	18
7	Elm	20–25	25	12	19
8	Hawthorn	10	11.5	7	9
9	Maple/sycamore	17–24	20	9	12
10	Cherry/plum	8	11	6	7.5
11	Beech	20	15	9	11
12	Birch	12–14	10	7	8
13	White beam/rowan	8–12	11	9.5	11
14	*Cupressus macrocarpa*	18–25	20	3.5	5

*The ranking has been achieved by weighting the number of species-cases recorded by the ratio of species percentage of the sample to estimated species percentage of the total relevant tree population.

conditions on clay soil, the maximum recorded distance at which damage was attributed to tree root action and the distances within which 75 and 90 per cent of cases occurred. It is debatable whether the ranking should be on the basis chosen rather than, say, on the distance within which 75 per cent or 90 per cent of cases occurred.

Various alternative tree ranking in descending order of damaging influence have been compared by Biddle (1979). Culter and Richardson's analysis is based on 2600 data cards which makes their sample by far the largest. It could be argued that to preclude 75 per cent of cases of damage caused by root action is more realistic than total preclusion and Table 10.4 suggests that this could be achieved by specifying a minimum distance between tree and building of about half the expected tree height. This specification appears to work well except for rowan/whitebeam, cherry/plum and willow. It is evident that Ward's original rule-of-thumb is appropriate if all cases are to be excluded, except for willow and poplar.

The reverse of the tree-induced subsidence problem, ground heave following tree removal, has been documented in two cases: Samuels and Cheney (1975), and Cheney and Burford (1975). In the former, investigation of damage to a terrace of single-storey houses on 1.2 m strip foundations in London Clay (PI: 45–60%; % < 2 μm: 60–70) revealed that large elm trees had been felled almost directly beneath the building before construction. Observations indicated that heave continued for about 20 years, with total movements estimated to be about 100 mm. Desiccation close to the sites of the trees was detected to a depth of about 5 m. Swelling was predicted both from assuming that a desiccated soil moisture profile close to tree positions attained an unaffected profile and from oedometer swelling tests. In these tests on carefully prepared samples, swelling pressures were first measured at constant volume and then progressively reduced, allowing the samples to swell to equilibrium under total overburden pressure. Both methods gave predictions ranging from 65 to 70 mm.

In the other case, Cheney and Burford monitored the heave movements of a three-storey office block built on 1.3 m deep strip footings in Boulder Clay (PI: 22–29%). Large oak trees had been removed from the immediate vicinity before construction. A maximum heave of 22 mm was recorded at one location 8 years after felling, causing minor cracking.

In view of the variability of clay behaviour, differences in hydrological conditions between individual sites and differences in tree species behaviour, it is difficult to offer comprehensive guidance on the likelihood of future damaging tree root action and the extent and degree of existing root action without recourse to specific soil laboratory tests or complex *in-situ* suction measurements which are, at least in the house construction industry, outside normal site investigation practice. In the case of existing tree influence the problem becomes one of trying to determine the magnitude of any desiccation from readily obtainable soils data, in the absence of *in-situ* suction measurements.

The suction versus moisture content curves shown in Fig. 10.2, which are for the soils the properties of which are shown in Table 10.5, offer a simple, if perhaps crude, practical means of assessing the likelihood of desiccation in a soil. Desiccation could be presumed to commence where the rate of change of moisture content, w, (and hence volume) with increasing suction increases noticeably; this point could be taken to be approximately at a suction of $pF = 2 (10 \, kN/m^2)$. Similarly, significant desiccation could be taken to have occurred when the suction is sufficiently large that on its release a low-rise building could be lifted. Typically, this value would be about $pF = 3 (100 \, kN/m^2)$. A plot of values of w at these suctions, taken from Fig. 10.2, against Liquid Limit (LL) is shown in Fig. 10.4. It is

Table 10.5 Index properties for expansive clays of figure 10.2.

Clay	Plastic limit (per cent)	Liquid limit (per cent)	Plasticity index (per cent)	w (pF = 2)	w (pF = 3)
Lias	24	60	36	30	26
Weald	25	68	43	31	27
London	26	78	52	32	28
Kimmeridge	24	77	53	34	31
Oxford	24	72	48	36	31
Gault A	25	70	45	37	31
Gault B	29	81	52	41	36
Gault C	34	102	68	51	43
Cotton soil	38	82	44	51	41
Gault D	32	121	89	59	50

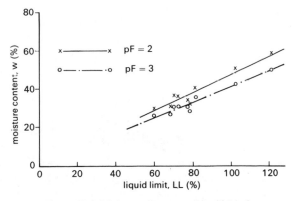

Figure 10.4 Moisture content v. Liquid Limit.

preferable to use LL rather than PL (Plastic Limit) or PI (Plasticity Index) since PL (and hence PI) values depend on a less reliable test procedure than that for determining LL. Figure 10.4 suggests the following relationship:

$$w = 0.5\,LL\,(pF = 2) \tag{10.1}$$

$$w = 0.4\,LL\,(pF = 3) \tag{10.2}$$

Concern about the possible influence of tree roots on clay soils has resulted in extreme caution being exercised, especially where trees are growing at distances from proposed buildings of less than expected maximum tree height. The first priority when

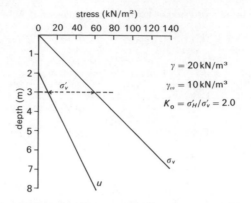

Figure 10.5 Total vertical stress and pore-water pressure distribution.

investigating such sites should be to demonstrate the existence of tree roots. However, the fact that they exist does not guarantee significant desiccation. Reynolds (1979) has suggested that sections of tree root systems may lie dormant until conditions requiring moisture supply in excess of normal activate these sections of the root system. If specific swelling tests are conducted as part of a site investigation some care is required in the interpretation of the results since a measurement of sample swelling need not mean that heave of the ground will occur. Many of the stiff clay deposits in the S.E. of England are overconsolidated and possess *in-situ* horizontal stresses which, especially near the surface, greatly exceed the vertical stresses. Figure 10.5 shows the distribution of total vertical stress (σ_v) with depth below ground level; also shown is a hydrostatic pore-water pressure distribution (u) below a water table at 2 m from the ground surface. Suppose an oedometer sample is taken from a depth of 1 m below the water table, so that no desiccation can exist in the sample. For the soil properties listed in Fig. 10.5 (which are typical of London Clay, for example), at the sample depth of 3 m:

$$\sigma'_v = 60 - 10 = 50 \, \text{kN/m}^2,$$

$$\sigma'_h = 2.0 \times \sigma'_v = 100 \, \text{kN/m}^2.$$

On sampling, the suction (u_s) initially set up in the sample due to the release of *in-situ* stresses will be approximately equal to $(\sigma'_v + 2\sigma'_h)/3$, $u_s \simeq 83 \, \text{kN/m}^2$. If the oedometer swelling test is conducted by applying the total overburden pressure $\sigma_v (= 60 \, \text{kN/m}^2$ in this example) then sample swelling will occur because u_s exceeds σ_v, despite the presence of a saturated sample. The author has come across just such an example where expensive foundations were designed to resist heave identified from sample swelling.

10.4 The response of low-rise buildings

In view of the large conurbations that exist in the S.E. of England where both the soils of greatest volume change potential and greatest summer soil moisture deficits occur, there has been a long history of foundation movement and damage in this area. The Building Research Station has been examining such cases since the nineteen thirties. Bartholomew (1841) refers to damage ocurring to buildings during drought. Pryke (1975) cites many

Table 10.6 Classification of visible damage to walls with particular reference to ease of repair of plaster and brickwork or masonry (after BRE, 1981).

Category of damage	Degree[1] of damage	Description of typical damage	Approximate crack width mm
0	Negligible	Hairline cracks of less than about 0.1 mm width are classed as negligible	Up to 0.1[2]
1	Very slight	Fine cracks which can easily be treated during normal decoration. Perhaps isolated slight fracturing in building. Cracks rarely visible in external brickwork.	Up to 1[2]
2	Slight	Cracks easily filled. Re-decoration probably required. Recurrent cracks can be masked by suitable linings. Cracks not necessarily visible externally; some external repointing may be required to ensure weathertightness. Doors and windows may stick slightly.	Up to 5[2]
3	Moderate	The cracks require some opening up and can be patched by a mason. Repointing of external brickwork and possibly a small amount of brickwork to be replaced. Doors and windows sticking. Service pipes may fracture. Weathertightness often impaired.	5 to 15[2] (or a number of cracks up to 3)
4	Severe	Extensive repair work involving breaking-out and replacing sections of walls, especially over doors and windows. Window and door frames distorted, floor, slopping noticeably[3] Walls leaning[3] or bulging noticeably, some loss of bearing in beams. Service pipes disrupted.	15 to 25[2] but also depends on number of cracks
5	Very severe	This requires a major repair job involving partial or complete re-building. Beams lose bearing, walls lean badly and require shoring. Windows broken with distortion. Danger of instability.	usually greater than 25[2] but depends on number of cracks

Notes: 1. It must be emphasized that in assessing the degree of damage account must be taken of the location in the building or structure where it occurs, and also of the function of the building or structure.
2. Crack width is one factor in assessing category of damage and should not be used on its own as a direct measure of it.
3. Local deviation of slope, from the horizontal or vertical, of more than 1/100 will normally be clearly visible. Overall deviations in excess of 1/150 are undesirable.

cases of house damage in S.E. England. Unfortunately, it is only recently, following the 1975/76 drought, that an objective method of assessing the scale of building damage has been adopted (Tomlinson *et al.*, 1978). This assessment utilizes a damage classification which is reproduced in Table 10.6. The author has applied it to many low-rise masonry buildings and has concluded that about 75 per cent fell within and below Category 2, 20 per cent in Category 3 and the remainder in Categories 4 and 5. In most cases trees could have been involved though their contribution was rarely fully substantiated. Although it is not possible to be sure, it is to be expected that the proportions of damage severity experienced following the drought reflect the general levels which have occurred in past dry spells of weather. When considering the scales of damage that occur, it is necessary to consider again the monetary cost of the drought of 1975/76. The average repair bill for a case of subsidence damage has been estimated (Reece, 1980) at about £4000 at 1980 prices. This high figure reflects the tendency on the part of those investigating the cases (rarely geotechnical specialists) to specify expensive underpinning of foundations. In many cases little or no attempt was made to determine causes of damage or details of the ground conditions. An attitude appears to have built up amongst owners and those involved in the buying, selling and insuring of houses in which underpinning is regarded as a panacea. In most cases the underpinning has been applied mainly to that part of the foundation which has been felt to have moved and the author has encountered numerous cases where the stabilizing of only a part of building, all of which was probably moving slightly, has resulted in new, minor cracking elsewhere.

In an attempt to improve the standard of both assessment and diagnosis of low-rise building damage due to foundation movement, the Building Research Establishment (BRE, 1981) published guidance on the detection of building distortion and movement, and on the many reasons for minor cracks in masonry structures. Emphasis is placed on identifying conditions leading to progressive movement, such as heave following tree removal on clay soils, which could lead to increasing damage and for which the greater expense of a more thorough investigation and more comprehensive remedies is justified.

Probably the most difficult aspect of the interaction of structures with the ground is that of predicting how much distortion the structure can accept before both the function of the building and its structural integrity are adversely affected. Various authors have attempted to identify the onset of structural cracking in terms of movement. Burland and Wroth (1975) have discussed these attempts and have proposed the 'deflection ratio' (defined in Fig. 10.6) as being the most appropriate parameter to consider. Their work clearly identified the different responses of a building to either a sagging or a hogging mode of distortion, with the latter being more critical since bending tensile strains at the top of the building are less restrained than are sagging strains owing to the tying effect of the foundations. For typical house length/height ratios of 1–2 Burland and Wroth suggest that, to prevent the onset of cracking, Δ/L should not exceed $0.4 - 0.5 \times 10^{-3}$ for load-bearing walls in a sagging mode, and 0.25×10^{-3} in a hogging mode. For a 10 m-long wall these values represent deflections of about 4–5 mm in sag and 2.5 mm in hog. These values seem at first sight to be excessively small but it should be borne in mind that they apply to walls without openings. The author has observed 2.5 mm-wide cracks in a gable wall where a hogging distortion of $\Delta/L = 0.4 \times 10^{-1}$ was measured.

Until such time as a large body of data on low-rise masonry building distortions is established it will not be possible to offer more confident guidance on allowable distortions. It can, however, be stated that the structural response to any movement depends on the shape, flexibility and ductility of the structure and these factors vary enormously across the building population. Small building protuberances such as bay

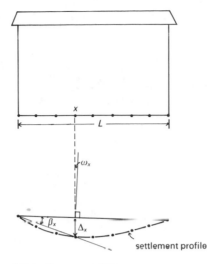

Figure 10.6 Definitions of distortion parameters.
ω = tilt.
$\beta \equiv \delta/l$ = 'angular distortion'[1].
  = 'angular deformation'[2].
  = 'relative rotation'[3].
Δ/L = 'deflection ratio'[3].
Δ = 'relative deflection'[3]–vertical displacement relative to line connecting two points L apart.
1, Skempton and MacDonald (1956); 2, Bozozuk (1962); 3, Burland and Wroth (1975).

windows and extensions are known to be especially vulnerable to foundation movement, while the brittleness of the hard mortars and bricks generally used in modern construction causes a sensitivity to small movements which is much less in properties constructed before World War II in softer bricks and ductile lime mortars. Approximately 25 per cent of all new house construction in Britain is now in timber-frame. While the loadbearing frame is inherently more flexible and ductile than lightweight concrete blockwork, the British insist on brick exterior finishing which is brittle. Consequently the potential scope to relax limiting distortion criteria in timberframe construction may be negated by the need to minimize cracking in the cladding.

The position outside the U.K. in terms of guidance on structural response offers little if any additional help. In Canada, Bozozuk (1962) published a damage classification comprising four categories in terms of crack width and differential settlement (Table 10.7a). He also reported data on distortions, choosing the term 'angular deformation' (δ/l) (*see* Fig. 10.6) to assess the performance of 77 houses in Ottawa. The results of this study are shown in Table 10.7(b) indicating quite large average differential settlements. It can be shown (Burland and Wroth, 1975) that by considering a structure to be idealized to an elastic beam deforming in both shear and bending about a neutral axis at mid-height, the maximum value of Δ/L can be related to δ/l by

$$\Delta/L = \frac{(\delta/l)}{3} \frac{1 + 3.9(H/L)^2}{1 + 2.6(H/L)^2} \quad \text{where } H/L \text{ is the height/length ratio}$$

Applying this equation to Bozozuk's minimum δ/l value of 1/180 gives a Δ/L values

Table 10.7(a)   Damage classification by Bozozuk (1962).

Degree of damage	Description
Slight	Some hairline cracks occurring in exterior brickwork
Moderate	Cracks in brickwork $\leq \frac{1}{8}$ in. wide
Heavy	Cracks in brickwork $\geq \frac{1}{4}$ in. wide; window and door frames distorted
Severe	Cracks ≤ 2 in.; badly distorted window and door frames; walls out of plumb.

Table 10.7(b)   Damage related to angular deformation

Degree of damage	No. of houses	Differential settlement		Angular deformation
		Max.	Average	
None	23	3	1.3	1/180
Slight	20	3.4	2.0	1/120
Moderate	17	5.3	3.1	1/90
Heavy	13	6.2	4.7	
Severe	4	14.0	9.4	1/50

(assuming $H/L = 0.7$) of about 2.3×10^{-3} which is 5–10 times the limiting values suggested by Burland and Wroth. This large difference may reflect the inclusion of tilt in Bozozuk's figures as well as the departure of his observed buildings from the idealized continuous walls of Burland and Wroth.

10.5   Foundation and structural design to minimize damage

In the U.K. foundation standards were greatly improved from the 1950s onwards following work at the Building Research Station (1949). It is apparent (Tomlinson et al., 1978) that the minimum depth requirement of 1 m identified in the Building Research Station work afforded satisfactory protection during the drought of 1975/76 to most of the houses so constructed; that is, where tree influence was absent. It is certainly the case that the majority of properties examined by the author following the drought possessed foundations less deep than 1 m. In view of the relatively minor levels of damage that occurred in 70 per cent of these cases it is quite probable that modern 1 m-depth foundations will continue to perform satisfactorily unless tree effects become substantial. Where trees are known to present a real threat to new buildings when constructing close to them, the recommendation (Building Research Station, 1949) has always been to use deep foundations comprising small diameter piles with capping beams (Ward and Green, 1952). In the case of ground heave following tree removal, piles should be reinforced for possible tensile stresses, precast beams suspended or cast *in-situ* on compressible or collapsible materials and floors suspended, to accommodate the heave (Tomlinson et al., 1978).

Abroad, especially in Australia (Walsh, 1975) and the U.S.A. (B.R.A.B., 1968), considerable resources have been applied to the design of reinforced rafts to resist the distortions arising from the long-term heaving of deeply-desiccated clays. No allowance is made for superstructure stiffening and it would appear that these rafts may generally be more expensive than the deeper pile foundations.

Figure 10.7 Hogging damage arising from corner subsidence due to tree root activity (Crown copyright).

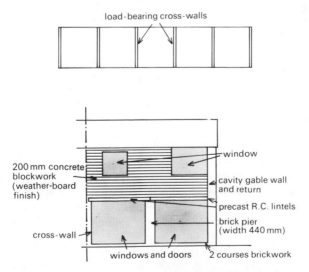

Figure 10.8 (Top) plan of terrace housing in 'cross-wall' construction; (bottom) rear elevation of end-of-terrace house illustrating lack of torsional and bending stiffness.

Choice of the most appropriate combination of foundation and superstructure design depends on the mode of distortion anticipated. Usually this is one of hogging where either tree activity causes corner-down subsidence (Fig. 10.7) or long-term ground swelling causes centre-heave. In these cases it is probable that stiffness is better incorporated near the top of the building than in the ground. While there has been some use of brick reinforcement in Britain to resist damage due to mining subsidence there does not appear to be any such use on clay soils where hogging movements are possible. In Britain much new public housing and cheaper private housing takes the form of terraces in 'cross-wall' construction (Fig. 10.8a) in which load-bearing walls run transverse to the length of the terrace. The usual specification of large openings on the long sides makes the structure relatively flexible in both longitudinal bending and torsion. Figure 10.8b shows the large structural openings in the rear elevation of a dwelling in which substantial torsional distortions (local slopes of 1/60) occurred following differential heave of the gable wall after tree removal.

It is evident that the designer of a building on clay soil has a clear choice: either he provides sufficiently deep foundations to ensure insignificant building movements or he acknowledges some foundation movement and designs the structure to tolerate the movements. There is little experience, at least in the U.K., of this latter option.

10.6   Conclusions

U.K. experience indicates that clay soils can be broadly classified by the degree of volume change potential using simple index properties. However, accurate assessments of actual ground movements are difficult to make because the changes in soil moisture suction cannot practically be measured. Based largely on London Clay data, it is known that in the U.K. vertical surface movements can vary from about 50 to 100 mm depending on the proximity of trees. Away from tree influence, ground movements are insignificant at a depth of 1 m and are known to extend laterally to about the height of the tree, for the more damaging species at least. Evidence exists to suggest that restricting the distance between tree and building to half the tree height would eliminate 75 per cent of cases of tree-induced damage.

For design purposes it is important to be able to identify soil desiccation within the zone of potential tree influence. A simple, though crude, means to detect desiccation in U.K. overconsolidated clays is offered whereby moisture content is compared with a proportion of the liquid limit.

Limiting values of structural distortion to prevent cracking are suggested, depending on the mode of distortion and building dimensions. However, these make no allowance for structural openings or complex plan forms.

Foundation techniques exist for preventing movements in buildings. Whether the designer chooses to adopt these techniques or to design the structure to tolerate a conservative estimate of distortion will depend largely on economic considerations.

Levels of damage observed in adverse conditions in the U.K. indicate, at least in structural terms, that most low-rise buildings respond satisfactorily to clay volume changes. However, market sensitivity to low levels of damage, which are usually only of aesthetic significance, has dictated that disproportionate amounts of money are frequently spent on underpinning existing houses and that very conservative new foundation designs are often adopted for marginal conditions.

Notation

H/L structure height-to-length ratio

K_0 coefficient of earth pressure at rest

Δ/L deflection ratio

pF   soil moisture suction (log scale)

u positive soil pore-water pressure

u_s negative soil pore-water pressure (suction)

w natural soil moisture content

Δw soil moisture deficit

β angular distortion

δ/l angular deformation

Δz ground surface vertical movement

γ soil bulk density

γ_w density of water

σ'_h effective horizontal soil stress

σ_v total vertical soil stress

σ'_v effective vertical soil stress

ω tilt or rotation

References

Avery, B.A. and Bullock, P. (1977) Mineralogy of clayey soils in relation to soil classification. *Soil Surv. Tech. Monogr. No.* 10, Soil Survey, Harpenden.

Bartholomew, A (1841) *Specifications for Practical Architecture.* 1st edn., J. Williams & Co, London.

Biddle, P.G. (1979) Tree root damage to buildings; an arboriculturist's experience. *Arb. J.*, **3** (6), 397–412.

Bozozuk, M. (1962) Soil shrinkage damages shallow foundations at Ottawa, Canada. *Eng. J.*, **45** (7), 33–37.

Building Research Advisory Board (1968) *Criteria for selection and design of residential slabs-on-ground.* Nat Acad. Sci. Rep. No. 33 to Federal Housing Agency, Washington, D.C.

Building Research Establishment (1980) Low-rise buildings on shrinkable clay soils: Part 1. *BRE Digest* **240**, H.M.S.O., London.

Building Research Establishment (1981) Assessment of damage in low-rise buildings, with particular reference to progressive foundation movement. *BRE Digest* **251**, H.M.S.O., London.

Building Research Station (1949) *House Foundations on Shrinkable Clays. BRE Digest* **3**, H.M.S.O., London.

Burland, J.B. and Wroth, C.P. (1975) Settlement of structures and associated damage. *Proc. Conf. Settlement of Structures, Cambridge*, Pentech Press, London, pp. 611–654.

Cheney, J.E. and Burford, D. (1975) Damaging uplift to a three-storey office block constructed on a clay soil following the removal of trees. *Proc. Conf. Settlement of Structures, Cambridge*, Pentech Press, London, pp. 337–343.

Croney, D. (1977) *The Design and Performance of Road Pavements*. H.M.S.O., London.

Cutler, D.F. and Richardson, I.B.K. (1981) *Tree Roots and Buildings*. Construction Press.

Driscoll, R.M.C. (1979) *Response of simulated house foundations to ground movements in shrinkable clay soils*. Unpublished BRE note.

Earl Jones, D. (1979) The expansive soil problem. *Underground Space* **3** (5), 221–226.

Flora, T. (1978) Treeless towns? *J. Landscape Arch.* **121**, 10–12.

Holtz, W.G. and Gibbs, J.J. (1957) Engineering properties of expansive clays. *Trans. ASCE* **121**, 641–663.

Johnson, L.D. (1977) *Evaluation of laboratory suction tests for prediction of heave in foundation soils*. Tech Rep S-77-7, U.S. Army Engrs. Waterways Stn, Vicksburg, Miss.

Meteorological Office (1979) *Estimated Soil Moisture Deficit and Potential Evaporation over Great Britain—explanatory notes*. Met. Office, Bracknell.

O'Neill, M.W. and Poormoayed, N. (1980) Methodology for foundations on expansive clays. *J. Geotech. Eng. Div., ASCE*, **106** (GT12), 1345–1367.

Penman, H.L. (1963) *Vegetation and Hydrology*. Tech. Commn. No. 53, Commonwealth Bureau Soils, Harpenden.

Pryke, J.F.S. (1975) Differential foundation movement of domestic buildings in S.E. England. *Proc. Conf. Settlement of Structures, Cambridge*. Pentech Press, London, pp. 403–419.

Reece, R.A. (1980) Insurance: the cost of mistakes. *Proc. Sem. 'Trees in Relation to Construction'*. British Standards Institution, London, pp. 28–36.

Reynolds, E.R.C. (1979) *A Report on Tree Roots and Built Development*. Dept. Environment, H.M.S.O., London.

Royal Society (1978) Scientific aspects of the 1975/76 drought in England and Wales. *Proc. Roy. Soc. (London)*, **A.363**.

Samuels, S.G. and Cheney, J.E. (1975) Long-term heave of a building on clay due to tree removal. *Proc. Conf. Settlement of Structures, Cambridge*. Pentech Press, London, 212–220.

Skempton, A.W. and MacDonald, D.H. (1956) Allowable settlements of buildings. *Proc. Instn. Civ. Engrs.* **II** (5) 727–768.

Tomlinson, M.J., Driscoll, R.M.C. and Burland, J.B. (1978) Foundations for low-rise buildings. *Struct. Engr.* **56A** (6) 161–173.

Walsh, P.F. (1975) The design of residential slabs-on-ground. *Div. Bldg. Res. Tech. Paper No. 5*, CSIRO, Australia.

Ward, W.H. (1947) The effect of fast growing trees and shrubs on shallow foundations. *J. Land Archt.* **11**, 7–16.

Ward, W.H. (1953) Soil movement and weather. In *Proc. 3rd Int. Conf. Soil Mechanics and Foundation Eng.*, **4** (31), pp. 477–482.

Ward, W.H. and Green, H. (1952) House foundations: the short bored pile. In *Proc. Pub. Wks. Mun. Serv. Cong.*, Instn. Civ. Engrs., London.

Weaver, C.E. and Pollard, L.D. (1973) *The Chemistry of Clay Minerals (Developments in Sedimentology)*, **15** Elsevier, New York.

Williams, A.A.B. and Donaldson, G.W. (1980) Building on expansive soils in South Africa: 1973–1980. *Proc. 4th Int. Conf. Expansive Soils, ASCE, Denver*, 834–844.

11 Settlement and stability of embankments on soft subsoils

R. MURRAY and I. SYMONS

11.1 Introduction

The location of road embankments, flood control levees and earth dams on soft and compressible subsoil can cause serious problems of instability during construction and of long-term settlement during the life of the structure. Moreover, in recent years such problems have become more frequent because of changes in design requirements, economics and construction practice. For example, in highway engineering, considerations of traffic engineering and environmental factors have played an increasing role in route selection. In many areas high land costs and agricultural or environmental considerations have meant that the only acceptable location for new roads is on ground where deposits of soft fine-grained or organic soils or even man-made fills exist to considerable depth. For all types of earthwork on soft foundations these problems have been exacerbated by the more rapid methods of mass earthmoving now employed.

At the design stage the engineering risk associated with these problems can only be minimized if reliable information is available on the probable magnitude and rate of settlement, and on the likelihood of instability developing in the subsoil under the embankment loading. Improvements in field and laboratory testing and in equipment for monitoring and controlling construction have enabled these problems to be studied in greater depth. However, for schemes involving long lengths of embankment of varying height the main uncertainties stem from the inherent variability of the alluvial deposits that comprise the soil profile in many areas of soft ground.

Grading of these recent alluvial deposits often becomes increasingly coarse with depth. Thus a typical profile might consist of a firm basal stratum at a depth of about 10 to 15 metres, overlain by gravels and sands and passing into silts, clays and peat towards the ground surface. In addition, variations in the depth and character of the deposits frequently occur over short distances. A detailed site investigation is therefore necessary to establish the longitudinal changes in soil properties.

11.2 Engineering appraisal

The imposition of an embankment on soft soil induces compression which is manifest by settlement of the embankment: in addition, instability in the subsoil supporting the embankment can be induced during construction and this is considered at the conclusion of this section. Two aspects of settlement are of concern to the engineer:

(i) The magnitude, which may range from a few centimetres to in excess of one metre.
(ii) The duration over which the movements continue.

Clearly, if the total movements are small the problems are unlikely to be serious however extended is the settlement period. Equally, if the time scale for large movements is small such that they are completed well within the construction period, there will be little difficulty in dealing with the problems as long as stability is maintained.

These are extreme examples of embankments founded on soft soil which will not present any particular engineering difficulties or incur much extra cost. Between these two extremes exist a large range of situations involving embankment construction on soft soils where significant amounts of settlement may occur over long periods. For example, an embankment forming part of a road structure will only tolerate a limited amount of post-construction settlement before the riding quality and structural adequacy of the road and performance of drainage systems are adversely affected.

Associated with settlement there is often a lateral movement of the subsoil which produces spreading of the embankment and may cause failure of any structure such as a road pavement which has been constructed before the movements are complete.

It is differential settlement which produces the main difficulties. Factors tending to reduce differential settlement include uniformity of the subsoil strata, similarity of loading over the embankment length and a construction programme which provides for a uniform rate of filling over the scheme. Conversely, problems of differential settlement are accentuated as a result of abrupt changes in the ground condition, the type of construction, or loading sequence. One example of an abrupt change in construction is provided by the usual procedure of founding a bridge on piles taken through the compressible subsoil, thus virtually eliminating vertical movement of that structure while the adjoining embankment settles directly on the compressible material.

Of equal concern to the engineer is the avoidance of instability of the embankment. For embankments on soft and compressible subsoils the most critical condition of stability is likely to develop during or shortly after the construction period, prior to a significant gain in shear strength in the subsoil resulting from its consolidation by the weight of the embankment. At the design stage, information is therefore required on both the initial strength of the subsoil and on the likely rate of consolidation. Embankment stability is also associated with a number of other factors including the embankment geometry, the depth of the compressible layer in relation to the embankment width, the strength properties of the fill material and the likely time available for construction.

11.3   Soil testing

The preliminary geotechnical investigations should establish the general characteristics and extent of soft and compressible subsoils so that an initial assessment of the likely stability and settlement problems can be made to determine the most suitable location for the works. A detailed geotechnical investigation will then normally be carried out to provide data for design which will enable the most economic alignment and form of construction to be determined. This phase of the site investigation will normally comprise in-situ tests to determine the subsoil profile at intervals, together with borings to extract samples for examination and testing. The extent of the investigation will depend

Table 11.1 Quality classes (after Rowe, 1972).

Quality class	Properties	Purpose	Typical sampling procedure
1	Remoulded properties Fabric Water content Density and porosity Compressibility Effective strength parameters Total strength parameters Permeability* Coefficient of consolidation*	Laboratory data on in-situ soils	Piston thin-walled sampler with water balance
2	Remoulded properties Fabric Water content Density and porosity Compressibility* Effective strength parameters* Total strength parameters*	Laboratory data on in-situ insensitive soils	Pressed or driven thin or thick-walled sampler with water balance
3	Remoulded properties Fabric A* 100% recovery Continuous B* 90% recovery Consecutive	Fabric examination and laboratory data on remoulded soils	Pressed or driven thin or thick-walled samplers water balance in highly permeable soils
4	Remoulded properties	Laboratory data on remoulded soils. Sequence of strata	Bulk and jar samples
5	None	Approximate sequence of strata only	Washings

* Items changed from German classification

on the scale of the earthworks and likely problems, and on the uniformity and depth of the soft subsoils. For schemes involving long lengths of embankment the investigation will normally concentrate initially on areas where the subsoils are poorest, the heights of embankment are greatest and where the embankments interface with structures (such as approach embankments to bridges).

Drilling equipment and soil samplers are described by Cole (1979, see also BS 5930, 1981). Rowe (1972) considers the importance of soil fabric in sampling and laboratory testing. Tables 11.1 and 11.2 show the classification of sample quality and required sizes of test specimen which he suggests. For soft subsoils, continuous sampling enables detailed examination of the stratigraphy before the location of high quality samples for laboratory testing is selected.

11.3.1 Laboratory testing—compressibility

The earliest and still most commonly employed method for assessing the compressibility characteristics of cohesive soils was developed by Terzaghi (1943). The method is

Table 11.2 (after Rowe 1972).

Clay type	Macro fabric	Mass k (m/s)	Parameter[†]	Specimen size (mm*)
Non-fissured, sensitivity < 5	None	10^{-10}	$c_u, c'\phi'$ m_v, c_v	37 76
	Pedal, silt, sand layers, inclusions Organic veins	10^{-9}–10^{-6}	c_u $c'\phi'$ m_v, c_v c_v	100–250 37 75 250
	Sand layers > 2 mm at <0.2 m space	10^{-6}–10^{-5}	$c'\phi'$ m_v	37 75
Sensitivity > 5	Cemented with any above		$c_u, c'\phi', m_v, c_v$	50–250[‡]
Fissured[+]	Plain fissures	10^{-10}	c_u $c'\phi'$ m_v, c_v	250 100 75
	Sand or silt-filled fissures	10^{-9}–10^{-6}	c_u c_v $c'\phi'$ m_v	250 100 75
Jointed	Open joints		ϕ'	100
Pre-existing slip			$c'_r\phi'_r$	150 or remoulded

* 75 mm samples for continuous quality 2–4 samples for fabric examination, strength as index test, c_u and $c'\phi'$ for intact low sensitivity
+ Size and orientation dependent on fissure geometry
‡ Tube area ratio 4%, sample diameter 260 mm
† See list of symbols for definition of listed parameters.

essentially a modelling approach whereby a small specimen of the soil in the shape of a disc is tested in the oedometer apparatus (BS 1377: 1975, p. 92). Typical dimensions for the specimens are 75 mm diameter by 20 mm in thickness. These dimensions are selected to reduce wall friction in the oedometer ring by maintaining the diameter-to-thickness ratio in the range 3–4.

The test procedure involves applying a series of increments of load to the specimen to cover the stress range anticipated from the embankment loading on the subsoil. Each increment of load is normally maintained constant for 24 hours to permit the relation between deformation and time to be established. With most soils and this thickness of specimen the 24 hour period is sufficient to permit the primary consolidation phase to be completed, that is, all excess pore-water pressures will be completely dissipated. As will be discussed later, in special circumstances the duration of the testing programme may be extended to permit data on the secondary phase of consolidation to be obtained.

In the usual testing procedure, the dissipation of pore-water pressure occurs only in the vertical direction to drainage platens above and below the specimen. The resulting coefficient of consolidation, designated c_v, therefore relates to this direction only. Clearly, to obtain the coefficient of consolidation for a particular direction in the subsoil, the specimen would have to be prepared from the sample such that the required drainage directions formed the vertical axis in the oedometer ring. Apart from the need to use more of the original undisturbed sample of soil when preparing specimens from sample tubes for other than vertical drainage directions, no alteration to the test procedure is required.

Disturbance of the specimen during its preparation should be minimized, as this tends to produce a reduction in apparent compressibility (Terzaghi and Peck, 1967, p. 76) and can also seriously affect the drainage characteristics.

A fundamental assumption of the method is that the percentage change in thickness produced by a stress increment or decrement in the laboratory test corresponds to the percentage change occurring in the field under the same stress conditions. It should be recalled, however, that the conditions in the oedometer apparatus correspond to a one-dimensional situation as the specimen is not permitted to strain laterally, a condition not necessarily true in the field. The data from the consolidation test should strictly be applied to predictions of the behaviour of soil beneath the centre of a wide, long embankment where conditions approximate closely to the one-dimensional situation. This problem has been recognized by soils engineers and, because the consolidation test has proved convenient and relatively easy to perform, analytical techniques have been developed to permit the one-dimensional test data to be corrected for field situations that are other than one-dimensional.

In some cases the dimensions of the consolidation specimen typically employed and referred to above are too small to provide an adequate representation of the field drainage characteristics. Studies by Rowe (1972) indicate that the specimen sizes given in Table 11.2 are required to provide more representative assessments of the consolidation parameters.

11.3.2 *Laboratory testing—shear strength*

To determine the strength of soil for use in total stress analyses of short term stability, undrained tests are normally carried out in the triaxial apparatus (Bishop and Henkel, 1962; BS 1377: 1975, p. 108). A minimum of three tests is normally required for each zone of soil identified. The minimum required size of test specimen is dependent on the structure and fabric of the soil (Table 11.2). Unless the soil layer does not have a micro fabric, and is relatively insensitive to disturbance induced by the sampling and specimen preparation processes, the larger the size of specimen tested the greater the confidence that can be placed in the results. For unsaturated cohesive fill materials, tests on specimens at a number of different cell pressures will be required to determine the parameters c_u, ϕ_u, for use in the analysis. Because of curvature in the failure envelope these tests should be carried out at stress levels representative of the field situation.

Other techniques which have been employed for determination of undrained strength include unconfined compression tests, direct shear tests, Swedish fall cone tests, laboratory vane tests and pocket penetrometer tests. These involve a smaller volume of material and/or less control of the stresses and drainage conditions than in the triaxial test, and can lead to a lower quality of assessment of the shear strength profile. For normally consolidated soils the relation between undrained strength, vertical effective stress and plasticity index suggested by Skempton (Terzaghi and Peck, 1967, p. 117) can provide a useful check on the measured strength.

To determine the strength parameters in terms of effective stress, consolidated undrained triaxial tests with the measurement of pore-water pressure or consolidated drained triaxial tests are customarily carried out. The former tests provide information on the pore-water pressure response to stress change required for predicting the pore-water pressures developed in the subsoil by the embankment loading. Tests should be carried out on a minimum of four specimens, preferably at stress levels appropriate to the field situation. Since the effects of disturbance are reduced by reconsolidation, tests are normally performed on 37 mm diameter samples (see Table 11.2). However, testing of larger diameter specimens can lead to improved definition of the cohesion c' for lightly overconsolidated soils. It is essential that the tests are carried out sufficiently slowly to allow full equilibration or dissipation of pore-water pressures. The difficulty of determining reliable values of cohesion c' demands a conservative approach to the selection of the design value.

11.3.3 Field testing

The main advantage of field testing is that disturbance to the soil caused by sampling, transportation, storage and preparation of specimens for laboratory testing is avoided. The principal field tests of relevance in establishing the strength and compressibility properties of soft subsoils are penetration tests, field vane tests, in-situ permeability tests and pressuremeter tests. A description of these tests is given in the U.K. Code of Practice on Site Investigations (BS 5930: 1981) and they are therefore only briefly outlined in the following subsection.

Penetration or probing tests are a traditional means of rapidly and economically determining the composition of the ground. The European Sub Committee of the International Society of Soil Mechanics and Foundation Engineering has established standards for penetration testing (Report of the Sub-committee, 1977). A comprehensive review of penetration testing is given by Sanglerat (1972) and a state of the art report on practice in the U.K. is given by Rodin et al. (1974).

The standard penetration test, (BS 1377: 1975, p. 103), is used mainly as an indication of the relative density of sands and gravels. Correlations between penetration resistance and the unconfined compressive strength of cohesive soils are given by Terzaghi and Peck (1967, p. 347) and with the angle of shearing resistance of cohensionless soils by Peck *et al.* (1974). The soil parameters obtained from such correlations are very approximate but the test, together with inspection of the disturbed soil from the sampler, provides a useful preliminary indication of the subsurface profile.

Static penetration testing involves the measurement of the resistance of a cone pushed into the soil at a constant rate of penetration and generally provides a higher quality assessment of soil properties than is available from the standard penetration test. Measurements are often made of the force on the cone and the lateral friction on the side of the penetrometer, which enable an assessment to be made of soil type (Sanglerat, 1972, pp. 190–214; Schmertmann, 1967). The undrained strength of clays can be determined from the measured cone resistance using formulae based on bearing capacity principles. These formulae depend on the dimensions and shape of the particular penetrometer. Assessments can also be made of the compressibility of cohesive soils, based on empirical relations between compressibility and cone resistance (Meigh and Corbett, 1970).

The testing procedure for determining the undrained strength of clays *in situ* (BS 1377: 1975, p. 99) comprises the measurement of the torque required to rotate a vane of cruciform shape about its vertical axis. The undrained shear strength is determined from the measured torque assuming that failure of the soil occurs along the surface of a cylinder with dimensions equal to those of the vane. The tests can also be used to measure the remoulded strength of clays for assessing sensitivity. Because in variable deposits the results can be invalidated by the presence of stones, shells, sand and silt layers, and root fibres field vane testing is normally restricted to fairly uniform cohesive soils.

In assessing the rate of embankment settlement on soft alluvial deposits, the measurement of *in-situ* permeability generally provides a more reliable determination of the natural drainage characteristics than can be obtained from conventional laboratory consolidation tests. This is because the test involves a much larger volume of subsoil. Various techniques are available for carrying out such tests but the method described by Gibson (1966) appears to be the most efficient. The test is generally carried out by applying a constant excess head to hydraulic piezometers installed within each of the main soil layers and measuring the rate of flow of water into or out of the piezometer with time. Once the value of field permeability coefficient (k_f) has been determined for a particular layer it is combined with the coefficient of volume compressibility (m_v) and unit weight of water (γ_w) to obtain the coefficient of consolidation as follows:

$$c_v = \frac{k_f}{\gamma_w m_v} \qquad (11.1)$$

The value of m_v in the above expression is usually obtained from standard consolidation tests, with the m_v chosen at the mean effective stress for the anticipated stress range in the field. As the field permeability value will generally be obtained at the over-burden effective stress before the start of construction it can be seen that the data employed in the above expression are somewhat inconsistent and moreover are likely to overestimate the true field value. Ideally it would be better if the permeability test could also be carried out at the mean effective stress. Some increase in effective stress can be achieved by carrying out a test in which water is drawn from the soil under suction, but the magnitude of such a gain in stress is then quite small.

If the subsoil is normally consolidated, a more consistent approach would be to employ the (larger) value of m_v corresponding to the initial overburden stress and to assume that the ratio of k_f/m_v remains constant over the stress range. However, most soils at their overburden stress show some evidence of over-consolidation and are much less compressible (with a correspondingly small value of m_v). With increasing effective stress induced by embankment loading, such soils approach a normally consolidated condition with a consequent increase in compressibility. Thus although the combined use of field permeability data with laboratory values of m_v have frequently produced good assessments of field behaviour (Lewis *et al.*, 1975), there is a danger that occasionally this approach will grossly overestimate the rate of settlement of an embankment founded on soft soil. Moreover, the problems of overestimation are accentuated if fissures or other drainage channels, which can remain open under the overburden stress, close when subject to the embankment loading. Evidence of such discontinuities may be apparent in undisturbed sampling cores and in these circumstances the best approach for obtaining the consolidation characteristics may be to carry out large diameter consolidation tests as well as field permeability tests. The field values of c_v can then be adjusted over the appropriate stress range to correspond with the proportional change in c_v evident in the laboratory tests.

A review of pressuremeter testing is given by Baguelin *et al.* (1978). In its simplest form the test consists of inserting a probe into a prebored hole, and expanding it laterally against the borehole sides. An approximate estimate can also be made of the *in-situ* lateral stress in the ground. The disturbance to the soil caused when boring holes for the installation of the probe is likely to have a considerable effect on the measured soil parameters. To overcome this problem self boring versions of the pressuremeter have been developed more recently (Baguelin *et al. op. cit.* pp. 482–573).

11.4 Settlement assessment

The compressibility of soft soils is generally considered as comprising the following three components:

(a) Immediate or 'elastic' settlement.
(b) Primary consolidation settlement.
(c) Secondary consolidation settlement.

Various procedures are available for calculating the magnitude of settlement for these three components and the more common of these are described below. Chapter 1 also provides an insight into settlement components, including 'undrained creep' which is particularly difficult to predict.

11.4.1 *Immediate or 'elastic' settlement*

Although the terms 'immediate' or 'elastic' settlement are often used synonymously, very little recoverable 'elastic' settlement actually occurs with normally consolidated soft soils. The main component of the settlement which occurs instantaneously following the application of a load, arises from shear distortion. In the oedometer apparatus, where such shear distortion is prevented by the confining ring, immediate settlement may in part be due to elastic behaviour but is more likely to be associated with compression of any gas bubbles present in the pore-water.

The shear distortion occurring initially in the subsoil beneath an embankment corresponds to the movements observed in a triaxial undrained compression test where the specimen contracts vertically but expands laterally to preserve the volume sensibly constant under undrained conditions. Beneath an embankment the lateral expansion only occurs in one direction because of the plane strain nature of the problem. It is apparent that very large vertical movements will occur as the ratio of the minimum to maximum principal stresses approaches the failure condition, that is, as σ_3' approaches $K_a\sigma_1'$. However it has been pointed out that, for surface loading less than half the value required to induce failure, the immediate settlements are small (Simons, 1971).

In critical state soil mechanics the soil is recognized as behaving elastically provided that the stress regime is below that of a yield locus previously attained. The soil must therefore be in an overconsolidated condition. It should be noted, however, that the manifestation of pre-consolidation pressure can result from ageing as well as stress history (Bjerrum, 1973) and the absence of any direct geological evidence of prior loading does not necessarily ensure that the soil will behave as a normally consolidated material.

The calculation of immediate settlement is often carried out employing an equation derived from elastic analysis (Davis and Poulos, 1968). For the situation where a subsoil of undrained deformation modulus E_u is subjected to plane strain conditions under undrained loading the equation takes the form

$$\delta_i = \sum_0^H \frac{3}{4E_u}[\Delta\sigma_z - \Delta\sigma_x]\cdot\Delta z$$

(11.2)

The increments of total stress $\Delta\sigma_z$ and $\Delta\sigma_x$ for vertical and horizontal directions are obtained from elastic analysis. The stress calculations are greatly facilitated by the use of Jurgenson's charts (Jurgenson, 1934).

The value of the undrained deformation modulus, E_u, is best obtained from the results of field loading or *in-situ* tests. Frequently, laboratory compression tests are used to derive values of E_u, or alternatively such values are obtained empirically from shear strength tests. Neither of these latter two approaches is considered reliable, although a better estimate may be obtained from laboratory tests in which the field stress conditions are accurately represented (Simons, 1975, p. 505).

In applying the above equation, the subsoil is subdivided into a number of layers of thickness Δz and the increments of vertical and horizontal stress at the centre of each layer determined. The deformation is then calculated by using the associated E_u in the above equation. The total immediate deformation, δ_i, in a vertical profile is obtained from the summation of the deformation of individual layers.

The above method has the merit that it employs a familiar engineering approach to stress-strain analysis. Several other methods available for calculating 'elastic' settlements avoid the need to compute the stress increments by making use of influence coefficients. The more common of these methods are described in standard reference works (*see*, for example, Lambe and Whitman, 1969, p. 212).

The stress regime beneath embankments is very much more complex than can be represented by the conventional consolidation test. For example, beneath the side slopes the principal stress directions rotate from their initial overburden condition during embankment construction. Because of the inadequacy of the conventional method, therefore, the use of a stress path approach has been proposed (Lambe, 1964). In this method high quality specimens of the soil from selected locations beneath the embankment are subjected to the anticipated stress regimes at these locations for purposes of assessing the consolidation characteristics.

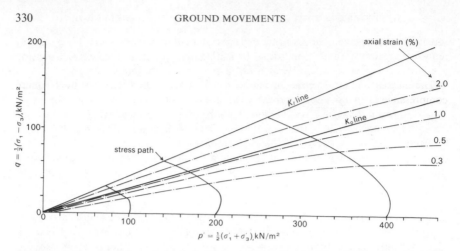

Figure 11.1 Stress paths in undrained triaxial tests and associated axial strains for remoulded clay.

The stress path approach can thus provide a useful insight concerning the behaviour of soft clays during undrained loading and can also be employed for predicting deformation behaviour. Fig. 11.1 presents the results of undrained triaxial tests carried out on a remoulded clay in which specimens were isotropically consolidated and then subjected to shearing under undrained conditions. As the soil was in a normally consolidated condition the stress path taken conforms to a yield locus which, according to the concepts of critical state soil mechanics, separates the elastic and plastic deformation behaviour of the specimen (Atkinson and Bransby, 1978).

During undrained shearing the specimens contract vertically while expanding laterally. Lines have been drawn on the figure which intersect the yield loci at vertical strains of equal magnitude. The limiting line conforms to the failure condition corresponding to about 6 per cent axial strain. It is therefore possible to obtain the change in axial strain, and hence the deformation, from the figure by locating the initial and final stress conditions (p' and q) on the appropriate yield locus. The results that are shown relate to triaxial tests, and clearly for assessing undrained behaviour beneath embankments, the testing procedure would more appropriately involve a plane strain approach.

11.4.2 *Primary consolidation settlements*

When an adequate factor of safety exists against undrained failure, primary consolidation will constitute the main proportion of the total settlement of the majority of soft soils with the exception of peat. It is to be expected that the primary compression of a soft clay and its water content are closely related and this hypothesis is generally supported by published data in Simons, (1975, p. 520).

The compression index C_c (the slope of the relation between void ratio and the logarithm (base 10) of pressure) also correlates well with liquid limit for quite a wide range of normally consolidated clays (Terzaghi and Peck, 1967, p. 72). For a preliminary assessment of the compression index of normally consolidated clays of low to moderate sensitivity the following approximate relation is available:

$C_c = 0.009(LL - 10\%)$ where LL is the liquid limit expressed as a percentage.

The classical equation for the prediction of primary consolidation settlement (δ_c) in a compressible layer having a total thickness of H is as follows:

$$\delta_c = \sum_0^H \frac{C_c}{1 + e_0} \log_{10}\left(\frac{\sigma_0' + \Delta\sigma_v'}{\sigma_0'}\right) \cdot \Delta z \qquad (11.3)$$

where e_0 is the original void ratio corresponding to overburden pressure σ_0' and $\Delta\sigma_v'$ is the increment of vertical stress occurring at the centre of a sub-layer of thickness Δz at depth z.

The above equation, although widely used, presupposes that the soil is in a truly normally consolidated condition. However, as pointed out previously, most soils exhibit some degree of overconsolidation which may have resulted from ageing (Bjerrum, 1973) rather than from stress history. Moreover, if the compression index is obtained from the ocdometer test, its direct application to the prediction of plane strain situations as occurs beneath embankments is incorrect.

The compression characteristics of a clay change radically at the preconsolidation pressure of σ_c' which represents the boundary between elastic and plastic behaviour of the soil (Fig. 11.2). A reliable estimate of σ_c' is therefore essential for assessing the primary consolidation. It has been proposed that in place of the usual procedure, in consolidation

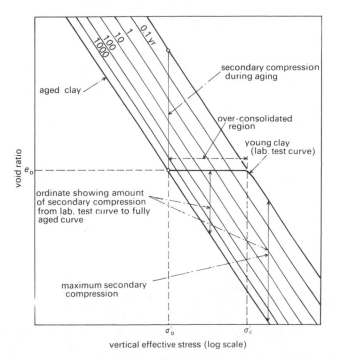

Figure 11.2 Relation between void ratio and effective stress showing influence of ageing on secondary compression (after Bjerrum, 1973).

testing, of doubling the load increment, relatively small increments are applied until the value of σ_c' is established (Bjerrum, 1973) by noting the point where a significant increase occurred in the time required for 90 per cent consolidation. Prior to attaining σ_c', these times would be very short, perhaps requiring only a few minutes. However, once σ_c' has been exceeded, the time required for 90 per cent consolidation might well be typically of the order of 24 hours for the conventional 2 cm thick specimen.

Several other methods are available for establishing the value of σ_c'. The most commonly-employed procedure is that due to Casagrande in which the value of σ_c' is determined from a graphical construction involving the minimum radius of curvature on a plot of void ratio versus the logarithm of effective pressure. However, this procedure may sometimes be rather artificial, as shown by plots of the same data to a natural scale whereby there appears to be no basis for selecting a particular point since the relation between void ratio and effective pressure (natural scale) produces a smooth curve. Details of the Casagrande method, as well as a method proposed by Schmertmann, are provided in standard references (Terzghi and Peck, 1967, pp. 77–78).

A relation between plasticity index and the over-consolidation ratio (σ_c'/σ_0') has been proposed by Bjerrum (1973) for assessing the influence of ageing on the preconsolidation pressure (Fig. 11.3) of soft soils. As shown in the figure the ratio is unity for young soils, that is, soils which have only recently been deposited and have not undergone any significant secondary consolidation. However, with time, such soils settle further under their self-weight to produce a more stable particle configuration as illustrated in Fig. 11.2. According to Bjerrum, this procedure may continue for thousands of years until the fully-aged condition is achieved. As shown in Fig. 11.3, this is manifest by an increase in the ratio of σ_c'/σ_0' above unity, particularly for the more plastic clays.

As the primary consolidation of clay occurs due to the dissipation of excess pore-water pressure it is apparent that the amount of compression will be related to the magnitude of excess pore-water pressure induced for a given soil. In the oedometer test, where lateral strains are prevented, the increase in pore-water pressure for saturated soils corresponds generally to the increment of stress applied to the specimen. However, beneath an

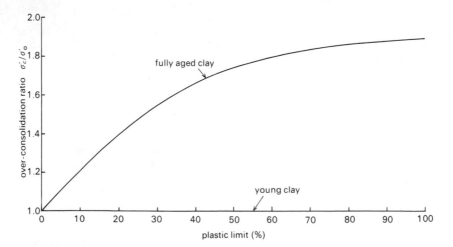

Figure 11.3 Relation between over-consolidation ratio and plastic limit for 'aged' clays (after Bjerrum, 1973).

embankment where lateral strains occur, the change in excess pore-water pressure as a result of a load increment is a function of both the increments of major and minor principal stress as well as the pore-pressure parameter A. To correct for this situation, while still employing the data from oedometer tests, Skempton and Bjerrum introduced a correction factor μ to apply to the equation for predicting primary consolidation settlement (Scott, 1963). The correction factor takes account of both the pore pressure coefficient A and the geometry of the consolidating region in relation to the loaded area. Charts are given which relate μ to A for different geometries, but as an alternative the following equation, which enables a reasonable approximation of μ to be obtained, has been developed by the present authors:

$$\mu = 0.33 + 0.728\,A + \frac{B}{H}(0.157 - 0.170\,A) \tag{11.4}$$

This equation relates to the conditions beneath a strip load of width B overlying a clay layer of thickness H. It can thus be used to provide an approximate assessment of the conditions beneath an embankment. The equation has been found to be satisfactory for A in the range 0 to 1.2 and the ratio of B/H in the range 0.25 to 3.

On the basis of the foregoing discussion, a more correct equation than equation (11.3) above for predicting primary consolidation for soil in a normally consolidated condition is therefore as follows

$$\delta_{c(NC)} = \mu_{NC} \sum_{0}^{H} \frac{C_c}{1+e_0} \log_{10}\left(\frac{\sigma_c' + \Delta\sigma_v'}{\sigma_c'}\right) \cdot \Delta z \tag{11.5a}$$

For soil in an overconsolidated condition, that is, when the stress is in the range σ_0' to σ_c', it is usually assumed that the stress-strain behaviour of the soil is linearly elastic. The equation therefore takes the following form:

$$\delta_{c(OC)} = \mu_{OC} \sum_{0}^{H} m_v(\sigma_0' + \Delta\sigma_v') \cdot \Delta z \tag{11.5b}$$

where $\sigma_0' + \Delta\sigma_v' \le \sigma_c'$.

The stress-path method of predicting primary consolidation settlement of embankments founded on soft soil (Lambe, 1964) ideally requires the data to be obtained from plane strain tests. In principle, the method involves the selection of a number of average points—the centres of layers, provided that these are not too thick, would frequently be most appropriate—to represent general settlement behaviour beneath the centre of an embankment and elsewhere. Having established the initial stress conditions and the stress changes as a result of the embankment loading at these average points, laboratory tests are carried out on undisturbed samples in which the stress regimes are made to correspond with those anticipated in the subsoil. The testing procedure is clearly required to replicate as closely as possible the duration of the construction works—appropriately scaled—such that the relation between field loading and subsoil consolidation is adequately represented by the test programme. The primary consolidation settlement of the layer can then be computed from the product of the observed vertical strain and the layer thickness. The total settlement in a vertical profile is obtained by summation of the values obtained for the individual layers.

To avoid the complex testing requirements of the stress-path method while benefiting from its conceptual advantage, a method has been developed by Simons and Som (1969) for adjusting the compression data from the one-dimensional oedometer test to account for three-dimensional conditions. The basis of the method is that the coefficient of volume compressibility appears to depend mainly on the increment of major principal effective stress and is not very sensitive to the test procedures (Simons, 1971). This situation applies, of course, provided that the soil is not close to failure. By employing elastic analysis to determine the ratio of major principal strain to volumetric strain the authors produced a correction factor λ, for different values of the increment of principal stress ratio, to apply to compression data obtained from oedometer tests. Essentially the method consists of adjusting the value of m_v from the oedometer test in proportion to the ratio of elastic axial strain to elastic volumetric strain, for different stress increment ratios. The analysis was developed by Simons and Som (1969) for application to three-dimensional conditions. As the conditions beneath embankments conform to plane strain, the following revised expression for the correction factor (λ_p) has therefore been developed for this condition employing a similar approach to that adopted by those authors:

$$\lambda_p = \frac{(1 - K_0 K')}{(1 - K_0)(1 + K')} \tag{11.6}$$

In the above expression K' equals the ratio of the increment of horizontal to vertical effective stress induced by the embankment loading, K_0 refers to the coefficient of earth pressure at rest, and for normally consolidated soils may be obtained from the empirical relation

$$K_0 = 1 - \sin \phi' \tag{11.7}$$

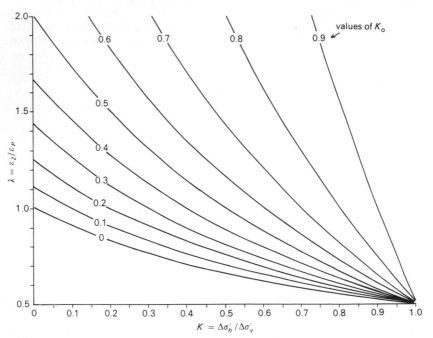

Figure 11.4 Correction factor in plane strain analysis for different stress-increment ratios.

Values of λ_p are plotted in Fig. 11.4 for a range of K' and K_0 where λ_p corresponds to the ratio of vertical strain (ε_z) to volumetric strain (ε_p) for plane strain conditions.

The modified forms of equations (11.5a) and (11.5b), taking account of the differences in the strain conditions between the oedometer test and the subsoil beneath a long embankment, are therefore as follows

$$\delta_{c(NC)} = \mu_{NC} \sum_0^H \lambda_p \left[\frac{C_c}{1 + e_0} \log_{10} \left(\frac{\sigma_c' + \Delta\sigma_v'}{\sigma_c'} \right) \right] \Delta z \tag{11.8a}$$

$$\delta_{c(OC)} = \mu_{OC} \sum_0^H \lambda_p m_v (\sigma_0' + \Delta\sigma_v') \Delta z \tag{11.8b}$$

where $\sigma_0' + \Delta\sigma_v \leq \sigma_c'$ in equation (11.5b) as previously.

The rate of primary consolidation settlement is most frequently determined by assuming that pore-water pressure dissipation occurs only in the vertical direction, which is virtually equivalent to an assumption that the conditions are one-dimensional. This may be considered to be a reasonable assumption beneath the centre of an embankment provided that the width of an embankment is at least three times the depth of compressible strata and the horizontal permeability is not significantly greater than the permeability in the vertical direction.

For deposits which are reasonably homogeneous, the following simple squared relationship is frequently employed

$$t_f = t_L (H_f/H_L)^2 \tag{11.9}$$

where t_f = time to achieve a specific degree of consolidation in the field,

  t_L = time to achieve the same degree of consolidation in a laboratory test,

  H_f = length of drainage path in the field,

and H_L = length of drainage path in the laboratory.

An advantage of the above method is that variation of the consolidation parameters is inherently taken account of and this may be of particular importance with peat and other highly-compressible organic soils. It is assumed in applying equation (11.9) that the stress range employed in carrying out the test corresponds to that occurring beneath the centre of the embankment.

More generally, the prediction of the rate of primary consolidation for one-dimensional conditions is carried out employing the coefficient of consolidation of the soil (c_v) in conjunction with the relation between time factor (T_v) and degree of consolidation ($\bar{u}$) given in standard references (Terzaghi and Peck, 1967, p. 181). The time to achieve a specific degree of consolidation is given by the relation

$$t = \frac{T_v H_f^2}{c_v} \tag{11.10}$$

where T_v is the time factor from the standard relation corresponding to a specified degree of consolidation. The primary consolidation settlement at any time t (that is, δ_{ct}) is therefore given by

$$\delta_{ct} = \bar{u} \cdot \delta_c \tag{11.11}$$

As the subsoil frequently consists of several layers with different consolidation properties, a simple procedure is available for computing the rate of settlement for one-dimensional conditions. This procedure involves the determination of an equivalent coefficient of consolidation $(\bar{c}_v)$ for the multi-layered strata. Once this value is established, the calculations take the same form as for a homogeneous deposit, utilizing the standard time-factor relation referred to above. The equivalent coefficient of consolidation is derived employing the following equation

$$\bar{c}_v = \frac{H_f^2}{\left[\displaystyle\sum_{i=1}^{i=N} \left(h_i/\sqrt{c_{vi}}\right)\right]^2} \tag{11.12}$$

where h_i = thickness of ith layer

c_{vi} = coefficient of consolidation of ith layer,

and N = number of layers.

The above equation provides only an approximate estimate of the degree of settlement at a given time and the errors increase with increasing differences between the compressibilities of the different layers.

A simplified method of dealing with consolidation in two dimensions, based on the standard one-dimensional solution, has been proposed by Carrillo (1942). In this method the average excess pore-water pressure $(\bar{u}_{xz})$ in a rectangular region subjected to two-dimensional consolidation is given by the following relation

$$\bar{u}_{xz} = \frac{\bar{u}_x \cdot \bar{u}_z}{\bar{u}_0} \tag{11.13}$$

where $\bar{u}_x$ = average excess pore-water pressures in the horizontal direction based on the standard one-dimensional solution,

$\bar{u}_z$ = average excess pore-water pressure in the vertical direction based on the standard one-dimensional solution,

and $\bar{u}_0$ = average initial excess pore-water pressure.

A more rigorous assessment of the rate of two-dimensional consolidation of the subsoil beneath embankments is provided by the solution to second-order partial differential equations. A discussion of those equations and their solution is beyond the scope of this chapter, but details are provided in the literature (Murray, 1978). Briefly, however, there are two systems of equations currently employed in consolidation analysis. These are referred to as the Rendulic and Biot systems, respectively. The former system utilizes the equations of diffusion theory. These are sometimes referred to as pseudo-consolidation equations as they do not fully describe the stress-strain characteristics of the soil during consolidation. The Biot system of equations are more complex but are generally recognized as providing a more accurate assessment of consolidation than is provided by the former system, although for many engineering applications the differences between the two solutions are not significant.

The solutions to these systems of equations are usually obtained by methods of numerical analysis, particularly for multi-layered soils and where the consolidation parameters vary with the effective stress. Both finite differences and finite elements have been used in this application and a number of computer programs for consolidation analysis based on these techniques have been described in the literature (Brady, 1982).

11.4.3 *Secondary consolidation settlements*

Secondary consolidation is commonly believed to be attributable to movement occurring in the adsorbed water layer which surrounds the particles. As the adsorbed water is several hundred times more viscous than free water the movements are extremely slow. Moreover, the forces activating these movements are associated with the effective stresses which induce the particle arrangement to form a more stable structure.

Because the deformations arising from secondary consolidation are associated with long-term effects, predictions usually involve an assessment of the rate of movement rather than an estimate of ultimate value. This rate of movement is usually defined by the coefficient of secondary consolidation, c_a. The most common procedure for determining c_a involves extended duration tests in the oedometer apparatus.

Following completion of the primary phase of consolidation it has been found that many soils exhibit a continuing deformation which appears linear when plotted versus the logarithm of time. When the deformation is expressed in terms of strain, the slope of the line on the semi-logarithmic plot is non-dimensional and is defined as the coefficient c_a. Typical values for c_a vary from about 0.002 for inorganic clay having a moisture content of about 20 per cent to values in excess of 0.1 for peaty soils with moisture contents of 1000 per cent. Factors known to influence secondary consolidation include pre-consolidation pressure, magnitude of applied load in relation to pre-consolidation pressure, disturbance and temperature. Field and laboratory assessments of c_a have been reported by Simons (1975, p. 515) in relation to their moisture content. The earliest method of predicting secondary consolidation appears to be that proposed by Buisman (Road Research, 1979, p. 75), and for a multi-layer system takes the form

$$\delta_s = \sum_0^H \left[c_a \log_{10}(t/t_{90}) \right] \Delta z \tag{11.14}$$

where t_{90} = time to achieve 90 per cent primary consolidation

and t = total elapsed time and must exceed t_{90}.

A variation of the above approach still widely used and which couples both primary and secondary consolidation in a single equation was proposed by Koppejan (Road Research, 1979, p. 75). Such methods are entirely empirical and the assumption that strain occurs linearly with the logarithm of time is not always correct.

A major contribution to understanding the deformations arising from secondary consolidation was provided by Bjerrum (1973). As pointed out previously it is assumed that, with increased age, the particles in soft normally consolidated clay deposits take up a much more stable configuration and the soil exhibits the characteristics of an over-consolidated material. Referring to Fig. 11.2 it was postulated by Bjerrum that a series of curves could be produced relating void ratio to the logarithm of effective stress, with each curve representing a particular age. To construct the figure it is necessary to first carry out consolidation tests to establish the compression curve for the young (24-hour) clay. A second curve is then drawn parallel to this through the value of *in-situ* void ratio corresponding to the initial overburden stress. The lower constructed curve gives the relation between void ratio and effective stress for the aged material. The vertical distance between the two curves provides a measure of the change in void ratio resulting from secondary consolidation which has occurred over the period from deposition to the present time. Clearly, for reasonable estimates, it is important that the soil samples used in testing are free from disturbance. Moreover, if the soil is young, either in terms of age or because it has been previously loaded in recent times, it will not be possible to construct the lower curve for estimating secondary consolidation.

The process of loading a specimen takes it back towards the void ratio-effective stress curve for a young clay (see Fig. 11.2). According to Bjerrum, the change in void ratio produced by secondary consolidation corresponds to the vertical distance between the laboratory compression curve and the lower constructed curve. The maximum value is attained when the increment of applied load increases the stress to σ_c' or beyond. The void ratio-stress path of the soil is now on the upper curve and will require a period of secondary consolidation corresponding to the present loading age of the specimen to attain the lower void ratio curve. As this void ratio change corresponds to that induced by an increase in effective stress on the young (24 hour) curve from σ_0' to σ_c', it is obvious that the magnitude of secondary settlement is given by

$$\delta_s = \sum_0^H \left(\frac{C_c}{1 + e_0} \log_{10} \frac{\sigma_c'}{\sigma_0'} \right) \Delta z; \qquad (\sigma_0' + \Delta\sigma' \geq \sigma_c') \qquad (11.15a)$$

It should be noted that the estimated secondary settlement given by equation (11.15a) relates to that for the existing age of the deposit, and may be many thousands of years. As the design life requirements for most structures is generally of the order of 100 years, the predicted secondary settlement on the basis of the above equation is likely to overestimate the value for the shorter period. However, because the secondary movements are frequently observed to continue linearly with the logarithm of time, the greater proportion of the movements will occur during the design period and may amount to more than 75 per cent of the total movements.

The form of the above equation has much in common with that used for predicting primary consolidation settlements. Some consideration must be given, therefore, as to whether the corrections applied to the former equation should be employed here. It appears inappropriate to apply the Skempton and Bjerrum correction factor (μ) associated with the development of pore pressure in view of the very slow nature of secondary consolidation. However, it is to be expected that the magnitude of secondary consolidation will be influenced by differences in the strain field between an oedometer and the plane strain conditions beneath an embankment. It is proposed, therefore, that a better estimate of secondary deformation is given by the following modified version of the above equation

$$\delta_s = \sum_0^H \lambda_\rho \left(\frac{C_c}{1 + e_0} \log_{10} \frac{\sigma_c'}{\sigma_0'} \right) \Delta z; \qquad (\sigma_0' + \Delta\sigma' \geq \sigma_c') \qquad (11.15b)$$

The application of the correction factor, λ_ρ, would apply, as previously, when data from the oedometer test are to be used for predicting the deformations in plane strain conditions.

11.5 Stability assessment and control

There are a number of methods of assessing the behaviour of an embankment which serve different purposes in the design process. These methods vary in their complexity and in the extent and quality of the soils information that is required. Their use in practice therefore depends on the scale and importance of the structure. The engineer needs to be aware of the assumptions inherent in the analysis and to clearly distinguish between the accuracy of the calculation and the overall reliability of the assessment. This is particularly so with the more complex and computer based methods. The most important limitation on reliability of any of the methods is almost certainly the accuracy of the soil parameters.

The finite element method is used principally for predicting the deformations and stresses under working conditions. A range of computer programs based on assumptions of linear or non-linear elastic or elasto-plastic behaviour of the soil is available. The method normally requires comprehensive information on the initial stress conditions and characteristics of the soil, and this may represent a limitation on its use for routine purposes. The method is of particular value in conducting sensitivity studies of the effect of variations in the input data on the predicted behaviour. For major structures, particularly when used in conjunction with centrifugal model tests, the finite element method provides a powerful analytical tool for the designer (Bassett, 1978).

Most of the other methods consider conditions at collapse and it is therefore essential that the most likely failure mechanism is taken into account in the analyses. A wide range of possible failure mechanisms can be considered using the limit analysis method. This approach aims at defining the failure load imposed by the embankment between close limits based on the upper and lower bound theorems of plasticity. However, good definition of the failure load by this method is not always easy to achieve.

More recently, methods which consider the probability of failure have been proposed (McGuffey et al., 1982). These are based on a statistical approach to the variability in the input parameters and therefore require a good data base for the soil parameters. Further development and evaluation of probabilistic methods are required since they appear to be potentially useful for structures such as road embankments and flood control levees where account has to be taken in design of significant longitudinal variations in the depth and composition of the soft subsoil and in embankment height.

The limit equilibrium method is the approach normally adopted for assessing stability and is therefore considered in greater detail in the subsequent sections (see also chapter 2). Reference should also be made to the U.K. Code of Practice on Earthworks (BS 6031:1981, p. 23). The main advantages of the method arise from its relative simplicity and well-proven record in practice. Although the method takes no account of compatibility between stress and strain in the soil, in most circumstances the solutions obtained are adequate for routine design purposes.

11.5.1 Conditions for the stability of embankments

Problems of embankment instability are generally associated with the presence of fine-grained soils of low permeability within the subsoil. For these soil types two limiting states of strength can be defined. The first applies before the application of the embankment loading and the second when complete equilibrium has been achieved. These two strength states lead to two corresponding conditions of stability; the 'short-term' and the 'long-term' condition. For an embankment constructed on soft subsoil the 'short-term', or end-of-construction condition, will almost invariably be the more critical, as the consolidation which occurs with time leads to an increase in soil strength.

11.5.2 *Shearing strength of soil*

The composition of soil can be simply envisaged as comprising a skeleton of solid particles surrounded by voids which may be partially or completely filled with air or water. Since neither air nor water can provide significant resistance to shear, the soil strength is governed entirely by the shearing resistance of the skeleton of solid particles. This resistance in turn depends on the magnitude of the stress acting between the particles. In addition, the soil skeleton is relatively compressible in relation to water. Thus, when a saturated soil is subjected to a rapid change in loading, the water within the soil voids will carry nearly all the normal components of the stress change while the soil skeleton carries the shear components. If drainage of water to or from the soil voids can take place, then the pore-water pressures and stress between soil particles will change with time, until all components of the change in applied stress are carried by the soil skeleton. It follows that under compressive loading, drainage will be accompanied by a reduction in pore-water pressures and a corresponding increase in the shearing strength of the soil.

For design purposes the shearing strength of soil is generally expressed in terms of either the total stresses carried by the soil system or the effective stresses which act only on the soil skeleton. In the former case the relation between the shearing strength τ and the applied total stress σ acting on the failure plane is given by Coulomb's empirical law

$$\tau = c_u + \sigma \tan \phi_u \qquad (11.16a)$$

where c_u is the apparent cohesion, and ϕ_u is the angle of shearing resistance, both being expressed in terms of total stress.

For saturated soils, the shear strength on any plane in the soil is a function of the difference between the total normal stress and the pore-water pressure. This may be expressed by an equation in terms of effective stress of the following form

$$\tau = c' + \sigma' \tan \phi'$$

and

$$\sigma' = (\sigma - u) \qquad (11.16b)$$

where c' is the effective cohesion intercept,

ϕ' is the effective angle of shearing resistance,

σ' is the effective normal stress,

and u is the pore-water pressure.

The 'short-term' condition of stability is usually assessed from a total stress analysis. For soft saturated soil, ϕ_u is equal to zero and the shear strength is unaffected by an applied stress change if no drainage occurs; the strength is then equal to the apparent cohesion c_u. The inherent assumption of this method is that undrained conditions will prevail in the subsoil during the period of embankment construction. The shear strength can therefore be determined before construction from undrained field or laboratory tests. The 'long-term' condition of stability is customarily assessed from an effective stress analysis using pore-water pressures based on the likely ground water conditions ultimately existing at the site. An effective stress analysis can also be used to assess stability during, or at any time after construction provided that the pore-water pressures present in the subsoil are known. The shear strength parameters required for this method of analysis can be obtained in the laboratory from drained or consolidated undrained shear tests with the measurement of pore-water pressures (section 11.3.2).

11.5.3 *Stability calculations*

The limit equilibrium method of analysis consists of examining the equilibrium of the proposed embankment cross-section along trial failure surfaces. The main disturbing force is provided by the weight of the soil above the assumed failure surface, and the principal resisting force is derived from the shear strength of the soil along this failure surface. The factor of safety can be defined as the ratio of the available shear strength of the soil to the shear strength required just to maintain the earthwork in a state of limiting equilibrium. Numerous trial failure surfaces are examined until the minimum factor of safety for the section being analysed is obtained.

For preliminary studies a rapid assessment of stability can sometimes be carried out using published stability charts or simplified analyses such as the wedge method for non-circular failures (Lambe and Whitman, 1969, p. 366). In general, however, the embankment and subsoil are not sufficiently homogeneous to satisfy the simplifying assumptions of the charts and analyses are carried out using computer programs based on the method of slices. This divides the soil mass lying above the failure surface into a number of vertical slices and, for the general case, determines the overall factor of safety by considering force and moment equilibrium for each slice as well as the overall equilibrium of the trial surface.

The procedure is an iterative one and requires assumptions concerning the forces acting between the slices to obtain a solution. Developments based on the method of slices for assessing stability using circular and non-circular failure surfaces are described in chapter 2, and in standard references (for example, Lambe and Whitman, 1969, p. 366).

11.5.4 *Reliability of the stability assessment*

The factor of safety must ensure that instability does not develop and that unacceptable deformations do not occur under working conditions. It must therefore allow for uncertainties in the input parameters (for example, shear strength and pore-water pressure data) and inaccuracies in the stability calculations. The factor of safety should also cater for occurrences not explicitly considered in the design and which may adversely influence the stability. These could include loading from construction plant, temporary stockpiling of fill material and realignment of drainage in the vicinity of the embankment toe.

In selecting the value of the factor of safety to use in a particular situation consideration should be given to the scale of the structure and the overall consequences of instability, including the type of failure and likelihood of a progressive collapse. The value adopted will also depend on the purpose of the stability assessment, whether for a preliminary or final design.

For road embankments on compressible subsoil the factor of safety selected for both total and effective stress analysis is typically of the order of 1.5, although values as low as 1.2 are sometimes employed when the soils data and site conditions are very well established (Road Research, 1979, p. 70).

The reliability of a stability calculation largely depends on the accurate assessment of the strength which can be mobilized along a potential failure surface in the field. In practice, and as demonstrated in chapter 2, the position of the failure surface will be affected by anisotropy in soil strength and inhomogeneity in the deposits. Because of

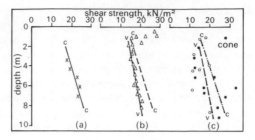

Figure 11.5 Profiles of undrained shear strength obtained from Canvey Island from (a) camkometer (b) vane (c) Dutch cone and triaxial tests. (After Windle and Wroth, 1977).

differences in strain conditions along the potential failure surface, the average strength mobilized in soils that exhibit strain softening is likely to be below the peak strength (soils such as sensitive clays and compacted fill).

The measurement of soil strength and particularly undrained strength is subject to a number of uncertainties. Disturbance is produced by the sampling process and preparation of specimens for laboratory tests and by the installation of equipment for carrying out *in-situ* tests. The measured strength is also likely to be affected by the rate of testing, the testing procedure and stress path followed. Significant differences can be expected in undrained shear strength profiles obtained by different testing methods (Windle and Wroth, 1977). This is illustrated in Fig. 11.5. The relevance of different shear tests to conditions along a potential failure surface has been considered by Bjerrum (1973), as illustrated in Fig. 11.6. Flaate (1966) has reviewed the factors influencing the results from vane tests and concluded that test data have to be calibrated against the type of problem in which they are to be used.

The best method of assessing the overall accuracy of any stability assessment procedure is from back analyses of failures occurring in the field. Parry (1971) undertook such a study using the case records of six embankments on soft ground and obtained factors of safety ranging between 1.3 and 2.0 from total stress analyses ($\phi_u = 0$). He drew attention to the factors that can influence undrained strength and recommended that both field and laboratory tests should be used in its determination and the lower values taken for stability calculations. Studies by various research workers have also shown that the assessment of shear strength using the standard field vane frequently overestimates the factor of safety at failure. On the basis of observed embankment failures, Bjerrum (1972) has therefore proposed the following corrections to the vane results to take account of the influence of time and anisotropy

$$c_u(\text{field}) = c_u(\text{vane}) \cdot \mu_r \cdot \mu_a \qquad (11.17)$$

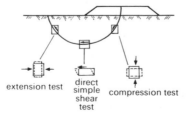

Figure 11.6 Relevance of laboratory shear tests of shear strength in the field (after Bjerrum, 1973).

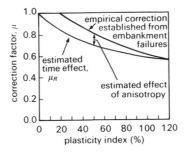

Figure 11.7 Empirically established correction factors to be applied to the results of a vane test in order to predict the shear strength which can be mobilized in the field over a period of some weeks to several months (after Bjerrum, 1973).

where μ_r is a factor correcting for the time effect,

and μ_a is a factor correcting for anisotropy, which varies according to the inclination (a) of the slip surface.

The forms of these correction factors are illustrated in Figs. 11.7 and 11.8 (after Bjerrum, 1972). Other failure studies have indicated non-conservative errors in these correction factors (Ladd, 1973; La Rochelle *et al.*, 1974), while studies by Dascal and Tournier (1975) suggest the need for an additional correction to take account of progressive failure in soft and sensitive clays.

The total stress method may overestimate the factor of safety, and its accuracy is limited by the difficulty of determining reliable values for the undrained strength of the soil. The method nevertheless forms an essential first step in assessing the stability of embankments on soft foundations, and if a reasonably conservative factor of safety that can be considered to take account of the above uncertainties can also be achieved economically, it provides a suitable basis for design.

The effective stress method has been extensively used for assessing long-term stability and results indicate that it is generally reliable. Moreover, the strength parameters in terms of effective stress are likely to be less affected by sampling disturbance because of

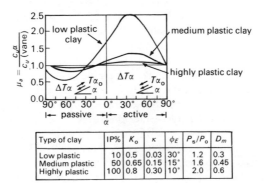

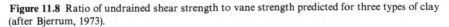

Type of clay	IP%	K_o	κ	ϕ_E	P_s/P_o	D_m
Low plastic	10	0.5	0.03	30°	1.2	0.3
Medium plastic	50	0.65	0.15	15°	1.6	0.45
Highly plastic	100	0.8	0.30	10°	2.0	0.6

Figure 11.8 Ratio of undrained shear strength to vane strength predicted for three types of clay (after Bjerrum, 1973).

reconsolidation of the test specimens. Use of the effective stress method for assessing short-term stability requires information on the pore-water pressures developed in the subsoil during embankment construction. These will vary with depth and lateral position from the embankment centreline and their magnitude will also change with time as a result of consolidation in the subsoil. In variable alluvial deposits accurate prediction of pore-water pressures is difficult and the effective stress method cannot be used reliably unless field measurements of pore-water pressure are made. There is therefore little to be gained from use of effective stress analyses if a satisfactory design can be achieved using total stress analyses. Pilot *et al.* (1982) have compared results of stability assessments using total and effective stress analyses for four embankments built to failure.

11.5.5 *Stability monitoring and control*

In the more critical situation where total stress analyses carried out at the design stage indicate an inadequate factor of safety for the completed works, techniques for improving stability must be considered (section 11.6). If sufficient time is available, consideration should also be given to modifying the construction schedule to take advantage of the improvement in stability which takes place as a result of the gain in strength in the subsoil during consolidation. Often this is likely to be the most economic solution but does necessitate close control over the construction. Where possible, an advanced earthworks contract should be let. This would include provisions for limiting the rate of placement of fill and for pauses in the work if required, when construction exceeds a specified (safe) height determined from the total stress analysis. This approach is likely to be particularly applicable where long lengths of embankment are required, such as for roads and flood control purposes. To provide an effective means of control during construction, the behaviour should be monitored at one or more key sections.

Probably the most widely used method of monitoring the stability and controlling the work has been the use of effective stress analyses linked to field measurements of pore-water pressure. An early example for the control of construction of an earth dam is given by Skempton and Bishop (1955). Lewis *et al.* (1975) and Cole and Garrett (1981) describe the use of this method in practice. In its simplest form the use of this approach involves undertaking calculations of the stability at intervals during the construction to determine the factor of safety based on the measurements of pore-water pressures. An alternative approach is for charts to be prepared in advance based on predictions of pore-water pressures. Because of the difficulty of accurately predicting the magnitude and distribution of pore-water pressures in variable alluvial deposits this can best be done when the embankment is constructed in stages. The predictions of pore-water pressure can then be based on measurements made during the first stage construction to a 'safe' height determined from total stress analyses. Using these predicted values the factors of safety can be calculated for a number of increased heights of embankment. The calculations are repeated using pore-water pressures above and below the predicted values to cater for changes in the rate of construction as well as changes in the pore pressure response to loading and dissipation. A chart can then be prepared for relating the pore-water pressures to the factor of safety for each height of embankment. As construction proceeds beyond the first stage the factor of safety can be estimated from the chart using the measured pore-water pressures and current height of embankment. A chart of this type was used to monitor the stability during construction of a road

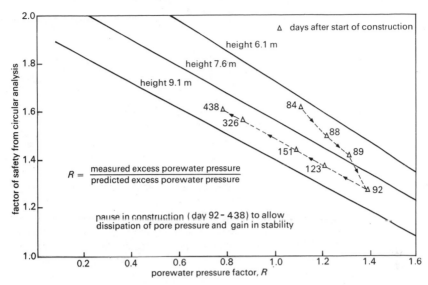

Figure 11.9 Relation between factor of safety and R for different heights of embankment at the over causeway site.

embankment on compressible subsoil, as shown in Fig. 11.9. Further details of this approach are given by Symons (1976).

In areas where the subsoil conditions are reasonably uniform and where previous failures or trial constructions have been closely monitored, some control over stability may be possible based on observations of ground movement. However, under normal conditions it is not possible to specify limits on either the magnitude or rate of movement which would constitute a generalized method of control. Routine observations on heave markers, profile boards, settlement gauges and inclinometers nevertheless provide a

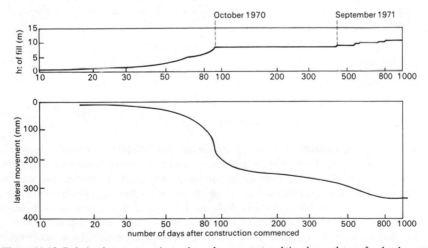

Figure 11.10 Relation between maximum lateral movement and time beneath toe of embankment.

useful qualitative assessment of performance. An example of the changes in maximum lateral movement recorded during construction on an inclinometer tube installed beneath the toe of an embankment is shown in Fig. 11.10.

11.6 Engineering solutions

At one end of the spectrum of possible engineering solutions to potential problems of settlement and instability of embankments is the complete removal and replacement of the soft compressible soils by excavation or controlled displacement during construction (Road Research, 1979, p. 101). This approach may well prove economic for relatively thin layers of very soft subsoil located close to the ground surface. Included in this category for highway work is the adoption of a structural solution in place of an embankment, such as a viaduct founded on piles that pass through the soft subsoil and terminate in a more competent stratum (Reid and Buchanan, 1983). Such a solution is likely to increase greatly the costs of construction, particularly where the compressible subsoil extends to an appreciable depth. At the opposite end of the spectrum of solutions, a comprehensive assessment of the engineering properties and likely behaviour of the subsoil is made at the design stage and construction is carefully controlled and monitored. This will usually require more extensive site investigations and may be supported by predictions of behaviour in advance of construction, based on results of model and analytical studies, or from trial constructions. The extent of such studies will depend both on the purpose of the earthworks (that is, an earth dam or a road embankment) and on the scale and economic consequences of the likely problems which may be anticipated from assessments carried out at the preliminary design stage. In general, such an approach is likely to produce a much more economic solution than methods of avoiding the problems, but may involve some time penalties if the duration of the construction period is extended.

Between these two extremes there are numerous construction expedients which directly or indirectly improve embankment stability or reduce the magnitude or time required for settlements to take place. The range of methods is too large for comprehensive review, and this section will therefore only consider the more common methods for practical use. These may be classified as involving one or more of the following:

(i) Modification of embankment loading or geometry.
(ii) Pre-consolidation of the soft soil.
(iii) Acceleration of consolidation.
(iv) Miscellaneous methods that increase stability and reduce settlement.

11.6.1 *Modification of loading or geometry*

The most direct method of reducing settlement and alleviating stability problems involves the use of lightweight fill. Many lightweight fill materials have been used for embankment construction in different countries. These include waste materials such as bark, sawdust, pulverized fuel ash, slag and clinker and manufactured materials such as lightweight aggregate and super lightweight expanded polystyrene. The most commonly-employed lightweight fill in the United Kingdom is pulverized fuel ash (PFA) which has a unit weight of approximately two-thirds that of normal fill. The cost of PFA is about the same as that of common fill when haulage distances are short. In contrast, the unit weight of expanded polystyrene is only one-twentieth that of normal fill and, although this material is expensive, it is receiving increasing application in Scandinavia and North America where there are extensive deposits of sensitive clays that require particularly careful treatment during construction.

Where the problems are mainly in terms of stability, a modification to the proposed design of cross-section to provide flatter side slopes, or alternatively to construct the embankment with berms, may provide the simplest solution.

The use of berms is generally preferable in terms of the volume of fill required for a given degree of stability. Moreover, since the main function of the berms is to provide a counterweight to offset the disturbing forces produced by the embankment, consideration can be given to the use of the cheapest available sources of material for the berm construction, including material that would be deemed unsuitable for the main earthworks. The designer must then ensure that the berms themselves have an adequate factor of safety in the short term. In some circumstances it may prove advantageous to erect the berms using normal density fill while lightweight fill is employed for the embankment. Such modifications to embankment geometry unfortunately necessitate greater land take.

11.6.2 *Pre-consolidation*

Where sufficient time is available, one of the most cost-effective procedures involves preloading the site in advance of the main contract such that shear strength is enhanced and settlements during and subsequent to the main works are reduced to within acceptable limits. Frequently the procedure will involve an advance earthworks contract whereby the embankment is constructed to the required height and cross-section together with additional fill to compensate for any subsequent settlement as well as perhaps an allowance for imposed construction such as, for example, a road pavement. Where stability problems are expected it may be desirable to use flatter side slopes than is planned for the final design, otherwise the construction may have to be delayed to permit the gain in shear strength of the subsoil as discussed previously.

Situations may arise where site access roads or embankment fills are unavailable and in these circumstances the technique of vacuum pre-loading may prove advantageous. In principle, an impermeable membrane is placed over the area to be treated and a pumping system is employed to reduce the subsoil pore pressure (Road Research, 1979 p. 89). Thus a differential surcharge pressure is created on the membrane between the atmospheric value and that of the pore pressure to induce consolidation. It should be noted that as consolidation is associated with a negative change in pore pressure, the danger of instability is avoided.

An alternative procedure, which may be considered as preconsolidation, is the technique referred to as dynamic consolidation, or more explicitly, impact compaction. In principle, the method involves the repeated dropping of large weights on to the ground surface in free fall. For soft soil applications, weights in excess of 100 kN dropped from heights of 30 m have been used (Menard and Broise, 1975). Such large dynamic forces are claimed to reduce the voids and create fissures which improve the rate of pore pressure dissipation. It is also to be expected that a considerable amount of displacement occurs with soft soils.

The compaction operation is carried out in a series of passes with delays to allow for pore-pressure dissipation. *In situ* test methods are employed during the treatment to monitor the ground condition and to establish when the required ground improvement has been attained.

11.6.3 Acceleration of consolidation

Extensive studies have demonstrated (Lewis *et al.*, 1975) that the rate of settlement of recent alluvial deposits is often very rapid as a result of natural drainage paths in these soils. The influence of such drainage paths is best evaluated by means of *in situ* permeability tests or large scale consolidation tests. The application of the results from small scale laboratory consolidation tests is not generally recommended for providing a reliable assessment of the rate of settlement, particularly where the need for special construction expedients such as described in this section is being considered.

The simplest and most direct method of accelerating consolidation is to construct the embankment to a greater height than is finally required. Assuming that the form of the relation between degree of consolidation and time is unaltered by the additional surcharge loading, then the time to achieve the ultimate settlement for the final height of embankment will be reduced by virtue of the fact that this value corresponds to only a proportion of the total settlement for the increased height. The surcharge is removed when the required ultimate settlement is attained. The use of surcharge can also be of value for alleviating secondary settlement.

Although the method is relatively low cost, it may not always provide a practical option, particularly if the problems are associated with instability since then the additional loading may not be feasible. Moreover, because the compressibility of the soil is related linearly to the logarithm of effective stress, very large additional loads may be required to significantly accelerate consolidation, particularly beneath high embankments.

To improve the effectiveness of the surcharge, it may be used in conjunction with lightweight embankment fill. Where it is proposed to use cohesive fills of low permeability for embankment construction it is essential to incorporate a blanket of high permeability granular material placed on the natural ground at the base of the embankment to allow vertical drainage to occur. More recently, geotextiles have been used in conjunction with granular soils for this application. However, care must be taken to ensure that the geotextile does not offer a preferential failure plane.

As discussed previously, the time to achieve a specific degree of consolidation is related to the square of the length of drainage path. It is thus possible to accelerate consolidation by also incorporating drains into the subsoil whereby the length of drainage path is significantly reduced. The most common procedure involves the installation of vertical drains through the subsoil at centres of typically about one metre.

A considerable range of vertical drainage systems is currently employed, including sand drains, cardboard wick drains, sand-wick drains and plastic drains (Institution of Civil Engineers, 1982). The design principles are the same for all types although their spacings may vary according to discharge capacity and surface area in contact with the soil. Horizontal drainage systems have also been employed in recent years whereby trenches are excavated in the compressible strata at intervals over the site and backfilled with free-draining granular fill. Currently the maximum depth of trench which can be excavated is about six metres. The use of such drainage systems can greatly increase the cost of the works and on many schemes this may add between 30 and 50 per cent to the earthworks costs.

Because drainage systems do not generally influence the rate of secondary consolidation, Bjerrum (1972) has proposed a method of assessing their effectiveness by means of an efficiency factor (η) defined as the ratio of primary consolidation settlement to total settlement. Where the proportion of primary settlement was small (η equal to 0.25 or less), the drainage systems were considered to be inefficient.

11.6.4 Miscellaneous methods

Several methods based on the use of stone columns or piles have had limited applications for reducing settlement and improving stability. One technique involves the formation of vertical stabilized columns within the subsoil by mixing unslaked lime (Road Research, 1979 p. 96). In addition to reducing the load carried by the soft ground, it is claimed that the increased permeability of the stabilized columns allows them to serve as drains.

Another method involves the use of relief piles installed beneath the loaded area. In some installations the piles extend upwards into the embankment, and have cappings of concrete, compacted granular material or fabrics to assist in the transmission of load from the fill to the piles (Reid and Buchanan, 1983). The method has an additional advantage for the approach embankments to bridges founded on piles, where a gradual transition in the longitudinal settlement profile between the embankment and the structure can be achieved by progressively increasing the length and reducing the spacing of the relief piles towards the abutment. Such a technique is likely to be effective but very expensive and should only be considered in special circumstances.

Reinforcement, comprising sheets of fabric, meshes of plastic or wire or strips of plastic or steel have been incorporated within the fill or at the subsoil-embankment interface to improve stability. The action of the reinforcement is to impart an increase in tensile strength through friction and adherence between the reinforcement and adjacent soil. For design purposes the reinforced zones can be considered to apply additional resisting forces against collapse although the actual reinforcing action is probably more complex, involving stress redistribution within the fill and foundation. The method has been used to a limited extent to contain shallow embankments on very weak foundations where otherwise very flat side slopes would be needed to prevent penetration and lateral spread of the fill.

Notation

A Skempton's pore-water pressure parameter

C_c compression index

c' cohesion

c'_r coefficient of primary consolidation

c_u undrained shear strength

c_a coefficient of secondary consolidation

c_r residual cohesion

E_u deformation modulus for undrained loading in terms of total stress

e_0 initial voids ratio

K_a coefficient of active earth pressure

K_0 coefficient of earth pressure at rest

K' ratio of increment of horizontal effective stress to increment of vertical effective stress

k coefficient of permeability

k_f coefficient of field permeability

LL liquid limit

PI plasticity index

T_v time factor

$\bar{u}$ average degree of consolidation

$\bar{u}_x$ average excess pore-water pressure in horizontal direction (x)

$\bar{u}_z$ average excess pore-water pressure in vertical direction (z)

$\bar{u}_{zx}$ average excess pore-water pressure for two dimensional consolidation

m_v coefficient of compressibility

$p' \quad = \frac{1}{2}(\sigma'_1 + \sigma'_3)$
$\left.\right\}$ stress parameters for $\sigma'_2 = \dfrac{\sigma'_1 + \sigma'_3}{2}$
$q \quad = \frac{1}{2}(\sigma_1 - \sigma_3)$

u pore-water pressure

γ_w unit weight of water

δ_c total primary consolidation settlement

δ_i total immediate settlement

δ_s total secondary consolidation settlement

η Bjerrum's drainage system efficiency factor

λ correction factor to compression index

λ_p correction factor to compression index for plane strain conditions

$\left.\begin{array}{c}\mu_r \\ \\ \mu_a\end{array}\right\}$ Bjerrum's correction factor for vane tests

μ correction factor for primary consolidation settlement .

σ total normal stress

σ'_0 initial overburden pressure in terms of effective stress

σ'_c effective preconsolidation pressure

σ_1 major total principal stress

σ_2 intermediate total principal stress

σ_3 minor total principal stress

$\Delta\sigma'_v$ increment in vertical effective stress

$\Delta\sigma_z$ increment in total stress in vertical direction z

$\Delta\sigma_x$ increment in total stress in horizontal direction x

τ shearing strength

ϕ' angle of shearing resistance in terms of effective stress

ϕ'_r residual angle of shearing resistance

References

Atkinson, J.H. and Bransby, P.L. (1978) *The Mechanics of Soils. An Introduction to Critical State Soil Mechanics.* McGraw–Hill Book Company Ltd, London.

Baguelin, F., Jezequel, J.F. and Shields, D.H. (1978) *The Pressuremeter and Foundation Engineering.* Trans Tech Publications, Clausthal, Germany.

Bassett, R.H. (1978) The centrifugal test as part of the design system for road embankments. In *Proc. Symposium Settlement and Stability of Earth Embankments on Soft Foundations.* Department of the Environment Department of Transport, TRRL Supplementary Report SR 399, Crowthorne Berkshire, England.

Bishop, A.W. and Henkel, D.J. (1962) *Measurement of Soil Properties in the Triaxial Test.* Edward Arnold Ltd., London.

Bjerrum, L. (1972) Embankments on soft ground. In *Proc. Speciality Conf. Performance of Earth and Earth Supported Structures,* ASCE, Vol. 2, pp. 1–54.

Bjerrum, L. (1973) Problems of soil mechanics and construction on soft clays. In *Proc. 8th Int. Conf. Soil Mech. Foundn Engng,* Moscow, Vol. 3, pp. 111–159.

Brady, K. (1982) *The application of programs ONEDIM and TWODIM to analysis of consolidation problems.* Dept. Environment Department of Transport, TRRL Supplementary Report SR 760, Crowthorne, Berkshire, England.

British Standards Institution, BS. 1377 (1975) Methods of Test for Soils for Civil Engineering purposes.

British Standards Institution, Code of Practice for Site Investigation (1981) BS5930:1981.

British Standards Institution, Code of Practice on Earthworks (1981) BS6031:1981.

Carrillo, N. (1942) Simple two- and three-dimensional cases in the theory of consolidation of soils. *Journal of Math. and Physics* **21** (1) March.

Cole, K.W. (1979) State of the art Review of Soil Sampling in the United Kingdom. In *Proc. International Symposium on Soil Sampling, Singapore,* pp. 171–186.

Cole, K.W. and Garrett, C. (1981) Two Road Embankments on Soft Alluvium. In *Proc. 10th Int. Conf. Soil Mech. Foundn Engng. Stockholm,* **1,** pp. 87–94.

Dascal, O. and Tournier, J.P. (1975) Embankments on soft and sensitive clay foundations. *Proc. ASCE, J. Geotech. Eng. Div. ASCE* **101** (GT3) 297–314.

Davis, E.H. and Poulos, H.G. (1968) The use of elastic theory for settlement prediction under three-dimensional conditions. *Géotechnique* **18** (1) 67–91.

Flaate, K. (1966) Factors influencing the results of Vane Tests. *Canadian Geotechnical J.* **3** (1) 18–31.

Gibson, R.E. (1966) A note on the constant head test to measure soil permeability in situ. *Géotechnique* **16** (3) 256–259.

Institution of Civil Engineers (1982) *Vertical Drains.* Thomas Telford, London.

Jurgenson, L. (1934) The application of theories of elasticity and plasticity to foundation problems. *J. Boston Soc. Civil Engrs.* **21** (3) 206–241.

Ladd, C. (1973) Invited discussion on Bjerrum (1973) In *Proc. 8th Int. Conf. Soil Mech. and Foundn. Engng. Moscow*, Vol. 4.2 pp. 108–115.

Lambe, T.W. (1964) Methods of Estimating Settlement. *J. Soil Mechanics and Foundation Division, ASCE*, **90**, (SM5), 47–22.

Lambe, T.W. and Whitman, R.V. (1969) Soil Mechanics. John Wiley and Sons Inc., London.

La Rochelle, P., Trak, P., Tavenas, F. and Roy, M. (1974) Failure of a test embankment on sensitive Champlain clay. *Canadian Geotechnical J.* **11**, (1), 142–164.

Lewis, W.A., Murray, R.T. and Symons, I.F. (1975) Settlement and stability of embankments constructed on soft alluvial soils. *Proc. Instn. Civ. Engrs.*, Part 2, **59**, Dec., 571–593.

Meigh, A.C. and Corbett, B.O. (1970) A comparison of in situ measurements in a soft clay with laboratory tests and the settlement of oil tanks. In *Proc. Conf. in situ investigation of soils and rocks*, British Geotechnical Society, London, pp. 173–179.

McGuffey, V., Grivas, D., Iori, J and Kyfor, Z. (1982) Conventional and Probabilistic Embankment Design. *Geotechnical Eng. Div. ASCE* **108** (GT10), 1246–1254.

Menard, L. and Broise, Y. (1975) Theoretical and practical aspects of dynamic consolidation. *Géotechnique* **25** (1), 3–18.

Murray, R.T. (1978) Development in two- and three-dimensional consolidation theory. In *Developments in soil mechanics*–1, Applied Science Publishers Ltd., London, pp. 103–147.

Parry, R.G.H. (1971) Stability analysis of low embankments on soft clays. In *Proc. Roscoe Memorial Symposium, Cambridge*, pp. 643–668.

Peck, R.B., Hanson, W.E. and Thorburn, T.H. (1974) *Foundation Engineering*. John Wiley, New York. p. 310.

Pilot, G., Trak, B. and La Rochelle, P. (1982) Effective stress analysis of the stability of embankments on soft soils. *Canadian Geotechnical J.*, **9**, 433–450.

Reid, W.M. and Buchanan, N.W. (1983) Bridge Approach Support Piling. In *Proc. Conf. on Advances in Piling and Ground Treatment for Foundations*. Instn. Civ. Engnrs, London, (in press).

Report of the Subcommittee on Standardisation of Penetration Testing in Europe (1977) In *Proc. 9th Int. Conf. Soil Mech. Foundn. Engng. Tokyo*, III, pp. 95–152.

Road Research (1979) *Construction of Roads on Compressible Soils*. Organisation for Economic Co-operation and Development, Paris.

Rodin, S., Corbett, B.O., Sherwood, D.E. and Thorburn, S. (1974) Penetration testing in the United Kingdom. In *Proc. European Symposium on Penetration Testing, Stockholm*, Vol. 1.

Rowe, P.W. (1972) The relevance of soil fabric to site investigation practice. *Géotechnique* **22** (2) 195–300.

Sanglerat, G. (1972) *The Penetrometer and Soil Exploration*. Elsevier Publishing Co., Amsterdam.

Schmertmann, J.H. (1967) Static cone penetrometers for soil exploration. *Civil Eng.* **37** (6) 71–73.

Scott, R.F. (1963) *Principles of Soil Mechanics*. Addison–Wesley Publishing Company Inc., London, p. 284.

Simons, N.E. and Som, N.N. (1969) The influence of lateral stresses on the stress deformation characteristics of London Clay. In *Proc. of 7th Int. Conf. Soil Mech. Foundn Engng, Mexico*, Vol. 1, pp. 369–377.

Simons, N.E. (1971) The stress path method of settlement analysis applied to London Clay. In *Proc. Roscoe Memorial Symposium, Cambridge*, pp. 241–252.

Simons, N.E. (1975) Normally consolidated and lightly over-consolidated cohesive materials. In *Proc. Conference on Settlement of Structures*, Pentech Press, London.

Skempton, A.W. and Bishop, A.W. (1955) The gain in stability due to pore pressure dissipation in a soft foundation. In *Proc. Cinquième Congrès des Grandes Barrages, Paris*, I, pp. 613–638.

Symons, I.F. (1976) *Assessment and control of stability for road embankments constructed on soft subsoils*. Department of the Environment TRRL Report LR 711, Crowthorne. (Transport and Road Research Laboratory).

Terzaghi, K. (1943) *Theoretical Soil Mechanics*. John Wiley and Sons, New York, pp. 265–296.

Terzaghi, K, and Peck, R. (1967) *Soil Mechanics in Engineering Practice*. John Wiley and Sons, New York.

Windle, D. and Wroth, C.P. (1977) The use of the self boring pressuremeter to determine the undrained properties of clays. *Ground Engng.* **10** (6) 37–46.

12 Seismic movements

N. AMBRASEYS and J. JACKSON

12.1 Introduction

The intention of this chapter is to present the background on earthquakes—their generation and characteristics—and to review the current state of knowledge with respect to seismicity assesment. Principal attention will be directed to the recent geological record, which is most valuable for estimating earthquake hazard, and to the superficial geology that controls to a considerable extent the vulnerability of engineering structures. Earthquake risk assessment requires knowledge of both the earthquake hazard and the vulnerability of the structure exposed to the hazard.

12.2 The nature of earthquake movements

12.2.1 *Faults and energy release*

Earthquakes result from the catastropic release of elastic strain energy stored in the Earth's crust. This energy is released by brittle failure on dislocations, known as faults, and thus earthquakes give rise to two types of surface displacement: permanent offsets on the fault itself, and transient displacements resulting from the propagation of seismic waves away from the source. Both types are of importance to the engineer. The ultimate source of the energy released in earthquakes is the heat loss of the Earth, which is responsible for the large-scale horizontal and vertical movements of the crust, including continental drift, that in turn create the major topographic features of the Earth's surface. Elastic strain energy builds up on a fault, which is held static by friction, until eventually slip is initiated at a point (known as the hypocentre or focus) and rapidly spreads out over the rupture surface. Because it is energetically more favourable to re-use a pre-existing surface than to initiate a new one, the same faults move repeatedly in successive earthquakes. Furthermore, old faults inherited from some earlier episode of continental deformation can be reactivated in present day earthquakes. Thus in New York State small earthquakes are occurring at the present day on the Ramapo Fault, a structure that first formed at least 600 million years ago (Sykes, 1978). Most hypocentral depths on the continents are in the range 5–20 km, though depths down to 650 km are found in areas within and bordering on the major ocean basins.

12.2.2 *Types of fault*

The three basic types of fault motion are called thrust, normal and strike slip, and involve shortening, extension and lateral movement of the crustal surface respectively (Fig. 12.1).

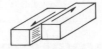

strike-slip fault

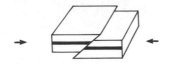

thrust fault (compressional)

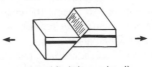

normal fault (extensional)

Figure 12.1 Basic types of geological fault motion.

Although one of these three types of motion is usually dominant in a single earthquake, both normal and thrust faults do sometimes move with a strike slip component as well. Within the very short part of the geological time scale of interest to engineers, the sense of motion on a particular fault is observed to be the same in successive earthquakes. Repeated movement on the same fault will therefore create a topography in active regions characteristic of the type of faulting: thrust faults are common at the base of mountain slopes, normal faults often form the steep-sided edges of lakes or coastlines, and strike slip faults give rise to long linear valleys. This allows the geomorphology associated with the sense of motion to be used as an important guide in identifying active faults.

12.2.3 *Quantification of earthquake size*

Attempts to quantify earthquake size by the amount of damage caused to man-made structures at the epicentre (the point on the earth's surface vertically above the hypocentre) are not satisfactory. Not only are such surface effects dependent on local conditions, but a particularly high level of damage (or intensity) can be achieved in either a small earthquake nearby or a large one further away. The magnitude scale was devised in an attempt to describe earthquake size according to the amount of energy released at the focus, and is logarithmic, varying from about -2 to about 9 (see equation 12.1, and Richter, 1958). Magnitude determination is in fact made by measuring the amplitude of seismic waves at particular frequencies and correcting for distance from the epicentre. Measurements are commonly made with either surface waves at 20 seconds period or body waves at 1 second period and the corresponding magnitudes called M_s and m_b, respectively. Local magnitude (M_L) was originally measured from the maximum amplitude recorded on a Wood–Anderson seismograph. Although easy to make and widely used, magnitude determinations have several shortcomings: their relation to total energy release is only empirical and it is difficult to relate magnitude to other important source parameters such as rupture length, fault displacement, stress drop and radiated seismic energy, all of which can be easily related to each other but can only be related to magnitude if the spectral description of the source (that is, the complete space and time history of faulting) is known. In particular, the greater the size of an earthquake and its causative fault, the more efficiently long period waves are generated. Thus the position of the 20 second (for M_s) or 1 second (for m_b) period within the emitted seismic spectrum will vary with the size of the earthquake. As long as the wavelength at which the magnitude is measured (roughly 80 km for M_s, 6 km for m_b) is long compared to the length of the radiating fault, the logarithmic dependence of the seismic energy (E_s) on magnitude is fairly good. Where the length of the fault is comparable to the wavelength involved, the magnitude is said to 'saturate', that is, the slope of the magnitude v. energy curve starts to flatten. This happens at about 5.5 for m_b and about 6.5 for M_s and is responsible for the observed upper limit to the magnitude scale. In place of the empirical relation $\log E_s = 11.8 + 1.6 M_s$ given by Gutenberg and Richter (1956), Kanamori and Anderson (1975) suggest the following three approximate relationships:

$$\left. \begin{array}{l} \log E_s \propto 3M_s \quad \text{for very large earthquakes} \\ \log E_s \propto 1.5M_s \quad \text{for most moderate earthquakes (up to about } M_s 7) \\ \log E_s \propto M_s \quad \text{for very small earthquakes.} \end{array} \right\} \qquad (12.1)$$

Because of these drawbacks, seismologists generally prefer to use seismic moment (M_0) as the measure of earthquake size. This is related to the rigidity of the medium (μ, $\approx 3 \times 10^{-11}$ dyne-cm), the fault area (A) and the average fault displacement ($\bar{u}$) by

$$M_0 = \mu \cdot A \cdot \bar{u} \qquad (12.2)$$

As well as being directly measurable in the field, M_0 can also be measured from the long period (theoretically zero frequency) level of the radiated seismic spectrum. Observational data from large earthquakes show a remarkable consistency in the fault displacement to length ratio, which typically has a value of about 10^{-4}. This can be shown to imply a roughly constant average stress drop ($c \cdot 10$–100 bars) in earthquakes within particular tectonic environments, and leads to a theoretical relationship between E_s and M_0 whereby they are roughly proportional to each other. This allows relations between M_0 and M_s to be estimated. Empirical relations also exist between M_s and m_b. Typical values for these parameters, for earthquakes of a variety of sizes, are shown in Table 12.1. From this it can be seen that large earthquakes account for most of the seismic motion and energy release in the Earth, and that smaller earthquakes, although much more frequent in occurrence (see section 12.5), collectively account for much less. Almost all shallow earthquakes of M_s greater than about 5.0, on close examination, are found to be associated with faulting visible at the surface.

12.2.4 *The engineering problem*

For the engineer assessing the risk due to fault-related ground movement at a site, the main questions are: (1) where are the major and minor faults in the area? (2) are they active? (3) what are the likely displacement amplitudes on any active or potentially active faults that cross the site itself? (4) what are the likely durations and amplitudes of the transient accelerations the site might experience from both near and distant earthquakes? The remainder of this section is directed at these questions.

Table 12.1 Some earthquake data. Surface wave (M_s) and body wave (m_b) magnitudes, moments (M_0), fault areas (A), average displacement ($\bar{u}$), fault length (L) and strain drop (u/L) for shallow earthquakes of a variety of sizes. These data are from various sources and are intended to illustrate the wide range of M_0 values, the saturation of M_s and m_b at high magnitudes, and the roughly constant value of u/L.

Place	Date	M_s	m_b	M_0 ($\times 10^{27}$ dyne-cm)	A ($\times 10^3$ km^2)	$\bar{u}$ (m)	L (km)	u/L ($\times 10^{-4}$)
Alaska	1964.3.28	8.5	6.4	820	130	21.0	650	0.3
Iran	1968.8.31	7.3	6.0	1.0	1.6	2.0	80	0.2
Turkey	1967.7.22	7.1	6.0	0.83	1.6	1.7	40	0.4
Algeria	1980.10.10	7.3	6.5	0.25	0.38	2.2	25	0.8
Greece	1981.2.24	6.7	5.9	0.08	0.15	1.1	15	0.7
California (Truckee)	1966.9.12	5.9	5.4	0.0083	0.1	0.3	10	0.3
California (San Fernando aftershock)	1971.2.10 (11 31 34 G MT)	—	4.2	0.000001	0.00002	0.02	0.1	2.0

The decision to undertake measures to protect engineering structures from ground movements produced by earthquakes is usually based on assessments of the risks to the community and to its economy as well as judgement as to whether these risks are acceptable. The data needed for the quantification of risks are at present insufficient to support predictions with a high degree of certainty. However, we do have a conceptual basis for progress in this field.

Earthquake risk may be defined as the probability of the loss of property or loss of function of engineering structures, life, utilities, and so on. The factors entering into the assessment or qualitative estimation of earthquake risk are, the *earthquake hazard* (the probability of occurrence of ground motions due to an earthquake), the *value* of the elements exposed to the hazard (property and lives), and the *vulnerability* of these elements to damage or destruction by ground motions associated with the hazard (Fournier d'Albe, 1982).

In their simplest functional form the factors entering into the assessment of risk may be expressed by the following relation:

$$(\text{risk}) = (\text{earthquake hazard})*(\text{vulnerability})\ (\text{value})$$

or

$$(\text{risk/unit value}) = (\text{earthquake hazard})*(\text{vulnerability}).$$

This relation may be applied to estimate the risk, in financial or economic terms, to any object or set of objects—buildings and their contents, a structure of any kind, a utility network—as well as to assemblies of such objects.

Value, or damage, may be taken as either in the sense of loss of capital or loss of production. *Vulnerability* is a measure of the proportion of the value, as defined above, which might be expected to be lost as a result of a given earthquake. It can be calculated in a deterministic manner for individual structures, if their dynamic behaviour to various earthquake ground motions is known, and can, by summation, be estimated for a group of buildings, a city or a region. *Earthquake hazard*, in this context, is the probability of occurrence, at a given place or within a given area and within a given period of time, of ground motion due to an earthquake capable of significant loss of value. It may normally be expressed by sets of figures indicating the probability of occurrence of certain ground accelerations, velocities and displacements, of ground movements of various duration, or any other physical parameter which is of significance for the vulnerability of a structure.

The assessment of risk therefore involves the assessment of three independent factors: Value, Vulnerability and Hazard.

(1) The assessment of the economic *value* of structures and assemblies of objects is beyond the scope of this chapter.
(2) For the assessment of *vulnerability* as a function of siting, building type and construction quality, we rely mainly on field observations. Every damaging earthquake affords an opportunity to observe and study the damage caused by strong ground motions, and there is already an abundant literature on the subject. These studies, while making it possible to test the validity of certain earthquake engineering hypotheses and techniques, have so far not yielded much information permitting one to forecast the extent of damage likely to be caused in the event of future earthquakes. The principal reason for this is that only in very few cases has the ground motion during earthquakes been recorded by appropriate instruments. It has thus been generally difficult to correlate damage directly with the parameters of ground motion

and thence to derive relationships of predictive value (see Ambraseys and Jackson, 1981).

(3) There remains the problem of making estimates of the *earthquake hazard*, that is, probability of occurrence, within a given period of time (usually determined by the useful life of the structure in question), at a given site or within a given area, of earthquake ground motion capable of causing significant damage. This problem is a complex one, since it involves: (a) estimating the probability of occurrence of earthquakes of various magnitudes at various distances from the site, (b) identifying the probable source, fault and mechanism of such earthquakes, (c) determining the attenuation factors affecting the propagation of seismic waves between the probable earthquake source areas and the site, and (d) estimating the influence of the local subsoil properties and of the structures themselves on the characteristics of strong ground motion at the site. It is these factors that provide answers to the questions posed at the start of this section.

12.3 Permanent ground displacement on faults

12.3.1 *Faults that generate major earthquakes*

It is clear from the discussion in section 12.2.3 that, because of the observed relation between displacement and length, large fault offsets only occur in major earthquakes on large faults. However, along the length of the fault-break, surface ruptures are neither continuous, nor do they follow precisely the surface outcrop of pre-existing faults. They seem rather to follow planes of weakness within a comparatively broad shear zone from a few metres to a few kilometres wide, shifting laterally from a shear plane in one part of the zone to another elsewhere, in most cases with well-developed large-scale *en échelon* patterns. Tension features and grabens are often connected with *en échelon* shears (Fig. 12.2 and 12.3) (see Ambraseys and Zatopek, 1969; Tchalenko and Ambraseys, 1970.)

These details will show if the fault zone is mapped on a large scale. Mapping on a smaller scale would tend to obliterate the details of the actual shear pattern and the fault would appear as a continuous and smooth trace. Widths of shear zones associated with strike slip faulting range up to 2 km, while widths in normal faulting may be twice as broad, up to 5 km. In thrusting earthquakes, shear and fracture zones can be much wider, depending on the dip of the fault plane. For shallow dips, the width of the fracture zone may reach up to 10 km, a typical example being the earthquake of 16th September 1978 at Tabas in Iran (Berberian, 1979; Ambraseys, 1981).

Both horizontal and vertical displacements on individual ruptures within the fault zone can be large, and lead to the formation of grabens and pressure ridges. Usually the width of the fault zone is narrowest where strike slip fault-breaks occur, becoming wider for normal and thrust faults. All ruptures and other deformations associated with a fault zone will clearly have serious consequences for engineering structures within these zones. As well as seismic movements occurring during the mainshock, both post-seismic creep and fault movement in aftershocks may cause additional damage to structures.

12.3.2 *Small faults*

Although only large faults, which are comparatively easy to recognize, are responsible for major earthquakes, small faults are often of great importance to the engineer, particularly

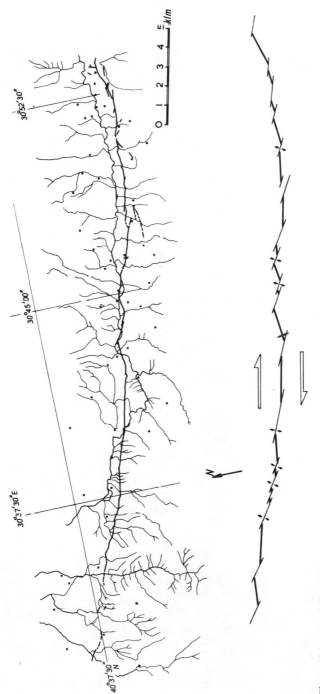

Figure 12.2 Map of main fractures in central part of the Mudurnu (Turkey) fault zone associated with the earthquake of 22 July 1967, showing generalized en echelon pattern of main fractures (small arrows show tensional fissures associated with en echelon shear).

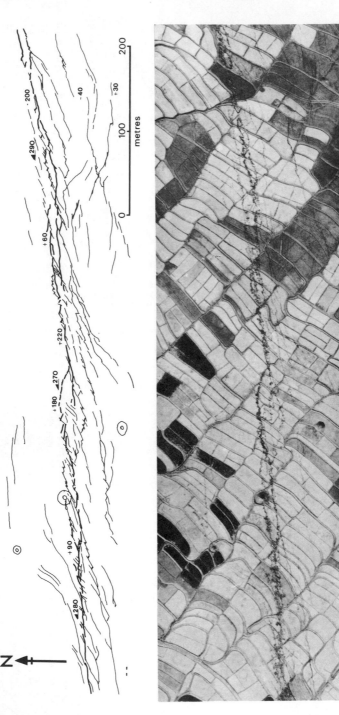

Figure 12.3 Aerial photograph of a small part of the 80 km long fault-break associated with the earthquake of 31 August 1968 in Iran with mapped ground deformations drawn on the same scale. Figures on the map marked with plus or minus indicate relative vertical ground displacements in centimetres; arrows indicate direction of motion (left-lateral) and magnitude of relative horizontal displacement in centimetres. Location of Figure 12.3,800m SE of the village of Dasht-e-Bayaz. (For exact location of the features shown in this figure, and for photographs and descriptions of other types of ground deformation, see Ambraseys, N and Tchalenko, J. 1969. "Dasht-e-Bayaz earthquake of 31 August 1968". UNESCO publ. No. 1214/BMS/RD/SCE. Paris).

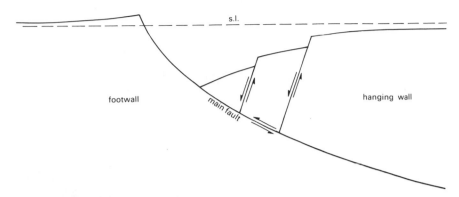

Figure 12.4 Examples of small faults resulting from internal deformation of blocks bounded by major faults. Although these small fault are not responsible for generating large earthquakes, displacements on them resulting from nearby earthquake notion on major faults can be large; so they are of importance to the engineer..

(a) (Above) Cross section of a normal fault, showing a curved (listric) geometry in which the fault is concave upwards. This is a common geometry of normal faults and requires severe internal deformation of the hanging wall block as it moves down the main fault. This internal deformation is seen as numerous minor faults in the hanging wall (see Jackson and Mckenzie, 1983).

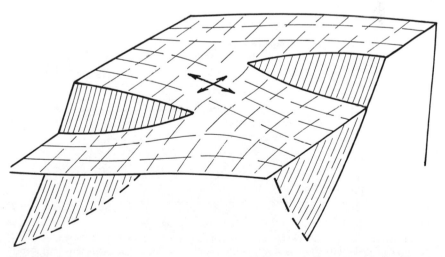

Figure 12.4 (b) Cartoon to show the warped surface between two offset normal faults. The area between the ends of the major faults is often broken up by numerous small faults (see Jackson *et al.*, 1982 and Figure 12.4c).

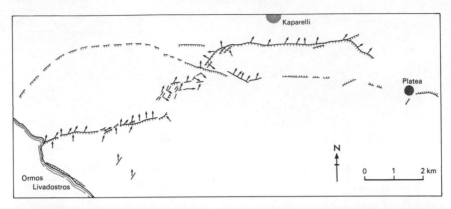

Figure 12.4 (c) Sketch of the normal faulting associated with the 1981.3.4. earthquake in the Gulf of Corinth (Greece). Faults marked with heavy lines moved at the surface in this earthquake. Other faults, which did not visibly move at the surface, are shown in lighter lines. Ticks are on the downthrown side (hanging wall). Open fissures have no ticks. Arrows show the horizontal direction of displacement across the faults or fissures. The major faulting occurred on two offset east-west segments, each 6 kilometres long and downthrown to the south. On these segments the vertical displacement was about one metre and the horizontal projection of the slip vector was consistently in a northern direction. In the area between the two offset segments displacements were smaller (typically 2–30 cm) and widely varying in direction (cf. Figure 12.4(b)).

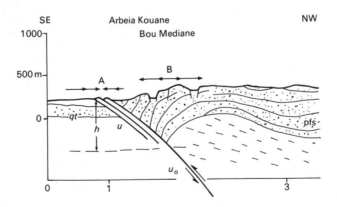

Figure 12.4 (d) Cross section of the thrust fault (A) responsible for the El Asnam (Algeria) earthquake of 1980.10.10, showing severe normal faulting in the hanging wall above the thrust plane (B). Displacements on these normal faults was up to to one metre and they were continuous for distances of 1 km or more along the strike (see Figure 12.4(e)). Displacement on the thrust in places reached 2 to 4 metres (see Figures 12.4(f) and 12.4(g)).

Figure 12.4 (*e*) Normal faulting corresponding to features 'B' in Figure 12.4, located between Oued Fodda and Arbeia Kouane, about 1.4 km NNW of Oued Fodda R. Station; looking SW.

Figure 12.4 (*f*) Compression ridges along surface trace corresponding to thrust zone 'A' in Figure 12.4, running SW, from north-west of Oued Fodda towards Zabadja, looking SW.

Figure 12.4 (*g*) Effect of compression on irrigation channel crossing thrust zone 'A', about 0.6 km NNW of Oued Fodda R. Station, looking NE. Channel was telescoped with a total shortening of 1.9 m over a length of 90 m.

if he plans to build structures actually across them. In this case, motion of only a few centimetres may be critical in deciding the relative merits of different types of structure. In many cases, small faults have been observed to move in response to motion on large faults nearby. This can happen during the mainshock, in aftershocks, or as the result of post-seismic creep. Although these small faults cannot themselves generate serious earthquakes, they can have substantial displacements on them. The reason for this is that their presence is usually the result of internal deformation in the blocks either side of the main seismogenic fault, resulting either from the geometry of the main fault surface or from non-uniform slip on it. Examples of such effects are shown in Fig. 12.4. When the fault surface is not vertical, internal deformation is usually most severe in the block overlying the fault plane (the hangingwall block); as seen in the numerous aftershocks that occur within this wedge, and its often very broken-up character at the surface. The amplitudes of these deformations accompanying the main faulting are observed to be highly variable. They need not extend to any great depth, do not follow the 10^{-4} ratio of u/L (displacement/length) seen for seismogenic faults, and their displacements are often exacerbated by local conditions, especially variations in soil, lithology and topography.

12.3.3 *Fault creep*

Creep movements, too slow to radiate seismic energy, have been observed on faults of all scales in many parts of the world, and may occur before or after major earthquakes. In some places, such as the central part of the San Andreas Fault in California, creep probably accounts for motion of up to 1–2 cm/yr (see for example Steinbrugge and Zacher, 1960). In the Alpine–Himalayan mountain system the seismic moment release can account for less than half the overall motion of India and Africa northwards into Eurasia (North, 1974; Chen and Molnar, 1977). The rest of this motion is assumed to take place by a seismic creep. Fault creep can only be monitored by numerous instruments installed close to the fault, and, where it has been detected, is usually found to be episodic. Creep is thus difficult to monitor on a worldwide basis, whereas earthquakes of m_b 4.6 or greater can be detected and located with the existing worldwide network of seismograph stations. Faults which exhibit creep usually have earthquakes as well, and, like other active faults, tend to produce geomorphological features which aid greatly in their recognition.

12.4   Recognition of active faults

12.4.1 *Degrees of activity*

Using geological, historical and seismological data it is often possible to assess the relative activity of a recognizable fault, and classify it in one of the following categories: (a) active, (b) potentially active, (c) uncertain activity, and (d) inactive.

(a) Active faults: these show historical or recent surface faulting with associated strong earthquakes; fault creep or geodetic indications of fault movement; geologically young deposits displaced or cut by faulting; fresh geomorphic features characteristic of active fault zones along the fault trace; physical groundwater barriers (spring lines) in geologically young deposits; offset streams. Seismologically, earthquake epicentres arc associated with individual faults with a high degree of confidence, including historical earthquakes that were felt over a wide area.

(b) Potentially active faults: no reliable report of historic surface faulting; faults are not known to displace or cut the most recent alluvial deposits; geomorphic features characteristic of active fault zones are subdued, eroded and discontinuous; water barriers may be present in older materials; the fault may be in a geological setting in which its geometric relationship to nearby active or potentially active faults suggests similar levels of activity. There may by an alignment of some earthquake epicentres along the fault trace, but locations are assigned with a low degree of confidence.

(c) Faults of uncertain activity: available information is insufficient to provide criteria that are definitive enough to establish fault activity. If the fault is considered critical to the project, additional studies are necessary.

(d) Inactive faults: no historical activity based on a thorough study of local sources of information. Geologically, features characteristic of active fault zones are not present, and geological evidence is available to indicate that the fault has not moved in the recent past. The lack of known epicentres on the fault traces is, of course, a prerequisite, but is *not* sufficient evidence to classify a fault as inactive.

12.4.2 *Data for assessing fault activity*

12.4.2.1 *Geological.* An active fault is indicated by young geomorphic features such as fault scarps, fault rifts, pressure ridges, offset streams, enclosed depressions, fault valleys, uplift and subsidence of shorelines and river terraces (where these are seen to be related to the surface outcrop of the fault), and ground features such as open fissures, 'moletracks', rejuvenated streams, folding or warping of young deposits, groundwater barriers in recent alluvium, *en échelon* faults and slickensides on recent surfaces. Usually, a combination of these geomorphic features is generated by fault movements at the surface. However, where there is a lack of young sedimentary material it may be impossible to assess the age of the latest movements on a fault system. Erosion features are not necessarily indicative of active faults, but may be associated with some active zones.

12.4.2.2 *Historical.* Useful information may be obtained from historical sources examined at first hand, local tradition and from sedentary people in the region. Fault movements, including creep, may be detected from displaced man-made structures such as early canals (on the surface or underground), roads, archaeological sites, railway and power lines.

12.4.2.3 *Seismological.* Clearly, if earthquakes can be positively associated with a fault, that fault is active. For this reason much attention is given to the evaluation of instrumental seismic data for earthquake risk. It is important for the engineer to appreciate the inherent inaccuracies and limitations of seismic data.

Epicentral and hypocentral locations are calculated from the arrival times of seismic waves recorded by seismograph stations. To do this, a model for the velocity of propagation of seismic waves in the Earth must be assumed. Although most earthquakes of $m_b > 4.5$ are routinely recorded and located using the existing worldwide network of seismographs, errors in the assumed Earth velocity model, as well as errors in reading arival times and the uneven distribution of recording stations, lead to considerable location errors. These errors are greatest for poorly-recorded small earthquakes, but even for large shocks epicentres can be wrong by up to 30–40 kilometres (see e.g. Jackson *et al.*, 1982). Epicentral errors before about 1960, when there were fewer seismograph stations and timing was bad, are much greater (Ambraseys, 1978). When the size of these errors (Fig. 12.5) is compared with the length of fault involved in the earthquake (Table 12.1) it can be seen that even a shock of M_s 6.0 could easily be assigned to the wrong fault on the basis of epicentral location alone. Errors in hypocentral depth are even larger, commonly being 50–100 kilometres for small shocks (Jackson, 1980). By installing a

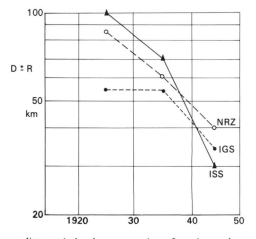

Figure 12.5 (*a*) Average distance in km between region of maximum damage and instrumental epicentre for earthquakes in Iran as a function of the year of occurrence. Figure shows large errors in teleseismic locations for early events (see Ambraseys, 1978).

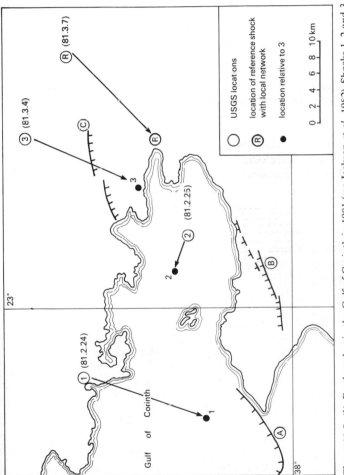

Figure 12.5 (b) Earthquakes in the Gulf of Corinth in 1981 (see Jackson et al., 1982). Shocks 1, 2 and 3 were the main shocks of magnitudes M_s 6.7, 6.4, 6.4 and associated faults A, B and C, respectively. Normal faults are marked as in Figure 12.4(c). R is a later aftershock of M_s 4.8. All four were located by the U.S. Geological Survey (USGS) using the arrival times of seismic waves at worldwide seismograph stations and a standard earth velocity model. These locations are not satisfactory: 1 is a long way from its fault break (A) and 3 is on the wrong side of its fault (C) (the epicentre should always be in the hanging wall), and R is 15 kilometres from its real epicentre, which was accurately determined by a local network of portable seismographs installed after event 3. By locating 1, 2 and R relative to 3, an accurate epicentre pattern is obtained, which can be placed geographically by positioning R over its known, locally determined, epicentre. These new locations are a great improvement: 3 is now the correct side of its fault, (C), and 1 is nearer to the fault A. Arrows link the USGS locations to the new locations and show the former to be wrong by up to 15 km.

network of local seismograph stations, in which the station spacing and the minimum distance from the epicentre is the order of the focal depth, these errors can be dramatically reduced, and in the best possible situations may reach 1–2 kilometres in epicentre and about 4 kilometres in focal depth (see for example Soufleris *et al.*, 1982). Most large earthquakes happen unexpectedly and not within local networks, but, in spite of this, improvements in their locations can be made. A group of earthquakes can be located *relative* to each other using the *difference* in their arrival times at world-wide stations; in this way a much improved pattern of locations is obtained which can be placed geographically if one of the pattern, such as an aftershock, was also recorded by a local network so that its *absolute* location is accurately known. This can reduce errors in epicentral position to about 5 kilometres (Fig. 12.5c and Jackson *et al.*, 1982). Focal depth estimates can be much improved by modelling the *shape* of the waveforms recorded at distant stations using synthetic seismograms generated on a computer. This can give depth errors of ± 4 kilometres (Jackson and Fitch, 1981), but, because a good signal-to-noise ratio is required, the technique is only routinely applicable to earthquakes larger than about M_s 5.5.

The pattern of seismic radiation emitted by an earthquake is strongly dependent on the orientation and type of fault movement. It is possible to make use of this to deduce such geometrical properties of the earthquake source from observations made at seismic stations with a varied azimuthal distribution from the epicentre (see for example Aki and Richards, 1980; Kasahara, 1981). Such studies yield 'fault plane' (or 'focal mechanism') solutions, which are estimates of the orientation of the fault plane and slip vector in space. Again, a good signal-to-noise ratio is needed, so that this technique is routinely applicable to shocks larger than about M_s 5.5 unless they are recorded by a dense local network. The best-quality fault plane solutions can easily distinguish between thrust, normal and strike slip faulting and can estimate the trend and inclination (strike and dip) of the fault plane to within about 20°. It is important to remember that fault plane solutions give the orientation and type of faulting at the point of rupture initiation only (for this reason they are also called 'first motion solutions') and that this may not be representative of later stages of seismic slip, since other nearby faults, with different orientations, may become activated in the same earthquake.

With these techniques and their errors in mind, let us consider the uses and limitations of instrumental seismic data. High-quality instrumental data have been available only since about 1960. Fault plane solutions and epicentral locations of the major earthquakes since that time will give some idea of the type and orientation of active faults in the region, and may be able to identify some of the causative faults themselves. It will generally not be possible to associate earthquakes of $M_s \leq 5.5$ with any particular surface structure. Installation of a dense local seismograph network may indicate whether suspicious structures, identified by historical data, fault plane solutions or geomorphological observations, are active. It must be borne in mind that microearthquakes may well not define a major dipping fault plane but may occur on numerous small faults in the blocks either side of it. Since most continental earthquakes have focal depths between 5 and 20 kilometres, the location errors associated with even the best local seismic networks exclude any hope of positively associating microearthquakes with particular small faults only a few hundred metres in length at the surface. As mentioned earlier, the significance of such faults is anyway much more connected with their proximity to major faults and possible activation in response to nearby large earthquakes. Finally it must be remembered that the apparent lack of earthquakes on a fault does not necessarily mean that the fault is inactive.

The following empirical relation (from Ambraseys and Melville, 1982) may be used to estimate the expected surface-wave magnitude of an earthquake (M_s) which might result from faults of known of inferred length (L) and co-seismic displacement (R):

$$M_s = 1.1 + 0.4 \log(L^{1.58} R^2) \qquad (12.3)$$

where both the maximum relative displacement at the time of the earthquake (R) and the length of rupture (L) are in centimetres. Maximum observed displacements are usually 2 to 4 times larger than average displacements and they may include a component due to aftershocks and creep. For shallow earthquakes (focal depth $h < 20\,km$) of magnitude greater than 6.0, the length of rupture in kilometres may be assessed from

$$\log(L) = 0.7(M_s) - 3.24 \qquad (M_s \leq 7.5) \qquad (12.4)$$

Other semi-empirical formulae relating length of rupture with magnitude for shallow earthquakes are given by Tocher (1958); Iida (1959, 1965); Housner (1969); King and Knopoff (1968); Brune et al. (1967) and Wyss (1979).

12.5   Transient ground motions

12.5.1 Acceleration, attenuation and duration of shaking

The earthquake-resistant design of an engineering structure requires knowledge of the probable ground motion which the structure must resist with or without damage. The instruments normally used for the recording of ground motions are strong-motion accelerographs*, and Fig. 12.6 shows the record obtained from the earthquake of Tabas (Iran) $M_s = 7.3$, at a distance of 27 kilometres from the focus, as well as records obtained from other, smaller shocks at different focal distances.

These figures of ground acceleration show that the two horizontal components (longitudinal and transverse) of ground motion have nearly the same intensity and frequency content, while the vertical component is usually less intense but of higher frequency.

The amplitude of the pulses, that is, the acceleration, is often used to indicate the severity of the ground motion, and maximum values in excess of or near to $100\%g$ are known to have occurred (San Fernando earthquake of 9 February 1971, $M_s = 6.6$, $125\%g$; Gazli earthquake of 17 May 1976, $M_s = 7.0$, $85\%g$; Imperial Valley earthquake of 15 October 1979, $M_s = 6.5$, $81\%g$.) Even small-magnitude earthquakes can produce large peak accelerations near the focus, but acceleration alone is not the most significant parameter for design. Rather, it is the maximum velocity and, to some extent, the number of pulses or duration of strong shaking involved. The acceleration has a significance but not an overriding one, and in design it is essential, therefore, to consider more of the aspects of the ground motion than are involved in a peak value of ground acceleration (Newmark and Hall, 1982).

* There are two widely used makes of accelerographs: the SM-A Kinemetrics series (25 Hz, 6% damping, recording to 70 mm film) and the MO-2 series made in New Zealand (30 Hz, 70% damping, recording on 35 mm film). For details see Hudson (1981).

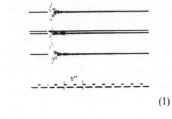

(1)

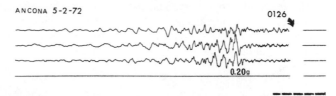

ANCONA 5-2-72   0126

0.20g

(2)

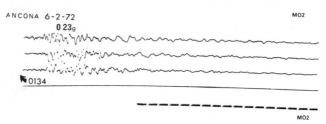

ANCONA 6-2-72   MO2

0 23g

0134

MO2

(3)

ANCONA 4-2-72   MO2

0242   0.13g

(4)

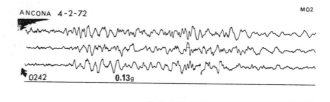

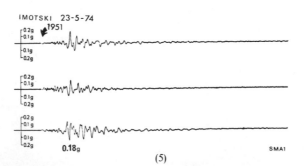

IMOTSKI 23-5-74

1951

0.18g   SMA1

(5)

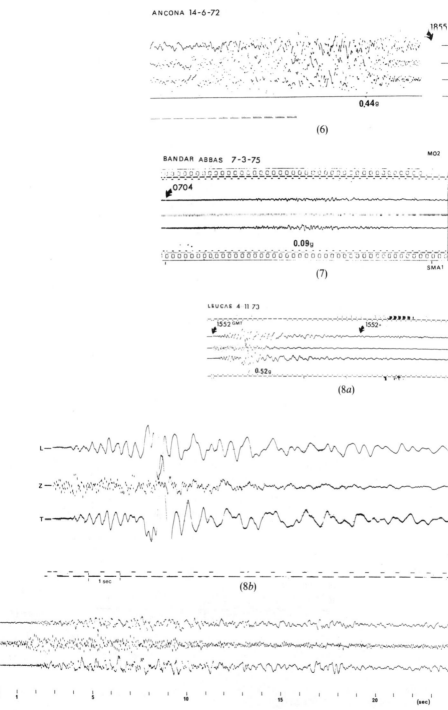

ANCONA 14-6-72

0.44g

(6)

BANDAR ABBAS 7-3-75

MO2

0704

0.09g

(7)

SMA1

LEUCAS 4 11 73

1552 GMT

1552+

0.52g

(8a)

L

Z

T

1 sec

(8b)

L

Z

T

0.5g

0   1   5   10   15   20   (sec)

(9a)

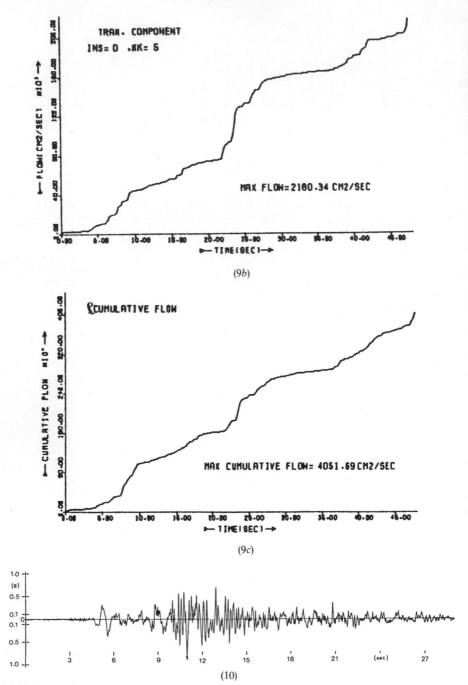

(9b)

(9c)

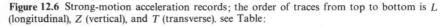

(10)

Figure 12.6 Strong-motion acceleration records; the order of traces from top to bottom is *L* (longitudinal), *Z* (vertical), and *T* (transverse). see Table:

Figure 12.6.1 to 12.6.10 (pp. 370–373). Various strong motion of ground acceleration records obtained by different instruments at different distances from earthquake sources of different magnitude. The order of traces from top to bottom is L (longitudinal), Z (vertical), and T (transverse).

Fig. No.	Date of Event	Origin time (GMT)	Magnitude M	Type of instrument	Epicentral Intensity (MM)I_0	Distance of instrument from focus D(km)	Maximum recorded acceleration a(g)	Location
12.6.1	1977 May 4	0611	3.1	OSM	—	3	0.19	Oroville (USA)
12.6.2	1972 Feb 5	0126	4.1	MO2	VIII	6	0.20	Ancona (Italy)
12.6.3	1972 Feb 6	0134	4.2	MO2	VIII	9	0.23	Ancona (Italy)
12.6.4	1972 Feb 4	0242	4.4	MO2	VIII	5	0.13	Ancona (Italy)
12.6.5	1974 May 23	1951	4.7	SM1	VI	15	0.18	Imotski (Jugoslavia)
12.6.6	1972 June 14	1855	4.7	MO2	VII	6	0.44	Ancona (Italy)
12.6.7	1975 Mar 7	0704	5.6	SM1	VII	45	0.09	Bandar Abbas (Iran)
12.6.8*	1973 Nov 4	1552	5.8	SM1	VII	16	0.52	Leucas (Greece)
12.6.9**	1981 Jul 28	1722	7.1	SM1	IX	38	0.36	Sirch (Iran)
12.6.10	1978 Sep 16	1536	7.3	SM1	IX	27	0.85	Tabas (Iran)

* *a* Full-scale size of strong-motion record
 b Enlarged record

 a Full-scale record of three components of ground acceleration, L = longitudinal, Z = vertical, T = transverse
 b Energy flux plot for T-component
 Cumulative flux plot for $L + Z + T$

$$\left(\text{Energy flux} = \int_0^t (\text{particle velocity})^2 \, dT\right)$$

As can be observed from the shape of the accelerograms in Fig. 12.6 the maximum accelerations of large amplitude occur only a few times and are of relatively short duration. Field observations and analysis show that these high-frequency pulses of acceleration do not have a significant influence on the response and behaviour of ductile structures. It is the part of the record that includes repetitive motion, which contains the bulk of the energy, that produces damage by accumulating non-linear deformations or developing large pore pressures in soils. However, for non-ductile, brittle structures the absolute magnitude of the maximum pulse is of great importance as it completely destroys their cohesive component of strength.

The predominant rise and fall time (that is, the half-period) of a single spike in the maximum horizontal ground acceleration near to the source of an earthquake varies between 0.05 and 0.25 seconds, and in the vertical motion is usually about half as long.

Ground motions decrease with distance from the fault zone, the rate of decrease being relatively small over distances comparable to the linear dimensions of the zone. Estimates of the ground motions in terms of acceleration and velocity can be made, taking into account observational and experimental data. There are several published attenuation laws of ground motion, expressed as a function of distance, magnitude and site conditions. All of them are empirical and hampered by the limited amount of strong-motion data at short distances and for large magnitude events. Also, the magnitudes used in the derivation of these laws are not uniform and the distance of the instrument from the hypocentre or causative fault is usually poorly known. Some of these laws are shown in Table 12.2, and in all cases great care and judgement are required in selecting the appropriate relations for use in design.

Table 12.2 Attenuation laws

(I)
$$\log a = -1.02 + 0.249 M_w - \log r - 0.00255 r + 0.26 P$$
$$r = (d^2 + 53)^{0.5} \text{ valid for } 5.0 \leq M_w \leq 7.7$$
$$\log v = -0.67 + 0.489 M - \log r - 0.00256 r + 0.22 P + 0.17 S$$
$$r = (d^2 + 16)^{0.5} \text{ valid for } 5.3 \leq M_w \leq 7.4$$

where a is the peak horizontal acceleration in gravitational acceleration (g) units, v is the peak horizontal velocity in cm/s, M_w is the moment magnitude, d is the closest distance to the surface projection of the fault rupture or source in km. S takes on the value of zero at rock sites and one at soil sites. P is zero for 50 percentile values and one for 84 percentile values (Joyner and Boore, 1981). Surface wave magnitudes (M_s) of less than about 7.5 are quite similar to moment magnitudes (M_w) (Hanks and Kanamori, 1979).

(II)
$$\log a = 2.00 + 0.39 M_L - 0.9 \log R + 0.24 P$$

Derived from north Italian earthquakes, where a is the peak horizontal acceleration in gravitational acceleration (g) units M_L is the local magnitude and R is the hypocentral distance in kilometres. P has the same meaning as in (I). (Chiaruttini and Siro, 1981; Boore and Joyner, 1982).

(III)
$$\log a = -1.80 + 0.377 M - 1.09 \log R^* + 0.16 P$$

where a is the *mean* of the peak values of the two horizontal components of acceleration in gravitational acceleration (g) units, $M = M_L$ if both M_L and M_S are less than 6.0, and $M = M_S$ if both are greater than 6.0. $R^* = D + 0.0606 \exp(0.7 M)$, where D is the shortest distance to the rupture surface in kilometres (Campbell, 1981; Boore and Joyner, 1982).

For other attenuation laws see Idriss (1978).

For soft foundation conditions the ground motions derived from attenuation laws must be modified to take account of the superficial geological conditions of the site (Idriss, 1978). For relatively stiff soils and for small accelerations the deposit will behave elastically and it may, therefore, amplify the bedrock motion. As the strength of the material decreases, or the bedrock accelerations increase, the response may bring about internal or near-surface failure, as a result of which accelerations above a certain amplitude will be prevented from reaching the surface and cause cracking and permanent displacements instead. It can be shown that the maximum acceleration in saturated, normally consolidated sands and coarse silts cannot exceed values of the order of c_u/p', where c_u is the undrained shear strength of the deposit at an effective consolidation pressure p'. Thus, the upper bound for the maximum acceleration in near-surface deposits is dictated by the undrained shear strength properties of the material (Ambraseys, 1973).

Duration of strong ground motion is one of the most important factors in producing damage. Prolonged cyclic loading in saturated foundation materials may cause excessive pore pressures to develop, leading to failure or progressive loss of cohesive strength and bonding in building materials.

The duration of shaking t_0 (in seconds) with ground accelerations in excess of 3% may be approximated (from Ambraseys and Sarma, 1967) by

$$t_0 = 11.5\,M_s - 53.0 \qquad (M_s \geq 5.0) \tag{12.5}$$

Bolt (1973) gives the following relations:

$$t_0 = 17.5 \tanh(M_L - 6.5) + 19.0 \qquad \text{for accelerations} > 5\% \tag{12.6}$$

$$t_0 = 7.5 \tanh(M_L - 6.0) + 7.5 \qquad \text{for accelerations} > 10\% \tag{12.7}$$

where M_L is the local or Richter magnitude.

12.5.2 Earthquake recurrence rates

Seismicity data can be used to establish earthquake recurrence rates for a particular area. These rates are useful for defining seismic hazard and establishing levels of acceptable risk.

The simplest recurrence law follows the empirical linear relationship

$$\log N = a - bM \tag{12.8}$$

where N is the number of earthquakes of magnitude equal to or greater than M occurring annually within a given area with b varying between 0.6–1.2 (Everden, 1970). The value of a is a function of the local seismicity and size of the area studied.

'Return period' is a term commonly used for the reciprocal of the annual frequency of occurrence $(1/N)$. Statistical studies of earthquake occurrence commonly assume a Poisson law, under which the likelihood of occurrence at any particular time within a particular area is independent of all preceding events. This would require excluding aftershocks in the derivation of the recurrence law. Thus, if there are n occurrences in t years, the mean annual rate of occurrence is $n/t = N$ and the annual probability of occurrence (P) is $P = 1 - \exp(-n/t)$, which is close to N if this quantity is small. The return period is simply $t/n = 1/N$, or approximately $1/P$. However, it is not true that a recurrence of the event is most likely after an interval approximately equal to this time. Thus, if a and b are, say, 4.0 and 0.9 respectively, then the return period of a magnitude 7.0 earthquake will be 200 years. For a structure having a useful life of say, 200 years, the

Table 12.3 Engineering assessment checklist.

	Task	Area	Sources
1. Preliminary assessment of regional seismicity	Identify important events, their location and magnitude. Assess accuracy of focal determinations and intensity distribution.	Depending on seismicity, a few square degrees ($2° \times 2°$) to ($10° \times 10°$) depending on kind of project.	Existing global and regional catalogues, earthquake data files (ISS, ISC, USCGS, BCIS, etc.)
2. Preliminary tectonic study of the area	Identify from existing maps all tectonic features of recent age, and throughgoing lineaments.	Somewhat larger	Existing geologic maps, reports and aerial photos.
3. Evaluation of regional seismicity	From (1) and (2) select most important events either because of their magnitude, large intensity or association with known fault and re-evaluate focal parameters, magnitude M_s or M_L, intensity distribution, and effects on ground.	Same as in (1)	Same as (1), station bulletins and recordings. Press reports and published technical literature. Field study of epicentral areas and local sources of information.
4. Evaluation of regional tectonics	From (2) and (3), field study of recent tectonics; identify relative capability of different faults and attempt to assess displacement history. Determine regional fault mechanism(s) (thrust, normal, strike-slip etc.)	Same as in (2)	Same as in (2), plus results from (3) and field studies. Re-triangulation of lineaments and C-14 dating.
5. Evaluation of regional earthquake hazard	Combining all available information historic, seismographic and geologic (3 & 4) determine the Operating Basis Earthquakes (OBE) for different types of engineering structures, using either a deterministic or probabilistic approach.	Same as in (1)	Cost-benefit analysis to data to determine economically acceptable (OBE) values for different degrees of damage (structural or non-structural)
6. Evaluation of local seismotectonic hazard	Check (OBE) values for local conditions. Detailed study of local, primary or secondary tectonics, extreme foundation conditions including creep.	A few square kilometres	Trenching, special soil investigations.
7. Evaluation of extreme hazard	For special engineering structures (dams, chemical plants, nuclear power plants etc.). Assess Maximum Credible Earthquake (MCE).	Same as in (1) or larger	From all available information.

probability (q) that the site will be subjected to ground motions corresponding to a magnitude 7.0 at least once in that time will be $q = 1 - exp(-200/200) = 63\%$ which is almost a certainty.

The probability of occurrence, q, may be used in defining design magnitudes for structures of known useful life. If, for economic reasons, we accept a probability of exceedance (or occurrence) of say 20% for an unimportant structure with a useful life of 25 years, this would correspond to an earthquake of magnitude 6.7 which has a return period of 112 years. This magnitude can be used in conjunction with the appropriate attenuation laws to derive design ground accelerations (Smith, 1982).

Equation 12.8 must be used together with a maximum magnitude value, M_{max}, at which this law must be truncated. The value M_{max} corresponds to what is usually referred as the magnitude of the maximum credible event, a value that must be assigned on the basis of all available information—historic, seismographic and geologic. Because the maximum credible earthquake is an exceedingly unlikely event that probably has not occurred within the historic record, and certainly not within the era of instrumental seismology, the primary evidence for its assignment must be geological. The length of faulting and the size of any recent displacements in the area may be used to derive upper bounds for M_{max}. For instance, the proximity of a site to a 100-kilometre-long fault of recent age may be used to infer through equation 12.4 a maximum M_{max} of 7.5.

It must be said that the Poisson process currently in practice is not ideal for earthquakes, but rather a first approximation to the real problem. Earthquake occurrence is dependent on the build-up and release of stress within the Earth's crust; a process which is not independent of preceding events. Great caution should therefore be exercised in deriving recurrence rates as there is a chance that all of the seismographic, and possibly the historic records, may be from a quiescent period in the seismic activity. This is the principal reason why statistics alone cannot answer the question of hazard evaluation.

For the use of recurrence relations in the statistical approach of hazard assessment, the reader is referred to Cornell (1968); Yogulap and Kuo (1974); Merz and Cornell (1973); and Lomnitz and Rosenblum (1976).

Notation

A Fault area

a peak horizontal acceleration in units of g

c_u undrained shear strength of any soil

D shortest distance to rupture surface

d closest distance to surface projection of fault rupture or surface

E_s seismic energy

h hypocentral depth

I_0 epicentral intensity

L fault length—length of rupture

M_L local or Richter magnitude of wave

M_{max} maximum magnitude value

M_0 moment

M_s magnitude of surface wave

M_w moment magnitude

m_b magnitude of body wave

N number of earthquakes of magnitude equal to or greater than M occurring annually

n_1 number of seismic occurrences

P probability of seismic occurrence

P parameter equal to zero (50% values) or one (84% values)

P' effective consolidation pressure of soil

q probability parameter

R co-seismic displacement of fault

R hypocentral distance

S parameter equal to zero (rock sites) or one (soil sites)

t number of years

t_0 duration of shaking

$\bar{u}$ average fault displacement

v peak horizontal velocity

μ rigidity

References

Aki, K. and Richards, P.G. (1980) *Quantitative Seismology*, Vols 1 + 2. W.H. Freeman and Co., San Francisco.

Ambraseys, N.N. (1973) Dynamics and response of foundation materials in epicentral regions of strong earthquakes. Invited Paper CXLVIII, *Proc. 5th World Conf. Earthq. Eng., Rome*, Vol. 1, p. c × lvii.

Ambraseys, N.N. (1978) The relocation of epicentres in Iran. *Geophys. J. Roy. Astr. Soc.* **53**, 117–121.

Ambraseys, N. (1981) The E1 Asnam earthquake of 10 October 1980. *Q. J. Engineering Geology* **14**. 143–48.

Ambraseys, N. and Jackson, J. (1981) Earthquake hazard and vulnerability in the northeastern Mediterranean. *Disasters* **5**, 355–368.

Ambraseys, N. and Melville, C.P. (1982) *A History of Persian Earthquakes.* Cambridge University Press.

Ambraseys, N. and Sarma, S.K. (1967) The response of earth dams to strong earthquakes. *Géotechnique* **17**, 187–213.

Ambraseys, N. and Zatopek, A. (1969) The Mudurnu Valley, west Anatolia, Turkey, earthquake of 22 July 1967. *Bull Seism. Soc. Am.* **59**, 521–589.

Berberian, M. (1979) Earthquake faulting and bedding thrusts associated with the Tabas-e Golshan (Iran) earthquake of 16 September 1978. *Bull. Seism. Soc. Am.* **69**, 1861–1887.

Bolt, B. (1973) Duration of strong ground motion *Proc. 5th World Conf. Earthq. Eng., Rome,* Paper no. 292.

Boore, D. and Joyner, W. (1982) The empirical prediction of ground motion. *Bull. Seism. Soc. Am.* **72**, S43–S60.

Brune, J. and Allen, C. (1967) A low stress drop, low magnitude earthquake with surface faulting. *Bull. Seism. Soc. Am.* **57**, 501–514.

Campbell, K. (1981) Near-source attenuation of peak horizontal acceleration. *Bull. Seism. Soc. Am.* **71**, 2039–2070.

Chen, W-P, and Molnar, P. (1977) Seismic moments of major earthquakes and the average rate of slip in central Asia. *J. Geophys. Res.* **82**, 2945–2969.

Chiaruttini, C. and Siro, L. (1981) The correlation of peak ground horizontal acceleration with magnitude, distance and seismic intensity for Friuli and Acona, Italy, and the Alpine Belt. *Bull. Seism. Soc. Am.* **71**, 1993–2009.

Cornell, A. (1968) Engineering seismic risk analysis *Bull. Seism. Soc. Am.* **58**, 1583–1606.

Fournier d'Albe, E. (1982) An approach to earthquake risk management. *Eng. Struct.* **4**, 147–152. (This paper formulates a conceptual basis for regarding earthquakes as a problem of economic and social risk.)

Gutenberg, B. and Richter, C.F. (1956) Earthquake magnitude, intensity, energy and acceleration. *Bull. Seism. Soc. Am.* **46**, 105–145.

Hanks, T. and Kanamori, H. (1979) A moment magnitude scale. *J. Geophys. Res.* **84**, 2348.

Housner, G. (1969) Engineering estimates of ground shaking and maximum earthquake magnitude. *Proc. 4th World Conf. Earthq. Eng.* Vol. 1, p. 1.

Hudson, D. (1981) *Reading and interpreting strong motion accelerograms.* Monograph, Earthquake Eng. Res. Inst.

Idriss, I.M. (1978) Characteristics of earthquake ground motions. *J. Geotech. Eng. Div. ASCE,* **3**, 1151–1265.

Iida, (1959) Earthquake energy and earthquake faults. *J. Earthq. Sci. Nagoya Univ.* **7**, 99–107.

Iida, (1965) Earthquake magnitude, earthquake fault and source dimensions. *J. Earthq. Sci. Nagoya Univ.* **13**, 115–132.

Jackson, J.A. (1980) Errors in focal depth determination and the depth of seismicity in Iran and Turkey. *Geophys. J. Roy. Astr. Soc.* **61**, 285–301.

Jackson, J.A. and Fitch, T.J. (1981) Basement faulting and the focal depths of the larger earthquakes in the Zagros mountains (Iran). *Geophys. J. Roy. Astr. Soc.* **64**, 561–586.

Jackson, J.A., Gagnepain, J., Houseman, G., King, G., Papadimitriou, P., Soufleris, C. and Virieux, J. (1982) Seismicity, normal faulting and the geomorphological development of the Gulf of Corinth (Greece): the Corinth earthquakes of February and March 1981. *Earth Planet. Sci. Lett.* **57**, 377–397.

Jackson, J.A. and McKenzie, D.P. (1983) The geometric evolution of normal fault systems. *J. Structural Geol.* **5**, 471–482.

Joyner, W. and Boore, D. (1981) Peak horizontal acceleration and velocity from strong-motion records. *Bull. Seism. Soc. Am.* **71**, 2011–2038.

Kanamori, H. and Anderson, D.L. (1975) Theoretical basis of some empirical relations in seismology. *Bull. Seism. Soc. Am.* **65**, 1073–1095.

Kasahara, K. (1981) *Earthquake Mechanics.* Cambridge University Press.

King, C. and Knopoff, L. (1968) Stress drop in earthquakes. *Bull. Seis. Soc. Am.* **58**, 249–257.

Lomnitz, C. and Rosenblueth, E. (1976) *Seismic Risk and Engineering Decisions.* Elsevier, Amsterdam.

Merz, H. and Cornell, C. (1973) Seismic risk analysis based on a quadratic magnitude frequency law *Bull. Seism. Soc. Am.* **63**, 1999–2006.

Newmark, N. and Hall, W. (1982) *Earthquake spectra and design.* Monograph, Earthquake Eng. Res. Inst.

North, R.G. (1974) Seismic slip rates in the Mediterranean and Middle East. *Nature* **252**, 560–563.

Richter, C.F. (1958) *Elementary Seismology.* W.H. Freeman and Co., San Francisco.

Smith, W.D. (1982) Pitfalls in estimation of seismic hazard. *Bull. New Zealand Ntal. Soc. Earthq. Eng.* **15**, 77–81.

Soufleris, C., Jackson, J.A., King, G., Spencer, C.P. and Scholz, C. (1982) The 1978 earthquake sequnce near Thessaloniki (northern Greece). *Geophys. J. Roy. Astr. Soc.* **68**, 429–458.

Steinbrugge, K.V. and Zacher, E.G. (1960) Creep on the San Andreas fault: fault creep and property damage. *Bull. Seism. Soc. Am.* **50**, 389–396.

Sykes, L. (1978) Intraplate seismicity, reactivation of pre-existing zones of weakness, alkaline magmatism, and other tectonism postdating continental fragmentation. *Rev. Geophys. and Space Phys.* **16**, 621–688.

Tchalenko, J. and Ambraseys, N. (1970) Structural analysis of the Dasht-e Bayaz earthquake fractures. *Bull. Geol. Soc. Am.* **81**, 41–60.

Tocher, D. (1958) Earthquake energy and ground breakage. *Bull. Seism. Soc. Am.* **48**, 147–152.

Wyss, M. (1979) Estimating maximum expectable magnitude of earthquakes from fault dimensions. *Geology* **7**, 336–338.

Yegulalp, T. and Kuo, J. (1974) Statistical prediction of the occurrence of maximum magnitude earthquakes. *Bull. Seism. Soc. Am.* **64**, 393–414.

13 Dynamic ground movements— man-made vibrations

B.O. SKIPP

13.1 Introduction

Dynamic ground movements may affect delicate production processes, damage structures and services and disturb people. Although there will be passing references in this chapter to ground vibrations associated with military activities, most attention will be paid to man-made ground vibration arising from peaceful activity.

The underlying mechanics of such man-made vibration or 'technical seismicity' has much in common with earthquake-generated ground vibration in that the transmission of energy through the ground is by elastic waves. The general background vibration at low level is usually of concern only to designers of foundations for sensitive installations. It is composed in the U.K. of microseismic surface waves, due probably to marine sources, at a few seconds period, and urban or cultural 'noise', again as surface waves (Skipp and Marriot, 1973).

The specificity of man-made ground vibration is linked to the usually small and shallow nature of the sources. With the exception of nuclear explosives, man-made sources rarely exceed 100 m in length and 500 m in depth, whereas earthquake sources are usually deeper than 1000 m and can have dimensions in the order of tens of kilometres. Excluding nuclear blasts, man-made explosions and impacts rarely radiate more energy than 10^9 joules. This compares with a moderate-sized earthquake radiating 10^{12} joules. There is, however, an overlap in that some small natural sources may generate vibrations similar to those from larger explosions and impacts.

The engineering significance of ground vibration is mainly in the response of structures or services coupled to the ground. There are, however, occasions where the ground itself may suffer 'damage'; for example, settlement, liquefaction, slope instability. This chapter does not deal explicitly with the response of structures except where the vibrations giving rise to concern are expressed in terms of structural response. Reference will however be made to texts dealing with such matters.

The engineer needs to know the levels of vibration which might give rise to problems. He then needs certain minimum information about a vibration so as to be able properly to characterize it and design measures to mitigate its effects. Herein lies a tautology, for in order properly to measure and characterize a vibration some assessment must be made as to the type of vibrations being measured. Simple vibration can be characterized by a little information using simple instruments. More complicated vibration calls for more and more difficult descriptions with correspondingly more elaborate instrumentation.

In applying engineering to mitigation of vibration effects it may also be necessary to understand the full chain: source–pathway–receiver. In the following sections an explanation of definitions, characteristics and measurement will therefore precede a consideration of 'levels of concern' and 'mitigation'.

13.2 Definitions

13.2.1 *Deterministic and random vibration*

The current vocabulary of vibration starts from a distinction between 'deterministic' and 'random' vibrations. A deterministic vibration is one where the instantaneous value at a certain time can be predicted from a knowledge of the instantaneous values at previous times. Deterministic vibrations may be periodic, aperiodic or transient. 'Shock' is regarded as a special case of transient vibration.

A random vibration is a vibration the instantaneous value of which at a certain time cannot be predicted from a knowledge of its time-history. A random vibration is called 'stationary' if its statistical properties remain the same at different times, that is, invariant with respect to translation in time. A vibration may depart from strict compliance with stationarity until its statistical properties are invariant with time. It is then called non-stationary. Detailed definitions are given in ISO 2041. The different types of vibration are classified in Fig. 13.1.

The instrumentation and duration of sampling required satisfactorily to characterize different types of vibration varies with the purpose of measurement as well as the type of vibration.

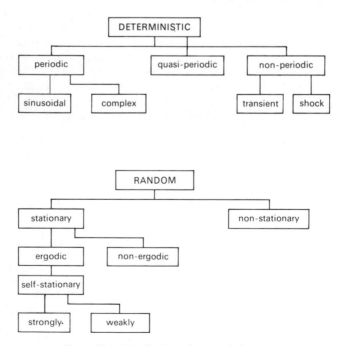

Figure 13.1 Classification of types of vibration.

Many types of vibration encountered do not fall easily into the categories outlined above and, for many practical purposes, especially when empirical 'acceptability' criteria are being used, vibrations are described as 'continuous', 'intermittent' or 'transient' without allocation to the formal classes of vibration.

13.2.2 Duration of vibration

It is recognized that the 'effective duration' of a vibration has a bearing upon the response of a structure and the damage to it. A structure which is lightly damped will require more cycles of excitation to reach its peak response than will one which is highly damped, although of course the maximum response in the latter case would be less than in the former.

The time constant of a resonant response is given by

$$T_R - \frac{1}{2\pi\xi_r f_r}$$

where $\quad f_r$ is the frequency of resonance r,

$\quad\quad\quad \xi_r$ is the damping ratio of resonances r

and $\quad\quad T_R$ is its time constant in seconds.

Three cases can be defined.

(a) *Continuous*
   If the forcing function (i.e. ground vibration) impinges on the structure continuously for more than $5T_R$ then the vibration is regarded as continuous.
(b) *Intermittent*
   If the forcing function 'switches on and off' at intervals comparable with $5T_R$ or longer, the response is then regarded as intermittent.
(c) *Transient*
   If the forcing function exists for a time less than $5T_R$, the response is regarded as transient.

This guide to the meaning of 'continuous', 'intermittent' and 'transient' in terms of input and response is of recent origin and has yet to be formally incorporated into codes.

So far vibrations have been regarded as variations of some quantity with time, that is, represented 'in the time domain'. Vibrations can be described in terms of their frequency content (i.e. in the 'frequency domain'). The transformation of time history information to the frequency domain and the converse process is a very important part of vibration analysis and is dealt with in a number of standard texts (Blackman *et al.*, 1958).

13.2.3 Kinematic quantities

A vibration is a time-varying quantity. For ground vibration that quantity may be presented as the variation with respect to time of the displacement (u) of a point, the first derivative with respect to time ($du/dt = \dot{u}$), usually termed 'particle velocity' or the second derivative ($d^2u/dt^2 = \ddot{u}$), termed particle acceleration*. 'Particle velocity' should not be confused with the propagation velocity of a travelling wave. In principle each of the

---

* Strictly an acceleration 'level' is expressed as micrometres/second squared on a decibel scale (ISO 2041)

parameters u, $\dot{u}$, $\ddot{u}$ can be derived one from another given information on frequency content. These quantities u, $\dot{u}$, $\ddot{u}$ are often referred to as 'kinematic quantities'. When a record is taken showing the variation of these quantities with time the 'amplitude'* in that record is taken from the mean or 'rest' position.

13.3 The chain of action

Most human activity generates vibrations which may be transmitted to and through the ground. Usually the source of the vibration can be indentified but the sources of low-level 'background' vibration which could affect sensitive installations are often difficult to pinpoint. Recent studies suggest that this background consists mainly of Rayleigh waves (Iyer and Hitchcock, 1976). Various sources and the types of vibration associated with them are classified in Table 13.1 (p. 388).

The complete vibration system consists of a source, a pathway and a receptor. The energy must be transferred to the ground, travel through it, and then be transferred from the ground to the structure, service, person or instrument affected. Each link in this chain alters the characteristics of the vibration. Mathematically the process can be described by the use of a 'transfer function' (ISO 2041).

13.3.1 Source

Man-made sources may be divided into those which are effectively continuous and those which are of relatively short duration. This may be exemplified by vibrations from, say, a compressor and vibrations from a single explosive blast. While it is easy to see the distinction at source, at some receptors the distinction may be blurred because the 'continuous', 'transient' or 'intermittent' nature of the vibration depends on what has happened 'in transit' and within the responding structure. Thus, the discrete blows of a piling hammer may set up effectively 'continuous' vibration in a structure. The signal from short-duration excitation is extended in time with greater distance from the source because of the phenomenon of 'pulse' broadening (Knopoff, 1956) and wave-train spreading, and because of dispersion (Ricker, 1977) through inhomogeneous and stratified ground.

The mathematical derivation of surface response to a pulsating force at the surface of an elastic half-space constitutes the classic 'Lamb's Problem' (Lamb, 1904) and has been extended to consider various geometries and stratified ground (Bycroft, 1956, 1957). Miller and Pursey (1954, 1955) showed that, for the special condition of an homogeneous, isotropic elastic half-space with Poisson's ratio equal to 0.263, there was a partition of energy between different wave types as follows:

	% of available energy
Surface Rayleigh waves	67
Body shear waves	26
Body compressional waves	7

---

* Strictly 'amplitude' should refer to a harmonic motion (see ISO 2041)

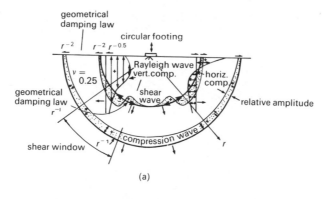

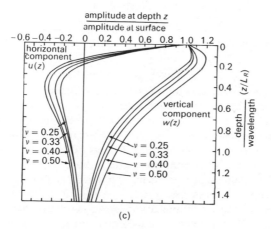

Figure 13.2 Radiation of elastic waves from a surface source (vertical oscillating force). (a) Distribution of displacement waves from a circular footing on a homogeneous, isotropic, elastic half-space (from Richart *et al.*, 1970). (b) Amplitude ratio *v.* dimensionless depth for Rayleigh wave.

Richart, Hall and Woods (1970) illustrate this condition (see Fig. 13.2). The mathematical treatment of a source within an elastic body (Pekeris and Lifson, 1957) has recently been further developed by Awojobi (1977).

In addition to the substantial literature devoted to fixed sources there are studies of vibrations generated in elastic continua by travelling loads (Fryba, 1970). Elastic waves are not radiated until the load travels at a speed equal to that of the velocity of the slowest propagating wave.

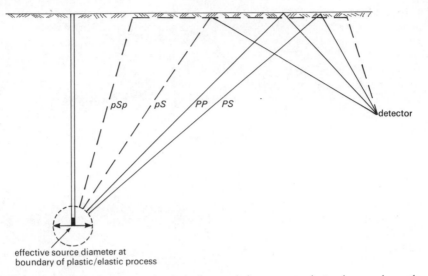

Figure 13.3 Waves from a pressure (explosion) source in homogeneous isotropic ground near the surface. PP—begins and ends as a P wave; PS—begins as a P wave, ends as a shear wave S; pSp—begins and ends as a P wave but travels most of its way as a surface wave; pS—begins as a P wave, ends as a shear wave S, having been reflected near the source. After Ewing *et al.*, 1957. (In real ground, anisotropy and charge geometry give rise to shear waves).

Theoretical studies of explosive sources postulate a sequence of zones, around the source, ranging from a 'hydrostatic' centre through zones of 'non-linear' behaviour to a radius at which 'elastic' processes dominate (Blake, 1942; Sharpe, 1942; O'Brien, 1969; see Fig. 13.3). There are now a number of advanced computer codes which permit more realistic modelling of explosive sources (Nelson *et al.*, 1971: Trulio, 1966; Cherry and Petersen, 1970). Generally with larger-source dimensions the proportion of longer-period waves increases, although experimental work with small charges does not show that the spectral character is very sensitive to charge size (O'Brien, 1969), even though the predominant spectral frequency dropped with increasing charge. Most mathematical models can only roughly approximate to the non-linear processes which are usually involved in 'intensive' man-made sources such as explosives, impacts on soils, and pile driving.

The proportion of total available energy radiated as an elastic seismic wave is called the 'seismic efficiency' of the source. Very large blasts may have an efficiency of less than 1%. Surface impacts can range up to 10% seismic efficiency.

13.3.2 *Transmission path*

Reference has been made to the different elastic waves which can travel through the ground (see Fig. 13.2). The most important are listed in Table 13.1. There are several standard texts dealing with elastic wave propagation (see, for example, Ewing *et al.*, 1957; Graff, 1975).

Except in the near field of intense sources (e.g. nuclear explosions) or in research on explosive breakage, elastic wave propagation is assumed. However, even at 10^{-4} to 10^{-6} levels of strain, departures from elasticity are needed to explain the full character of wave propagation in rocks and soils (Horton 1953; Brennan and Smyllie, 1981; Borcherdt, 1974). Some account of departures from strict elastic linearity of stress and strain may have to be allowed for when the vibrations are transmitted, for example, from a bedrock to an overlying soil (Seed, 1979). In soil, linear elasticity is an adequate assumption for strains of less than about 10^{-4} (Pyke et al., 1975). Piecewise linear approximations are often used in site response modelling (Lysmer, 1978).

For many situations where vibrations present significant problems, a source of finite dimensions is involved. Under these circumstances, close to the source there may be no clear separation of the different types of wave.

As waves radiate from their source, amplitudes decrease as radiated energy is spread over a greater length of wave front. Body waves travelling divergently in the solid attenuate by geometry according to a $1/r^2$ law except where they emerge, when the attenuation measured on the surface is proportional to $1/r$, where r is the radial distance travelled from source.

Surface waves attenuate according to a $1/\sqrt{r}$ law. In stratified ground this pattern may be distorted by emerging waves refracted from underlying stiffer layers. Wave 'guide' phenomena are also observed where a surface layer is stiffer than that of the underlying ground.

In addition to the above 'geometric' attenuation, anelastic processes degrade amplitude. A 'Quality' factor term, or 'Specific Dissipation Constant' is used to quantify this effect. This is a 'material' damping parameter which can be related for simple systems to a 'system' damping parameter D, such that $1/Q = 2D$ and $\dfrac{1}{Q} = \dfrac{\Delta W}{2\pi W}$

where ΔW = loss of energy per cycle,

$\qquad W$ = maximum stored strain energy,

and ΔW is the area of the hysteresis loop on the stress–strain curve for the soil.

Typical Q values for soils range from 2 to 20 (Abbis, 1981) and for rocks the range is from 20 to 200+ (Attewell and Ramana, 1966; Ambraseys and Hendron, 1969).

The radiation pattern from a real source is affected not only by source mechanism but also by anisotropy of ground and inhomogeneities of dimensions comparable to the wave lengths being propagated (Crampin, 1980). Variations of topography can have a significant effect (Trifunac, 1973; Bolt et al., 1974). In practice local magnifications arise when measurements are made on embankments, scarps and so on, at wave lengths similar to the dimensions of such features.

In addition to travelling waves, occasionally 'standing waves' have been observed. Such a standing wave could theoretically arise should a machine foundation, for example, go into resonance on a soil layer overlying a stiff bedrock (Bycroft, 1957) or a sustained horizontally polarized shear wave be radiated in a bedrock and propagate upwards through a less-stiff overburden. Travelling waves can excite 'standing' waves in layered systems. Such phenomena can explain the reported observations of a 'characteristic site period'.

Table 13.1 Principal elastic waves.

Wave	Designation	Remarks	Propagation velocity
compressional, dilatational	P	Particle motion in direction of propagation	$V_P^2 = E(1-v)/\rho(1-2v)(1+v)$ (infinite medium)
distortional, shear	S	Particle motion normal to direction of propagation	$V_S^2 = G/\rho$
vertically polarized	SV		
horizontally polarized	SH		
Rayleigh wave*	R	Retrograde elliptical motion at surface	$V_R^2 \simeq G_1/\rho$
Love wave	L	Particle motion normal to direction of propagation in plane of interface	Short λ: $V_1 = (G_1/\rho_1)^{1/2}$ Long λ: $V_2 = (G_2/\rho_2)^{1/2}$
Stonely wave (generalized Rayleigh wave)		Surface wave in elastic half space where two layers have similar shear wave velocities	$0.998\,V_P$

Conversions at boundary (solid/solid)

Incident	Transmitted	Reflected
P	P, SV	P, SV
SV	SV, P	SV, P
SH	SH	SH

V_P = velocity of propagation, P waves
V_S = velocity of propagation, S waves
G = shear modulus
ρ = density
λ = wavelength
h = distance
Subscripts refer to layers.
*Rayleigh waves do not appear at distances shorter than $V_R h / \sqrt{V_P^2 - V_S^2}$ or $V_R h / \sqrt{V_S^2 - V_R^2}$.

13.3.3 *Receiver*

The chain is completed by the receiver, which may be a sensor, a structure or a service. This aspect is dealt with briefly in a later section. For environmental problems the responding systems are usually linear elastic, and problems relate to serviceability rather than ultimate limit states (CIRIA, 1977).

13.4 Ground motion from various sources

The preceding sections have dealt with general aspects. The following section will deal with some of the more important sources of ground vibration. Steffans (1974) provides a review of vibration and building damage. Ground vibration from construction activities has been reviewed by Wiss (1981).

13.4.1 *Machine vibration*

Although most machines having rotating or reciprocating elements are designed to minimize out-of-balance forces and couples, they may generate some vibration. The levels of vibration on the bearings for rotating machinery are used to classify 'vibration severity' (ISO 2372). Many machines are mounted on isolating springs, so reducing the vibrations transmitted to the ground. When machines complying to their severity grades are running normally, the maximum source vibration would range from 2 mm/s to 20 mm/s. Some reciprocating machines such as compressors may generate vibrations with peak particle velocities of up to 30 mm/s on the machine foundation.

Where a rigid foundation block is used it may oscillate in six modes (three translational, three rotational), so imparting complicated motions to the ground. The levels are usually within the quasi-elastic range of behaviour of most soils, and surface waves predominate in the radiated energy. For both reciprocating and rotating machinery the vibrations generated are predominantly continuous and periodic, that is, deterministic in character. Reciprocating machinery will generate more than one frequency, depending upon the cylinders and their arrangement. When a foundation block rests on piles the coupling of vibrations to the ground is more complex. In practice a small gap often develops on the underside of the foundation slab, especially in granular soils, so the coupling actually is via the traction on the pile soil contact (skin friction) or, in the case of point-bearing piles, at the founding stratum. Since piled foundations are often used to cope with rocking and torsional behaviour, the coupling to the soil is via a normal load at the pile-to-soil contact. Depending upon the lateral soil-to-pile coupling conditions, the energy may be radiated from deeper levels along the pile-to-soil contact. The degree of embedment of a foundation influences the radiation of vibration and a friction piled foundation can be regarded as a 'deeply' embedded block foundation (Novak, 1974).

The continuous and periodic nature of much machine vibration can give rise to resonant behaviour in structures far removed from the source. A very lightly-damped element requires a longer time to reach the maximum response than does a highly-damped element, and maximum response may be many times greater than the input. Response therefore is a function of both damping and duration of vibration.

Not all machine vibration is periodic. Some machinery, such as crushers, ball mills,

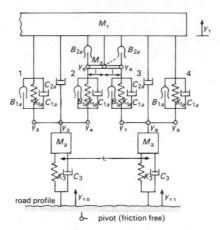

Figure 13.4 Mass-spring-damper representation of the Type A suspension. Symbols and values for Model A 'standard vehicle' (after Page, 1973):

M_1 = sprung (body) mass	= 9220 kg
M_3 = linkage mass	= 25 kg
M_2 = unsprung mass	= 225 kg
B_{2a} = linkage Coulomb friction	= 1000 N
K_s = spring stiffness	= 250000 N/m ($= 2K_{1a}$)
C_s = spring viscous damping	= 5000 N/m/s ($= 2C_{1a}$)
B_s = spring Coulomb friction	= 500 N ($= 2B_{1a}$)
C_{2a} = shock absorber damping	= 0
K_3 = tyre stiffness	= 1 800 000 N/m
C_3 tyre viscous damping	1000 N/m/s
l = half length of linkage bar	
L = axle spacing	= 1.33m (measured)
θ = angular displacement of linkage bar	
$y_1 - y_{11}$ = vertical displacements at those points.	

(measured) . (estimated)

weaving looms, and printing machines, contains impact and random vibrations along with periodic excitation from the rotating parts. The vibrations from such machines as rotary crushers and similar plant may be very difficult to characterize, being of an effectively 'non-stationary' variety with occasional 'spikes' in the near field and significant low frequency energy which may easily be transmitted considerable distances through the ground. Similar complex vibrations arise in rock-cutting tunnel machines (New, 1982).

The impact from forging hammers can generate troublesome ground vibrations, often with significant low frequencies (< 5 Hz) which travel far. It is now common practice to mount such equipment on isolated inertia blocks (Major, 1962). For large screw-hydraulic presses the source may involve both vertical impact and a torsional pulse.

13.4.2 Traffic vibration

Road traffic, with vehicles rolling along a paved highway, generates ground vibrations as the sprung mass is set in motion by the surface roughness. The analogue (Page, 1973) of a

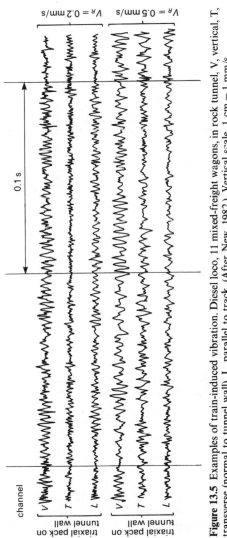

Figure 13.5 Examples of train-induced vibration. Diesel loco, 11 mixed-freight wagons, in rock tunnel, V, vertical, T, transverse (normal to tunnel wall), L, parallel to track. (After New, 1982.). Vertical scale, 1 cm = 1 mm/s.

rubber-tyred motor vehicle is given as Fig. 13.4. Inspection of this model would suggest that the character of the ground vibration generated would be periodic, with the amplitude a function of surface roughness and translational speed. Even on very soft soils the velocity of propagation of a surface wave is unlikely to be less than 80 m/s, so translational motion (that is, a travelling line load) will hardly generate a surface wave.

Examples of vibration from railways is given in Fig. 13.5. Both road traffic from multiple sources and railway-induced vibrations have the character of non-stationary random motion and may not easily be characterized by simple measurements and analysis.

Experimental studies have shown that the picture is more complicated in that the vibration generated from an individual vehicle often has periodic component frequencies and is not truly random in character.

There is some evidence (certainly as far as building response is concerned) to point to vibration energy air-coupled to the ground from the pulsations in the vehicle exhausts where they impinge upon the road surface and the building itself. Furthermore, when a vehicle travels at speed over a jump in a road surface some impact type of vibration can be generated and which has a broader frequency spread. Examples of traffic-generated ground vibration are given in several studies; see, for example, Whiffin and Leonard (1971), Lande (1974), Lande and Rundqvist (1981), Beam and Page (1976) and Martin

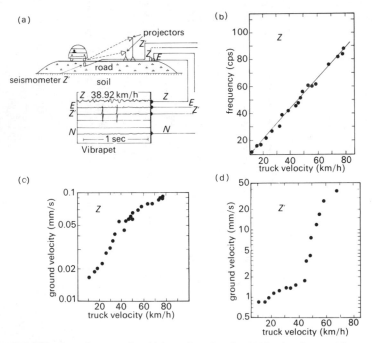

Figure 13.6 Vibrations from road vehicles (after Lande, 1974), showing rapid attenuation from under road position, and the frequency-speed relationship.

(1980). Rarely is a peak particle velocity of greater than 2 mm/s measured except directly under a pavement (see Fig. 13.6).

13.4.3 *Ground vibrations generated by impact*

Ground vibrations are generated by impact arising from demolition, ground treatment, hammer foundations and driven piling.

13.4.3.1 *Demolitions.* During the last decade, controlled explosive demolition of large structures has become an established procedure. Cooling towers, for example, have a mass of up to 6000 tonnes. The methods of demolition which have been developed involve kinetic energies of impact of up to 19^9 joules. Much of the potential energy is however dissipated in break-up of the tower itself. Empirical relationships between peak particle velocity and an energy-scaled distance have been used for prediction purposes (Skipp and Buckley, 1977). Such relationships are very dispersed and often dimensionally inconsistent (Fig. 13.7). Attempts have been made to take account of varying ground conditions and energy levels with a more prescriptive source model (Skipp and Buckley, 1977). Examples of time histories and duration of significant ground vibration and attenuation from a set of 40 cooling towers are given in Fig. 13.8. Response spectra (see section 13.6.1) have been derived from records at 20 sites with distances from 30 to 300 m from the zone of impact (Fig. 13.9).

The demolition of large chimneys (energy of impact ranging up to 5×10^9 joules) generates ground vibrations which tend to have a lower frequency content than that for cooling towers. They may be attributed to the source approximating to a travelling line load.

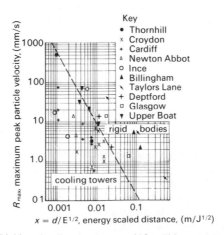

Figure 13.7 Rigid and collapsing impact. (After Skipp and Buckley, 1977).

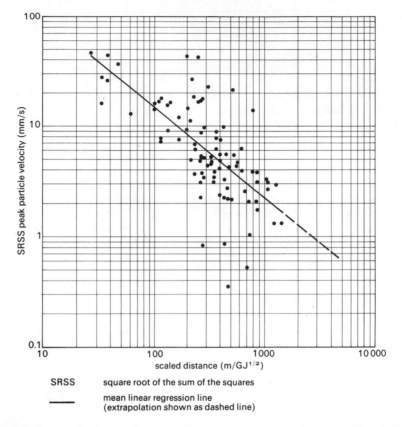

Figure 13.8 Ground vibration from demolition of cooling towers. SRSS particle velocity v, scaled distance. SRSS, square root of the sum of the squares; solid line, mean linear regression (extrapolation shown as dashed line); kinetic energy range 10^7–10^9 joules. Note: data from measuring points 30–200 m from position of impact, all ground conditions.

The source characteristics involved when buildings are demolished by explosives depends on the details of the method used. For buildings a sequential mechanism is often aimed for with ample delays between detonations in a pre-weakened structure (Kriebel, 1980).

The examples of demolition impact so far given involve what may be described as 'crumpling' structures, when energy is dissipated in many fractures giving a 'soft' impact as distinct from a 'rigid' impact. The transfer of kinetic energy from a falling body to the ground is a complex process (Scott and Pearce, 1975) which involves the stiffness of both the impacting body and the ground. Generally, the seismic efficiency for a 'stiff' body falling on firm ground is higher than that of a soft or yielding body (as long as there is no 'bounce'). The energy radiated from a stiff body falling on very soft ground would, however, be minimal if the dynamic pressures caused gross plastic yield in the soil. The

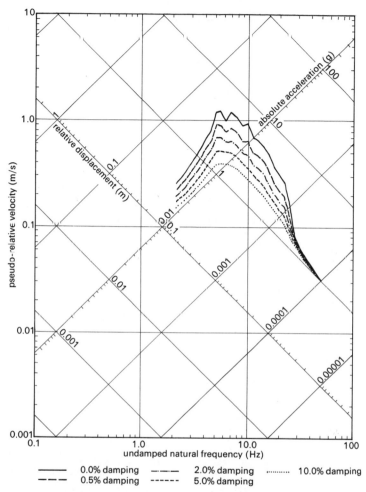

Figure 13.9 Ground vibration from demolition of cooling towers. Response spectrum, cooling towers, horizontal mean; 102 data points, 30 towers on firm ground.

high seismic efficiency of impact sources as compared with explosive sources has been exploited by some seismic exploration methods. Comparison between so called 'rigid' and 'collapsing' impact has been made by Skipp and Buckley (1977).

It should be noted that when explosives are used in demolition works they generate ground vibrations which in some cases (chimneys for example) may be of greater significance near a structure's (chimney) base than the vibrations generated by the impact of the structure (chimney) on the ground.

When explosives are used to weaken or break up a concrete foundation there is a coupling of explosive energy to the ground which may also be of vibration significance.

13.4.3.2 *Dynamic consolidation.* A widely-used method of ground improvement consists of dropping rigid blocks in an ordered pattern on the ground, thereby compacting it (Menard and Broise, 1975; Greenwood and Kirsch, 1983). The peak particle velocities of the ground vibrations generated can be related to the distance and kinetic energy of impact by a scaled distance law (Fig. 13.10). The seismic efficiency improves as the ground is stiffened by the treatment.

13.4.4 Ground vibrations from explosives sources

There have been extensive theoretical and experimental studies of the ground vibrations generated from small single explosion sources (Blake, 1952; O'Brien, 1969; Langfors and

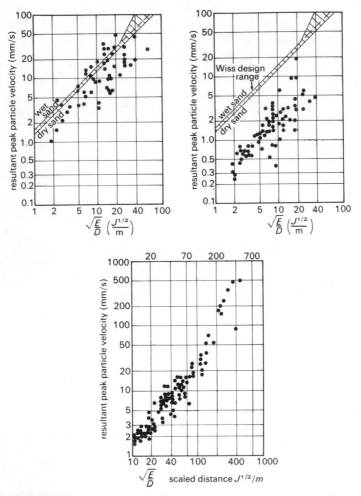

Figure 13.10 Vibrations associated with ground improvement (after Greenwood and Kirsch, 1983). Measurements on ground surface, vibroflot (top left); on structure footings, vibroflot (top right); and on ground surface, tamping (bottom).

Kihlström, 1978; Larsen, 1982). Near-field numerical studies are usually conducted in terms of shock-strain or displacement field, but most analytical work has been on elastic wave propagation from a pressure pulse acting on the surface of a spherical or cylindrical cavity and solution of the elastic wave equation.

13.4.4.1 *Site laws.* Experimental work for engineering purposes has concentrated in establishing so called 'site laws' which enable estimates of ground vibration (usually at ground surface) to be made knowing the size of explosive charge (expressed often as TNT equivalent), distance and some 'site factors'.

Such experimental investigations have been designed around the following suggested relationship:

$$PPV = KM^{\alpha}r^{-\beta}$$

where PPV is the peak particle velocity,

M is the mass (or weight) of explosive detonated per delay,
K is a 'site constant',

and α and β are exponents generally controlled by the site conditions.

Although a full determination of the various coefficients is occasionally warranted, it is more usual to develop a 'scaled distance' relationship, plotting some vibration parameter

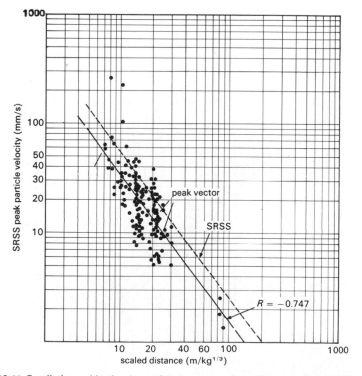

Figure 13.11 Small charge blasting in sandstone: attenuation relations using SRSS and peak vector and cube root scaling. R = correlation coefficient.

(usually peak particle velocity or a vector product) against a value r/W^n where r is the radial distance from source, W is the weight of explosive per delay, and n is a value which would be established by statistical fitting, but is usually chosen as $\frac{1}{2}$ or $\frac{1}{3}$.

Ambraseys and Hendron (1969) have shown that the value $\frac{1}{3}$ is dimensionally correct for a fully-confined explosion. Recent work points to departures from linearity and dependence on the porosity of the rock (Larssen, 1982). The value W is normally the instantaneous charge weight. When multiple sources are used separated by fractions of a second, ground vibration is often found by observation to be better related to an 'effective' charge weight, being some value between that of the instantaneous charge and the total (multiple delay) charge. This, however, is dependent on the distance, the velocity of propagation and the spectral content. At a large distance from a source the individual events within a multiple sequence become blurred and the relationship if developed for short intervals between blasts (millisecond delay) and 'effective' charge weight (Thoenen and Windes, 1942) may no longer apply (Ambraseys and Hendron, 1969; Skipp and Tayton, 1971).

A statistically better relationship is often found between the peak vector and a scaled distance rather than with an individual (directional) component of vibration.

It is sometimes more convenient when the ground vibration does not show signs of exhibiting clear separate phases to use the Square Root Sum of Squares (SRSS), $\bar{R}$, where $\bar{R} = (\dot{u}_x^2 + \dot{u}_y^2 + \dot{u}_z^2)^{1/2}$ (sometimes as in German practice called a 'resultant'),

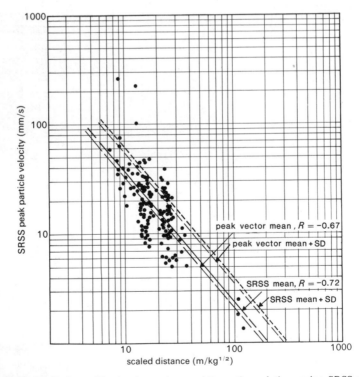

Figure 13.12 Small charge blasting in sandstone. Attenuation relations using SRSS and peak vector and square root scaling.

$\dot{u}_x, \dot{u}_y, \dot{u}_z$, being the peak particle velocities in the 'wave train' in three orthogonal (x, y, z) directions. In Figs. 13.11 and 13.12 a comparison is shown between scaled distance relationships using $n = \frac{1}{3}$ and $n = \frac{1}{2}$, SRSS, peak vector and individual peak components, and in Fig. 13.13 the distribution of the ratio between SRSS and the maximum single direction component is illustrated for the same study from which Figs. 13.10 and 13.11 have been obtained.

Ground vibrations from large sources such as quarries and opencast mines rarely arise from single events but from the interaction of waves generated from a number of blast holes delayed in time (milliseconds). The degree of interaction is a function of the geometry of the source, the delay between initiation at different positions in the source, the statistical variation in the time of detonation of holes normally designed to initiate at the same time, the mutual geometry of free surfaces and impedance mismatch surfaces, and the distance from the source region itself. The sequence of initiation of a line of holes affects the radiation pattern. Some examples of time history and frequency spectra of typical opencast coal mining blasts measured at varying distances are given in Fig. 13.14, and in Fig. 13.15 a square root scaling law from mineral and construction blasting is illustrated.

The classification of wave types from large quarry-type blasting has been studied using seismological terminology (Franetti, 1977). Three zones can be identified: the very near field where non-linear processes predominate, an intermediate field where wave conversions are taking place, and the far field. As for single sources, prediction of

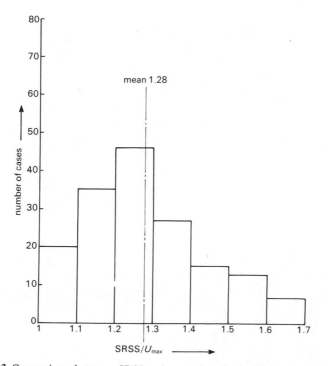

Figure 13.13 Comparison between SRSS values and peak particle velocity regardless of components.

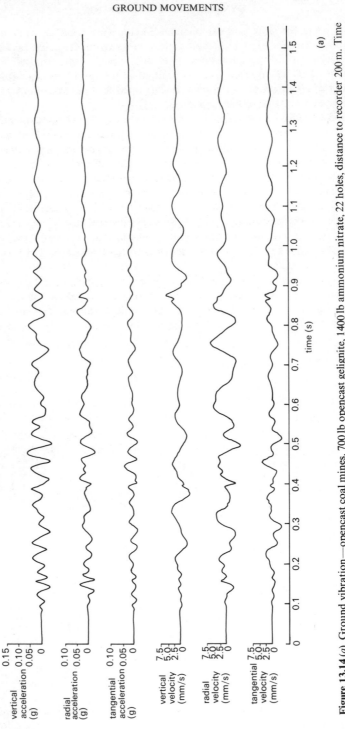

Figure 13.14 (*a*) Ground vibration—opencast coal mines. 700 lb opencast gelignite, 1400 lb ammonium nitrate, 22 holes, distance to recorder 200 m. Time domain traces.

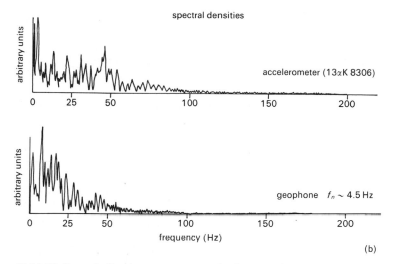

Figure 13.14 (*b*) Ground vibration—opencast coal mines. 7001b opencast gelignite, 14001b ammonium nitrate, 22 holes, distance to recorder 200 m. Spectral densities.

ground vibration from a knowledge of the weight of explosive to be used is usually based upon some 'scaled distance' law. Such laws are established by observation for a particular locality. The dispersion on 'global' relationships, not taking site specific factors into account, may be high with the standard error being similar to the mean. Since such laws imply 'point' sources they are properly limited to distances that are large compared with the dimensions of the source and those wave lengths carrying the energy which it is deemed may affect structures and sources. In practice this usually means distances in the order of hundreds of metres, even thousands of metres for quarries, and tens to hundreds of metres from construction blasting (tunnels, trenches and so on).

13.4.4.2 *Very large single explosions.* Studies of the ground vibrations set up by large single explosions (1000 tonnes), buried and at ground surface, have been carried out by various agencies (Johnson, 1971). Fully-coupled and contained explosions (that is, no ground surface breakout) generate a high proportion of direct P (Primary) waves, but surface waves are then generated by wave conversion (see Table 13.2). In Fig. 13.16 the ground vibrations expressed as peak accelerations are given for cratering explosions in terms of weight of explosive (TNT equivalent) and distance. Typically, the seismic energy is less than 1% of explosive energy. Such large explosions have the character of small earthquakes, and ground motions for design studies are usually expressed in terms of response spectra scaled to the effective peak ground acceleration.

13.4.4.3 *Other uses of explosives.* Explosive compaction of loose soils may generate troublesome vibrations. The technique has, however, been used without structural damage in urban areas (Docher, 1980; Ivanov, 1980). In Figs. 13.17a, b an example from explosive trials in dense saturated sand shows somewhat surprisingly that the charge scaling law is not dissimilar from that derived from blasting in rock (compare Fig. 13.16 with Fig. 13.17b).

13.4.5 *Ground vibration from piling*

The ground vibrations associated with piling have caused concern and are covered by legislation in several countries. In the United Kingdom ground vibration as well as air noise is regarded as 'pollution' and so is subject to the appropriate legislation (U.K. Control of Pollution Act, 1974). However, criteria are yet to be agreed for the U.K. (1983). Although the matter has affected the construction industry for many years, the quality of instrumental data on ground vibration from piling is generally poor. Recent studies (Palmer, 1983) have assembled and reviewed a mass of case history information. There are also several important papers in which an attempt has been made to understand how piling operations cause ground vibration (Attewell and Farmer, 1973; Mallard and Bastow 1980; Wiss, 1981).

Methods of piling in current use may be classified as either 'displacement' piling or 'bored' piling. There are many types of impact-driving piling systems having impact kinetic energies ranging from about 22kg.m (220N.m) a BSP 200 double-acting air hammer) to about 27 200 kg.m (272 000 N.m) (a Delmag D80-12 diesel hammer) energies per blow (220 to 272 000 joules, respectively). During the last two decades

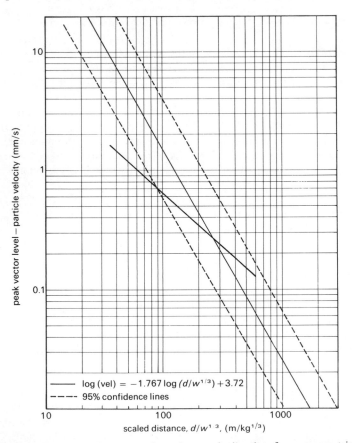

peak vector level – particle velocity (mm/s)

scaled distance, $d/w^{1.3}$, (m/kg$^{1/3}$)

—— $\log (\text{vel}) = -1.767 \log (d/w^{1/3}) + 3.72$

– – – 95% confidence lines

Figure 13.15 (*a*) Cube root distance scaling of ground vibration from quarry trial blasts. Confinement in soft rock. 60 data points.

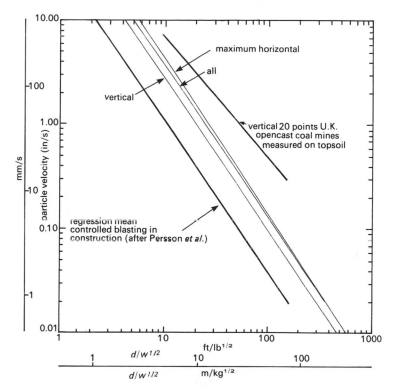

Figure 13.15 (*b*) Square root distance scaling of ground vibrations from mineral blasting and construction blasting. (*a*) Construction blasting. 1, mean regression, large quarry blasts, 500–300 m measurements on overlying glacial till, total charge 1–8 tonne; quarry in limestone, 20 data points. 2, mean regression, small excavation to open quarry in soft limestone, distance 30–300 m, charges 1–100 kg, 55 data points. (*b*) 1, vertical (Walker *et al.*, 1982) from 20 points UK opencast coal mines measured on topsoil. 2, regression mean, controlled blasting in construction (after Persson *et al.*, 1980).

vibrating pile drivers have become widely used in soil that is dominantly non-adhesive. They are also often used in conjection with bored piling systems to inset or extract casing. They are classified according to power and frequency and most operate in the frequency range 30–60 Hz. A useful development for reducing environmental vibration in built-up areas is the BSP ID 17 Impulse Pile Driver in which an air cushion impulse chamber is interposed between the piston and driving plate.

Bored piling systems use rotary or shell and auger methods to remove ground into which concrete is then placed. Bored piling systems generate ground vibrations when the casing is driven in or withdrawn (by impact or vibration) and if a chisel has to be used to clear obstructions. Those piles which are designed to be 'end bearing' are usually driven or socketed some short distance into that bearing stratum which may be rock. With driven piles there are transient shear forces on the pile/soil interface which stress the soil up to yield as well as an area loading at the toe which can be linked to the expansion of a spherical cavity in an elastoplastic material. The 'pile skin' source imparts vertically polarized cylindrical or conical body shear waves and surface waves to the ground, and the pile toe generates compression waves. Both by source

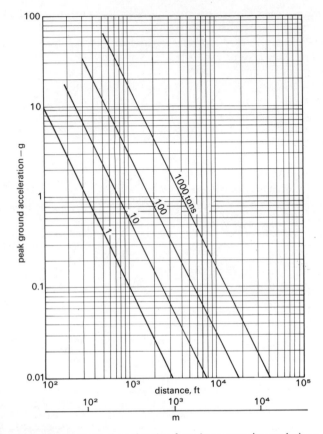

Figure 13.16 Ground acceleration from large cratering explosions.

mechanism and from conversions, surface waves then tend to predominate. When piles are driven to refusal on a bedrock, energy may be transmitted by interface waves and refracted body waves through the bedrock so as to affect structures, perhaps themselves founded on that bedrock at a distance from the source.

13.4.5.1 *Scaled distance relationships—piling.* There have been several studies in which 'scaled distance' laws have been developed. Attewell and Farmer (1973) suggested the relationship:

$$V = 0.76 \left(\frac{\sqrt{W_0}}{r} \right)^{0.87},$$

which may be approximated to

$$V = K' \frac{\sqrt{W_0}}{r}$$

to be applicable. The peak particle velocity V is expressed in mm/s, K' approximates to

0.75, W_0 is the energy of blow impact in joules (N.m), and r is the radial distance from the source in metres. Since most of the source energy arises from the pile toe, the radial distance will often be measured to that point. Similar relationships have been presented by Wiss (1981). These relationships are only convenient ways of presenting empirical data and are prone to dimensional inconsistencies.

Mallard and Bastow (1980) have also suggested that the distance used in this kind of relationship should be between source and detector rather than plan position (equivalent to the difference between an earthquake hypocentre and its epicentre). It would, however, seem to be more appropriate to use plan distance when the pile is to be a friction pile and the source can be treated as a line generating cylindrical or conical waves, taking account of the proportion of energy which would be generated from the pile tip as it is driven into stronger ground. In Fig. 13.18 the attenuation of SRSS velocity with distance in a selection of piling operations is given, and in Fig. 13.19 some examples of the use of scaled distance relationships are given for impact piling. An

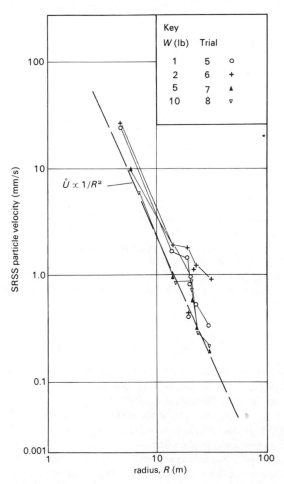

Figure 13.17 (*a*) Ground vibrations on 30° ray path from cratering explosions in dense sand.

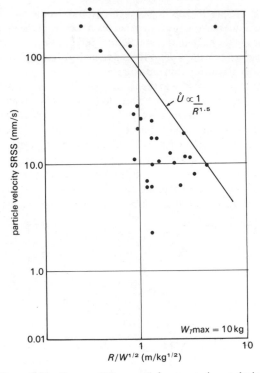

Figure 13.17 (b) Ground vibrations on 30° ray path from cratering explosions in dense saturated sand.

example of the attenuation of ground surface vibration associated with piling vibrators in a specific case is given in Fig. 13.18.

Although general relationships based upon energy scaling are very dispersed, site-specific relationships are much more usable.

There is some disagreement as to the significance of ground conditions, but recent work (Palmer, 1983) points to the importance of this factor along with variations during a driving history.

13.4.6 *Air to ground coupling*

The interaction between a vibration in air and the ground surface has long been referred to by exploration geophysicists as 'ground roll'. The interaction between pressures waves and buildings is well known (Siskind *et al.*, 1980b).

All pressure fluctuations travelling at the speed of sound at ground level can generate a Rayleigh wave in the ground if soils have a low stiffness and density. More energetic disturbances, such as shock waves from air blasts, may couple to the ground non-linearly (Hendron and Auld, 1967). Sonic boom from aircraft can generate a ground response which may initially confuse the seismologist. The impinging of powerful jets upon the ground gives rise to vibrations which in some cases may be measured at great distance, as

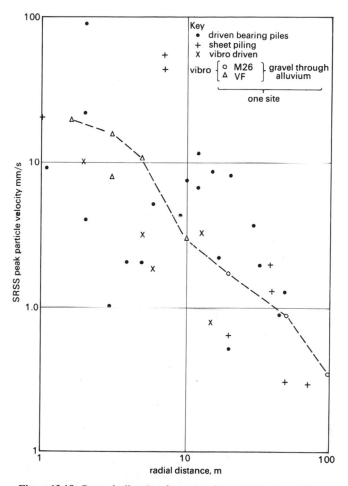

Figure 13.18 Ground vibration from a variety of piling operations.

for example from rocket launches at Cape Canaveral (Heller and Weiss, 1967). A similar phenomenon has been noticed when powerful jet aircraft take off. The maximum ground vibrations are generated as the aircraft rotates to take up its steep climb attitude. The mechanics of such vibrations are not clear but they could arise from coupling of the turbulent cells in the jet or an 'attachment' phenomena at critical angles.

13.5 Measurement of ground vibration

13.5.1 *General*

There are so many types of ground vibration and so many reasons for making measurements that a comprehensive guide on ground vibration measurement is difficult to assemble. There are some review papers (Skipp, 1978), and currently (1983),

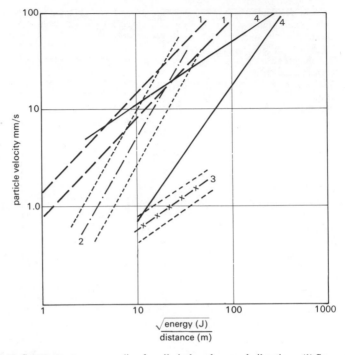

Figure 13.19 Square root energy scaling for pile-induced ground vibrations. (1) Suggested upper bound and design guide (Attewell and Farmer, 1973). (2) Driven bearing piles to sandstone/rock (SRSS) (Mallard and Bastow, 1980). (3) Driven steel bearing piles to gravel through soft alluvium (SRSS) (Soil Mechanics Ltd Report 7980). (4) Envelope for 7 cases with variety of piling and ground (Wiss Envelope, 1981).

a working group of the International Standards Organisation (ISO) is preparing a standard on the measurement of structural vibration which will also touch upon ground vibration measurement.

The fundamental principle is that the measuring methods must be able to characterize the vibration in a way consistent with the purposes of the measurement.

A simple periodic vibration is fully described by its period and maximum amplitude. The duration of the measurement need be relatively short. A complex non-deterministic vibration requires a longer duration of measurement if the spectral characteristics are to be adequately determined. For buildings where the frequencies of interest are usually above 0.5 Hz, several seconds duration of 'random' vibration may be needed to enable the frequency spectrum to be developed with a normalized error of 10%. Unless special signal analysis techniques are adopted, the use of procedures based on 'stationarity' may be in error if applied to short duration and 'non-stationary' signals. There are, however, more complex vibrations such as might emanate from a rock crusher, where in order to characterize motion for the design of isolation mountings much longer durations of measurement must be allowed. Selection of appropriate length of record and sampling intervals for analysis is a serious problem and compromises have to be made (Bendat and Piersol, 1971).

Measurement of vibration is carried out for the following purposes:

(a) Problem recognition: if there is some concern that vibration levels are unacceptably high in a building or on, say, a potentially unstable slope the first objective is to establish approximately whether or not that concern is warranted.
(b) Control, monitoring: where maximum vibration levels have been established by some agency.
(c) Documentation: where some dynamic loading has been recognized in the design of a structure, measurements are made to verify the production of response and provide new design parameters.
(d) Design diagnostic, mitigation: where an existing structure is to be exposed to a new source of vibration, or a new structure is to be constructed at a site when ground vibrations are to be expected, then ground vibrations must be described in a way adequate for designs to be implemented for mitigating their effects.

A measuring exercise may be a simple exploratory monitoring, a more substantial field survey, or a sophisticated project geared to an engineering analysis.

13.5.2 Instrumentation

13.5.2.1 The instrument chain. The instrumentation at its simplest may consist of a peak level recorder wired up to an accelerometer or geophone. This would be adequate for exploratory monitoring, and even for regulatory monitoring if the character of the offending vibration is known and the sensors (accelerometer, geophone) have the correct response properties. With the rapid development of microelectronics many of the constraints imposed on field instrumentation in the past have been removed. While analogue systems have not been displaced, there has been an upsurge of digital systems with great power and microcomputers are now often taken into the field.

Some vibration measuring systems are tailored to specific fields, for example, blast monitoring, and general purposes systems can be built around FM tape recorders, spectrum, and transient analysers.

13.5.2.2 Sensors. A wide range of sensors is available. In the frequency range 4–150 Hz, which encompasses most man-made ground vibration, geophones are still favoured, but there are now suitable piezo accelerometers and force balance accelerometers. These latter two devices may be needed for lower frequencies. The very sensitive range of seismological 'geophones' is usually unsuitable for the measurement of engineering vibrations which cause concern, except for very low-level background studies for special installations. Strong motion seismographs, such as are used in engineering seismology, may be useful for particular cases. They are usually self-triggered but can be operated remotely. With a frequency range of d.c. to about 40 Hz they do have limited applications in 'technical seismicity'. Their sensitivity may be low, however, when the human environmental aspects of ground vibration are involved.

The duration of recording is usually limited to a few seconds on most purpose-made ground vibration instruments. Recordings of longer than 10 seconds, with enough data points to enable frequencies of up to 150 Hz to be resolved, impose considerable demands on digital storage in a typical transient recorder. FM tape records are then favoured.

13.5.2.3 Location and coupling of sensors. The ideal sensor has the same density and stiffness as has the ground. This is in practice not achievable except where 'strain meters' rather than geophones or accelerometers can be used. Strain meters have far too limited a dynamic range and are not convenient for use in engineering practice.

Sensors used for general ground vibration measurement are stiffer and denser than soil. Unless coupled so as to minimize the effect of interaction with the surrounding ground they may have 'pseudo-resonances' which can distort the signal they transmit to the recording chain (Fail et al., 1961; Safar, 1978). It has been shown experimentally that, when measuring vertical vibration, distortion can be minimized by burying the sensor in the ground. This gives a result equivalent to mounting the unit on the surface with a light coupling plate of such dimensions as to ensure that no resonant interaction occurs. Where dense detectors are simply placed on soft surface soils pseudo-resonant amplification is more likely. It can in the worst case nearly double the measured vertical vibration amplitude. If the 'mass ratio', m, is kept below 2, where

$$m = \frac{m_0}{\rho \cdot r_0^3}$$

with m_0 being the mass of the sensor, ρ the soil mass density, and r_0 the radius of the base of the sensor, then the problem can be avoided.

Since the vibrations of concern at the ground surface are often in the form of Rayleigh waves there are dangers in indiscriminately burying transducers if the shorter wave lengths are of interest. Where, however, the foundations of a structure are taken down to a bearing stratum, then that is the proper level for the installation of sensors. Sensors may be clamped to the sides of plastic casing and even mounted on spikes, but the preferred way is to bed them in sand and withdraw the casing. This may be expensive. Sometimes the correspondence between the preferred option and a less expensive alternative (spikes, plastic casing) can be demonstrated and costs reduced.

13.5.3 Evaluation of ground vibration

The evaluation of ground vibration depends largely upon the purpose of measurement. Sometimes it will be sufficient to estimate a peak amplitude and some apparently predominant frequency—a procedure still widely used in blast monitoring (Bollinger, 1971). This requires access to a 'time history', preferably as a permanent record.

Random vibration may be analysed in the frequency or time domains (Bendat and Piersol, 1971; Blackman and Tukey, 1958). The most commonly-used quantity in the frequency domain and in general vibration analysis is the Power Spectral Density. For many purposes in ground vibration studies the Fourier amplitude spectrum is derived. When the prediction of structural response is envisaged, the Response Spectrum (see section 13.6.1) may be calculated. The Response Spectrum can be related to the Fourier amplitude spectrum (Pfaffinger, 1983).

The standard time domain analyses familiar to structural dynamicists (covariance, autocorrelation, cross correlation) are only rarely called up in ground engineering practice. In recent years, however, there have been attempts to develop transfer function analyses of ground to structural response, involving standard random vibration analysis techniques (Splittgerber, 1977; Walker et al., 1982).

13.6 Response of structures and services to ground vibration

13.6.1 Structures

Much of the theoretical analysis and design study of structural response to ground vibration has been directed towards the response of relatively large and important buildings during earthquake ground motion. This has necessitated attention to ultimate limit states and structural stability (CIRIA, 1977). Man-made vibration rarely imposes such loading on structures (except during war or civil disaster circumstances) and the serviceability limit state is more often of concern.

The basic principles of earthquake engineering hold, however, for structural response to man-made ground vibrations, and in recent years some progress has been made in transferring the response spectrum approach to what is often called 'technical seismicity' (Medearis, 1976; Walker et al., 1982). For an exposition of basic earthquake engineering reference should be made to standard texts (Newmark and Roseblueth, 1973; also see relevant Codes such as ATC 3, 1978).

It should be noted that the short duration of most man-made impulsive dynamic loadings through the ground (< 10s) should be taken into account and the adaptation of conventional earthquake engineering procedures would be inappropriate for small close sources of short duration radiating spherical and cylindrical waves.

The simplest approach is to adopt the concept of 'equivalent lateral force'. Horizontal base shear arising from horizontal acceleration is estimated from formulae giving this base shear. They take the form

$$V = C_s W$$

where C_s is a seismic design coefficient and W is the total gravity load of the building. It should be noted that the coefficient is not the maximum effective peak ground acceleration which is used in some semi-empirical analytical procedures to evaluate soil liquefaction potential (Seed, 1979). The 'base shear' approach can be extended to calculate an equivalent static lateral force at other levels of a structure.

The natural frequency of the structure may also be incorporated into the calculations either by modifying the seismic design coefficient or by using the 'response spectrum' approach (Newmark and Rosenblueth, 1973).

The response spectrum is the envelope of maximum response to base motion of a series of simple one degree of freedom mass-spring elements. One can visualize a bank of inverted pendulums each having its own natural frequency responding to horizontal motion of the ground to which they are fixed. For a given transient motion the maximum value of the relative displacement of a pendulum mass is denoted D. The maximum acceleration of that mass is approximately equal to a related quantity, 'pseudo-acceleration', A, and maximum velocity to a pseudo-velocity, $V = 2\pi f D$. For a system with zero damping, $A = (2\pi f)^2 D = 2\pi f V$. The maximum response (displacement, velocity, acceleration) is usually plotted on tripartite logarithmic paper and is usually normalized either to the maximum acceleration or velocity, or to a value of peak horizontal acceleration of 1 g.

If a building is thought of as a simple shear beam having natural frequencies in shear and torsion, the maximum 'whole' body response to a ground motion can be estimated by entering the design spectrum at the appropriate frequencies, reading off the maximum response, and then scaling according to, say, the peak ground acceleration measured in the time domain. 'Design' response spectra may be assembled from experimental data and the envelopes developed statistically (see Fig. 13.9).

Damping, up to relatively low values, can be incorporated in such design spectra. For high intensity loading into the plastic range, response spectra for elasto-plastic systems having ductility have been developed.

It should be noted that the quantity usually adopted in simple response spectra is not the true displacement, velocity or acceleration, but a 'pseudo' variety of these.

The assumption is made that the horizontal accelerations are applied at effectively the same instant in time across the base of the structure. This is a reasonable assumption to make where the time it takes for a travelling wave to cross the foundation is small compared with the natural period of that structure. This condition would not hold for high-frequency energy with low propagation velocity (for example, precision blasting in rock near to long buildings on superficial deposits). Some indication of the high frequency filtering potential of a simple foundation is given by Bycroft (1957). This restricts the applicability of some simple vibration peak particle velocity criteria. Recent work in Sweden (Holmberg, 1981) has attempted to include travelling wave factors in an estimate of permissible strains in foundations.

13.6.2 Earthworks

Concern is often expressed for the stability of embankments, cuttings, river frontages, retaining walls and so on when exposed to man-made vibration. Empirical data are sparse, although there is limited 'negative' information available in the form of earthworks that have sustained without damage the effects of intense vibrations from earthquakes. The use of analysis extends the already imperfect procedures of earthquake engineering into this field. It is important to take into account the 'strain softening' of the soils. An example would be the shaking of a modest earth embankment by quarry blasting.

If the main excitation is from surface waves in, say, a dense superficial soil then the wave lengths could be of the order of several tens of metres, and for small structures the time taken for the energy to travel across the base would be small compared with the time taken to travel its height. This permits relatively simple slip circle stability calculations to be made. For other geometries, account would have to be taken of the natural period of the embankment (Sarma, 1975, 1981).

If the embankment materials show no significant loss of strength with strain or cyclic softness, the vibrations, which with a combination of horizontal and vertical accelerations bring the embankment into a state of limiting equilibrium, can be accommodated with small deformation. If, however, the critical accelerations are often repeated, the accumulated deformation may then become unacceptable.

When soils are vulnerable to pore pressure accumulation and loss of strength it may be necessary to evaluate their reaction by one or other of the 'tuned analytical' procedures described by several authorities such as Sherif and Ishibashi (1982), Finn (1981), and Seed (1979). Recourse may have to be made to numerical modelling (Zienkiewicz et al., 1980).

13.7 Levels of concern

13.7.1 General

Ground vibration transmitted to structures may render them unserviceable or reduce their structural integrity. The amplitudes of ground motion at which such effects are manifest cannot rationally be laid down without reference to the response of the structure. There are, however, a number of codes and guides, to which reference can be made. Some are very conservative indeed. An often-used code is the German DIN 4150

(1970, 1975). This refers to measurements made at the foundations of structures and uses peak particle velocity as a single parameter index.

The use of a single kinematic parameter to indicate damage potential of a ground vibration is widely used, notwithstanding the fundamental objections to that approach. Empirical data bases are rarely transferable to different settings, that is, type of building, type of ground, type of source. Hence, there is a proliferation of 'national' codes reflecting national circumstances (see Table 13.2). Particle velocity has attractions in that it can be related on structural members to fibre stress (Gasch, 1967)*.

Peak stress, taking account of travelling waves, has been used by such workers as Dvorak (1962, 1970). Studies on actual strains in buildings have been made by Splittgerber (1969) and Fang and Kroemer (1977). There is very little data on strain limits under cyclic loading for ordinary building materials.

Although peak particle velocity is now the preferred index, numerous other measures have been suggested. Early criteria for blasting vibration used particle displacement (Teichmann and Westwater, 1957). Crandell (1949) suggested an 'energy ratio'.

Peak particle velocity criteria suggested by the United States Bureau of Mines have had considerable influence. They relate to surface mining. In Scandinavia the peak particle velocity criteria reflect the widespread use of blasting in hard rock for construction purposes (Langefors and Kihlström, 1978).

German work has lead to the DIN 4150: 1970 and 1975 (provisional) standards which

---

* The stresses in beams or plates vibrating close to resonance can be calculated from measurement of velocity, or displacement and frequency, if the measurement is performed at the points of maximum vibrating displacements.

For beams with full rectangular cross-section and constant stiffness and weight loading, the following relationship applies, independent of the length, height and width of the beam, between the largest bending stress σ_{max} and the vibration velocity $\hat{\omega}_{max}$:

$$\sigma_{max} \approx \sqrt{E_{dyn} \cdot \rho} \cdot \sqrt{\frac{3G_{ges}}{G_{beam}}} \cdot k_n \cdot \hat{\omega}_{max}$$

where

$\hat{\omega}_{max} = \hat{\omega}_{max} \cdot \omega_n$ is the maximum amplitude of the vibration velocity occurring at a point along the beam length,

ω is the forcing frequency approximately equal to ω_n (natural frequency of the beam),

E is Young's modulus of elasticity,

ρ is the mass density of the beam,

$\dfrac{G_{ges}}{G_{beam}}$ is the load coefficient, where the beam is stressed by other evenly distributed loads in addition to its own weight,

$G_{ges} = G_{beam} + G_{other\ loads}.$

and k_n is the mode coefficient (dimensionless) having values of 1 to 1.33.

The eigen-mode coefficient k_n is dependent on the boundary conditions and the degree of the mode, which only has a slight influence.

Gasch (op. cit.) has extended this approach to a consideration of random motion.

Table 13.2   A selection of national standards.

(a) *The DIN Specification* 4150 (1970 *draft*)
The standard recommends for intermittent and shock vibrations the measurement
of the resultant peak particle velocity v_R arising in the foundation of the building.
This entails the measurement of the three velocities on orthogonal axes $v_x(t)$, $v_y(t)$
and $v_z(t)$, and the resultant velocity is given by the expression:

$$v_{Rmax} = \sqrt{v_x^2(t) + v_y^2(t) + v_z^2(t)}$$

In practice since the three components v_x, v_y and v_z have maxima occurring at
different times, it is usual to make use of a 'substitute resultant' found from the
three maximum peaks.
Representative values of the resultant vibration velocity v_{Rmax} for use in assessing
vibrations from sudden shocks:

Class of building	Nature of building	v_{Rmax} permissible (mm/sec)
I	Ruins and damaged buildings, protected as monuments	2
II	Buildings with visible defects, cracks in masonry	4
III	Undamaged buildings in technically good condition (apart from cracks in plastering)	8
IV	Well stiffened buildings (e.g. industrial buildings)	10–40

(b) *The DIN provisional specification* 4150 (1975)
Damage criteria are given in Part 3 of the Specification. The 1975 provisional
specification criteria are similar to those in the 1970s draft, but differ in detail.
The guide values for maximum peak particle velocity $\bar{v}_R$, determined as given in
C3.5, are given for the frequency range 0–60 Hz:

Guide values for the resultant vibration velocity at the building foundation for the
assessment of the effects of shocks of short duration

Class of building	Type of building	Guide values for $\bar{v}_R$ (mm/sec)
1	Residential buildings, commercial premises and structures similar in their construction which are satisfactorily maintained in compliance with the generally accepted rules of good construction practice	8
2	Well-braced structures made of heavy structural elements and well-braced framed structures, which are satisfactorily maintained in compliance with the generally accepted rules of good construction practice	30
3	Structures other than those referred to in rows 1 and 2 and structures covered by a preservation order	4

Notes:
(a) If the blasting is more frequent than say twice a working day the guide values of the
velocities should be reduced to two-thirds of those given in the table.

(b) In general no damage is to be expected when the peak resultant velocity is less than 2 mm/sec.

(c) The guide values are not suitable when there is a possibility of ground compaction by vibration.

(d) The guide values in the table are suitable for impact hammer pile driving provided the blows are infrequent enough for the vibrations to dissipate before the next blow occurs.

(e) For vibrations in ceilings from blasting and pile driving guide values of up to 20 mm/sec in the vertical direction (v_z) may be permissible.

(f) The guide values are not suitable for piling driven by vibro-hammer which gives rise to continuous vibration.

It is recommended that one axis of measurement should be towards the source of vibrations or parallel to one of the side walls of the building. In large buildings the vibrations should be measured in several places simultaneously including the foundations, the ceiling and upper storeys.

(c) *Swiss Association for Standardization* (1978)

	Frequency bandwidth Hz	Blasting induced PPV mm/s	Traffic or machine-induced PPV mm/s
Steel or reinforced concrete			
structures such as factories,	10–60	30	
retaining walls, bridges,	60–90	30–40	
steel towers, open channels,	10–30		12
underground tunnels and chambers	30–60		12–18
Buildings with foundation walls	10–60	18	
and floors in concrete, walls	60–90	18–25	
in concrete or masonry, under-	10–30		8
ground chambers and tunnels	30–60		8–12
with masonry linings			
Buildings with masonry walls	10–60	12	
and wooden ceilings	60–90	12–18	
	10–30		5
	30–60		5–8
Objects of historic interest or	10–60	8	
other sensitive structures	60–90	8–12	
	10–30		3
	30–60		3–5

(d) *Standards Association of Australia ASC A23-1967*

Category of structure	Limiting value of v_R (minor, non-structural damage)
1. Buildings containing sensitive equipment	15 mm/s
2. Residential structures	25 mm/s
3. General commercial structures not included in 1 or 2	50 mm/s

Table 13.3   Older criteria for building damage and vibration.

(a) *Caution limits for displacement amplitude* (*blasting*)

Type of property	Allowable amplitude, μm
Civil engineering structures	760
Isolated properties	400
Closely congregated houses and properties	200
Structures of great value and frailty (e.g. ancient monuments and churches) and properties in poor condition	100

(b) *Crandell's* 1949 *energy ratios* (*blasting*)

Energy ratio (ER)		Likelihood of damage
Imperial ft^2/sec^2	Metric m^2/sec^2	
less than 3	less than 0.28	safe
3 to 6	0.28 to 0.56	requires caution—old structures in poor condition likely to be damaged
greater than 6	greater than 6	dangerous to all structures

(c) *Langefors criteria* (*blasting*) (*Langefors and Kihlström*, 1978)

Peak particle velocity (mm/sec)	Possible damage effect
less than 75	no noticeable cracks
75–100	insignificant cracking and fall of plaster
100–150	cracks
150–225	serious cracks

(d) *Sior damage criteria* (*piling*)

Sior's (1961) degree of vibration	Strength in vibrars	Possible damage effect
1	0–30	none
2	30–40	light damage such as plaster cracks but cracks in loadbearing structural units possible
3	40–55	damage in loadbearing structural units possible
4	greater than 55	damage in loadbearing structural units, destruction of entire structure

(e) *Zeller's criteria* (*continuous*)

Rating or degree	Zeller value mm^2/sec^3	Description
1–3	0–2 500	no damage
3–4	2 500–10 000	possibility of fine cracks in plaster
4	10 000–25 000	possibility of small cracks in plaster
6	$1-5 \times 10^5$	possible slight structural damage
7	$5-20 \times 10^5$	serious cracking
8	greater than 20×10^5	severe (destructive)

(f) *Damage criteria in vibrars* (*continuous*)

Strength of vibration (vibrars)	Classification	Possible damage effect
0–10	imperceptible	none
10–20	light	none
20–30	medium	none
30–40	strong	light damage (cracks in rendering)
40–50	heavy	severe (damage to main walls)
50–60	very heavy	destruction

(g) *Damage Figure criteria* (*continuous*)

Damage Figure (mm^2/sec^3)	Equivalent in vibrars	Extent of damage
0–50	0–26	none
50–500	26–36	small cracks in rendering
500–2000	36–42	occasional slight cracks in walls
2000–7000	42–47	cracks extend to main walls

where $R = 2\pi^2 A^2 f^3$ is the Energy Unit or Damage Figure.

use a combined peak particle velocity criterion, itself the square root of the sum of squares of the peak particle velocities in three orthogonal components measured at the foundation of the structure (SRSS).

Criteria for damage related to 'continuous' vibration have grown from the work on human perception. Zeller proposed an index 'Z' defined as acceleration²/frequency with units mm²/s³ which was further developed by Koch (1953; in Steffans, 1974) as the Vibrar ($S = 10\log(Z/10)$). These measures were related to human response, but it was suggested that there was a risk of slight structural damage when the value of the Vibrar S exceeded 40 (Sior, 1961). Koch (*op. cit.*) did however suggest another Damage Figure which was defined as the 'energy per unit mass vibrating for a time which is a quarter of the period of the vibration'. This is clearly a response-based criterion. These guides are summarized in Table 13.3.

Building Research Establishment 278 (1983) summarizes recent guides for building and human response.

Current codes do attempt to incorporate a frequency element* into the criteria, but often in a confused way. It is not always clear whether the 'frequency' refers to that of the excitation or the responding structure. Some current criteria in use are given in Tables 13.3, 13.4 and Fig. 13.20. Many agencies internationally are accumulating data on ground vibration and building damage. No universally accepted criteria are currently available. Recent work in Sweden (Holmberg *et al.*, 1981 and Fig. 13.21) contains many case history reports of damage caused by blast vibration and has shown that damage ratios are similar to those of Siskind *et al.* (1980a).

Recognition of the necessity to take the response of structures into account has led to the use of response spectra for the evaluation of blasting vibrations. Medearis (1976) has demonstrated for a set of low-rise structures that damage could be correlated best

* The DIN 4150 standard to be published in 1983 has a frequency classification.

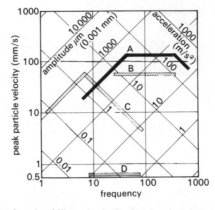

Figure 13.20 Damage and noticeability criteria (limit values) as functions of frequency. Line A represents the threshold for damage to normal buildings on hard rock. Line B represents the limit enforced by the Stockholm police authorities to prevent building damage. Line C is the upper limit for vibrations allowed in the supports of a large computer. Line D is the limit where vibration is just perceptible by a human being.

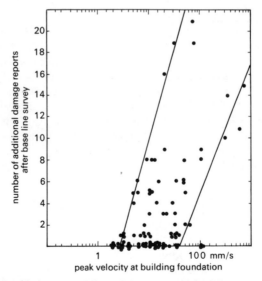

Figure 13.21 Relationship between additional damage at the final inspection and peak velocity (after Holmberg, 1981).

with the peak spectral velocity response. (see Fig. 13.22). Extensive studies (Siskind *et al.*, 1980*a*) have led to a probabilistic assessment of threshold damage in terms of a peak particle velocity measured at the foundations of low rise structures (Fig. 13.23). Very low levels have been known to propagate superficial cracks in plaster (Wall, 1967). This is unlikely to be a true threshold in terms of PPV for all circumstances.

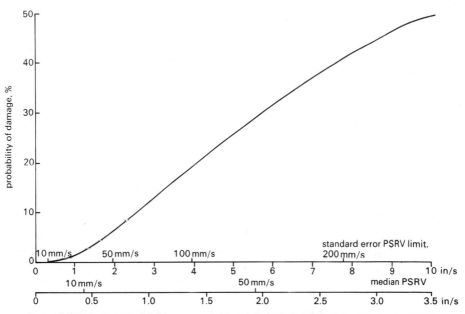

Figure 13.22 Probability of damage v, standard error PSRV limits (after Medearis, 1976).

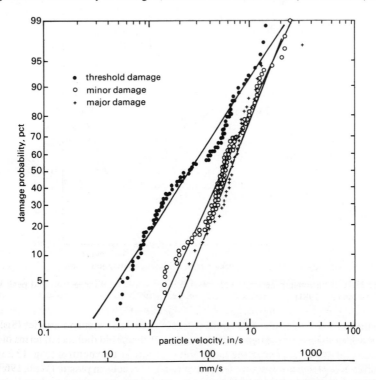

Figure 13.23 Damage probability and peak particle velocity (after Siskind *et al.*, 1976).

13.7.2 *Factors to consider in assessing potential damage from ground vibration*

In understanding the significance of ground vibration as it affects the serviceability or integrity of buildings it is important to take account of the following points in addition to measuring amplitudes.

    (a) Transient, intermittent or continuous character as related to the response of the building.
    (b) Frequency content of vibration input.
    (c) Duration of excitation.
    (d) Natural frequency of structure and its elements (walls, floors, chimney, windows).
    (e) Damping of the structure.
    (f) Proximity of the source as compared with building base dimension—is the travelling wave front planar, cylindrical, or spherical?
    (g) Wave length and propagation velocity of waves.
    (h) Type of construction and condition of the building.

In recent years there has been a process of critical assessment of empirical codes and guidelines and an extension of the database. Different countries have developed criteria based upon their own styles of construction, and social and economic climates. It can be unwise to use a national code out of context. In some circumstances the understandably conservative guidelines used in developed, environment-sensitive countries may be inappropriate. Structural dynamics analysis with criteria based upon response displacement, stress, fatigue and so on may be economically justified for important and sensitive structures. There are, however, areas of considerable uncertainty, especially where long-term effects are to be considered (Crockett, 1979).

13.7.3 *Buried structures*

The use of peak particle velocity in ranges similar to that adopted for low-rise buildings as an index for buried services has often been shown to be unduly conservative. As for buried structures there is no database to justify its use. Some estimate of peak stresses induced should be made.

Tunnels are generally capable of sustaining without distress more vibration than surface structures. Statutory authorities would normally accept PPV levels in the range of 15–30 mm/s for tunnels in reasonable condition.

13.7.4 *Buried services*

Public authorities are often concerned that construction-generated ground vibrations do not endanger their services. It has become conventional practice to use peak particle velocity as a limiting criterion. Recent research has shown this to be unduly conservative. The basis on which such low values as 10–15 mm/s have been arrived at is a calculation which assumes that the buried service deflects with the ground as a simple periodic ground wave is propagated through it. It probably includes a safety factor of around 10 if referred to maximum allowable strain in malleable cast iron pipes for example. Soil-structure interaction is neglected. Recent empirical relationships between pipeline stresses and explosive charge levels have been suggested by Esparza *et al.* (1981). Attewell (1983) and Attewell and Fry (1982) have also reviewed the problem. Pipes subjected to the effects of about 1 kg explosive detonated 2 m away are at significant risk. Recent theoretical earthquake engineering work indicates that neglect of soil-structure interaction can result in significant overestimates of induced strains, and hence pipe stresses (see Hwang and Lysmer, 1981).

Empirical experience suggests that buried services in good condition can sustain much higher vibration levels than current conservative levels would indicate. In the U.K. and other old industrial areas, pipes may be in poor condition. Nevertheless it must be recognized that under travelling loads (lorries on roads) local high-frequency vibration and a travelling 'flexure' could be damaging. This situation is not dynamically comparable to that of a travelling wave from a distant source. It sould also be noted that vibration may cause ground settlement and this may endanger a service.

13.7.5 *Vulnerability of ground*

The levels of vibration which would affect the ground itself range from the particle velocity criteria for the rupture of rocks—spalling effects in tunnels are seen at velocities of about 300 mm/s (Holmberg and Persson, 1979; Kiel *et al.*, 1979) to the threshold strains (10^{-4}) below which it appears that loose cohesionless deposits do not become densified (Pyke *et al.*, 1975). Loss of strength in saturated sands subjected to shaking is a well-studied phenomenon, usually referred to as 'liquefaction' and associated with earthquakes, but similar phenomena are seen when ground is subjected to dynamic (impact) consolidation or explosive compaction. Youd and Perking (1978) have shown Holocene materials, in which the velocity of propagation of shear waves is less than 400 m/s, to be vulnerable to liquefaction in earthquakes, whereas older Pleistocene deposits having a shear wave velocity greater than 400 m/s are relatively immune from liquefaction. Age sometimes has its advantages! As a very rough guide, young loose saturated sand or coarse silt-size deposits should be scrutinized if they are likely to be subjected to vibration with accelerations greater than 0.03 g for more than 5 significant stress reversals within the frequency range 0–50 Hz. Analysis of vulnerability of deposits to liquefaction is a well-developed subject (Seed, 1979), but is still controversial (Peck, 1979).

13.7.6 *Human sensitivity*

Most difficulties with man-made ground vibration relate to human perception and 'pollution' rather than to actual damage. Reference should be made to Steffans for criteria established prior to 1974. Subsequent work is described by Splittegerber (1977). The Reiher Meister Scale (Fig. 13.24) indicates zones of human sensitivity in a double logarithmic plot of frequency versus amplitude of displacement in microns.

Dieckmann introduced 'K' values which depend on frequency and also reflected findings that people responded to acceleration at lower frequencies in the range up to 100 Hz, to velocity in the middle range and displacement in the upper range.

German Standard 4150 (1939) introduced the Pale scale of intensity, with units by *S*, where

$$S = 20 \log 2.24(v_e/v_0)$$

where v_e is the RMS of particle velocity, and v_0 is the threshold value of 0.316 mm/s. The

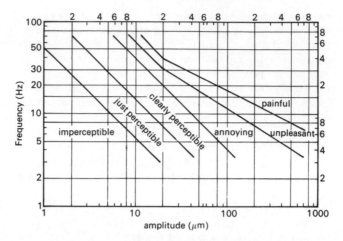

Figure 13.24 Human sensitivity: Reiher-Meister scale (vertical vibrations).

Table 13.5 Older criteria for human response to vibration

Effect on people (after Dieckmann)   (upper limits for each description):

K value	Description
0.1	Lower limit of human perception
1.0	Allowable in industry for any period of time
10.0	Allowable only for a short time
100.0	Upper limit of strain or endurance for the average man

Classification of K values (DIN 4025):

K value	Classification	Effect on work
0.1	Threshold value—vibration just perceptible	not affected
0.1–0.3	Just perceptible—easily bearable, scarcely unpleasant	not affected
0.3–1.0	Easily noticeable—bearable but moderately unpleasant if lasting for an hour	still not affected
1–3	Strongly noticeable—still tolerable but very unpleasant if lasting over an hour	affected, but possible
3–10	Unpleasant—can be tolerated for periods up to one hour, but not for longer	considerably affected, but still possible
10–30	Very unpleasant—cannot be tolerated for more than 10 minutes	barely possible
30–100	Extremely unpleasant—not tolerable for more than one minute	impossible
Over 500	Intolerable	impossible

Calculation of Dieckmann K-values:

Vertical vibrations		Horizontal vibrations	
Below 5 Hz	$K = 0.001\,Af^2$	Below 2 Hz	$K = 0.002\,Af^2$
5–40 Hz	$K = 0.005\,Af$	2–25 Hz	$K = 0.004\,Af$
Above 40 Hz	$K = 0.2\,A$	Above 25 Hz	$K = 0.1\,A$

where A is the displacement amplitude in microns and f is the frequency in Hz.

The Pal (DIN) scale of intensity:

Strength of vibration (Pal)	Sensation or effect
0–5	just perceptible
5–10	clearly perceptible
10–20	severely perceptible (annoying)
20–40	unpleasant

Vibrar Scale:

Vibrars	Power of vibration
0–10	imperceptible
10–20	light
20–30	medium
30–40	strong
40–50	heavy
50–60	very heavy

$$Z = \frac{a^2}{f} = 16\pi^4 A^2 f^3 = \text{Zeller value}$$

$$S = 10\log\left(\frac{Z}{10}\right)\text{vibrars}$$

$$S = 20\log 2.24\,v\ \text{Pals}$$

where v = peak particle velocity.

expression simplifies to $S = 20\log 2.24\,v$, where v is the peak particle velocity. It should be noted that these relations, and indeed the experimental data base, are for continuous periodic vibrations (see Table 13.5). The Vibrar has already been described. In the DIN 4150 (1975) the measure KB is introduced, which can be derived from acceleration, velocity or displacement amplitude. The criteria for human response are summarized in Table 13.6.

The International Standards Organisation (1978) have produced a basic document entitled *Guide to the evaluation of human exposure to whole body vibration*. Three limits are named: 'reduced comfort boundary', 'fatigue decreased proficiency boundary' and 'exposure limit'. The basic level curve in Fig. 13.25 is on a double logarithmic plot of RMS velocity versus frequency and is used in conjunction with various multiplying factors. An addendum suggesting a satisfactory magnitude for persons in buildings has been put forward (see Fig. 13.26 and Table 13.7).

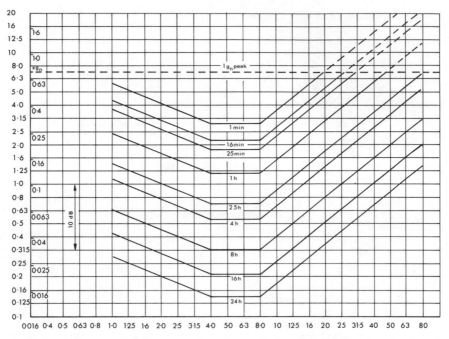

Figure 13.25 Human sensitivity to vibration—longitudinal acceleration limits as a function of frequency and exposure time (from ISO 2631: 1978). To obtain (1) 'exposure limits', multiply acceleration values by 2 (6 dB higher); (2) 'reduced comfort boundary', divide acceleration values by 3.15 (10 dB lower). Ordinate, acceleration (m/s^2) rms; abscissa, frequency or centre frequency of third octave band (Hz).

Table 13.6 National and international standards relating to human response
DIN 4150 1975 *Provisional:*
Guide values for the assessment of vibrations in dwellings or comparable rooms

Construction area	Time	KB-Guide	values
	By day (0600 to 2200 hr) By night (2200 to 0600 hr)	Continuous vibrations & those occurring repeatedly with interruptions	Rarely-occurring vibrations
1. Purely residential, general residential, weekend house, smallholding	by day	0.2 (0.15)	4.0
	by night	0.15 (0.1)	0.15
2. Mixed central	by day	0.3 (0.2)	8.0
	by night	0.2	0.2
3. Commercial, including offices	by day	0.4	12.0
	by night	0.3	0.3

4. Industrial	by day	0.6	12.0
	by night	0.4	0.4
5. Special area according to type of usage and proportion of housing	by day	0.1 to 0.6	0.15 to 0.4
	by night	0.1 to 0.4	0.15 to 0.4

Acceleration: $\qquad KB = a \dfrac{\alpha}{\sqrt{1 + \left(\dfrac{f}{f_0}\right)^2}}$

Velocity: $\qquad KB = v \dfrac{\beta f}{\sqrt{1 + \left(\dfrac{f}{f_0}\right)^2}}$

Displacement: $\qquad KB = A \dfrac{\gamma f^2}{\sqrt{1 + \left(\dfrac{f}{f_0}\right)^2}}$

where $\quad a =$ vibration acceleration in m/sec^2
$v =$ vibration velocity in mm/sec
$A =$ vibration displacement amplitude in mm
$f =$ frequency in Hertz
$f_0 = 5.6$ Hz (reference frequency)
α, β, γ are constants which for values of a, v and A are:
$\alpha = 20.2\,\text{m}^{-1}\,\text{sec}^{-2}$
$\beta = 0.13\,\text{mm}^{-1}\,\text{sec}^{-2}$
$\gamma = 0.80\,\text{mm}^{-1}\,\text{sec}^{-2}$

The KB value determined for the particular location should be compared with the table above. In the table:

(a) 'Continuous vibrations and vibrations occurring repeatedly with interruptions' are those which occur continuously for longer than 2 hours without, or even with interruptions of short duration.
(b) 'Rarely-occurring vibrations' are occasional occurrences of short duration, such as are caused by up to three blasting explosions per day.
(c) With extremely rare vibrations such as 1 or 2 blasting explosions per week the guide values given in Column 4, Row 4 are to be applied in Row 1 and 2 construction areas, if after prior notification the explosions take place in day time outside the rest periods 0600 to 0700, 1300 to 1500 and 1900 to 2200 h.
(d) With vibrations which are confined to a few days and take place only during daylight hours, such as pile driving and blasting performed during the course of construction work, guide values may be used up to double those in Column 4, provided the damage limits are not exceeded.
(e) The guide values in brackets should not be exceeded when structures are subjected to horizontal vibrations of 5 Hz and below.

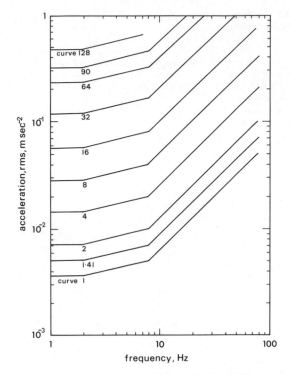

Figure 13.26 Human sensitivity to vibration in buildings (after ISO 2631 : DADI). See Table 13.7. Worst case combination of x, y, z axes.

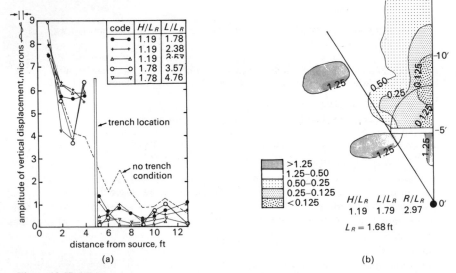

Figure 13.27 Mitigation by trenches (1). Physical models: surface wave generation (after Richart *et al.*, 1970). (a) Amplitude of vertical displacement *v*. distance from source for five tests (from Woods, 1968). (b) Amplitude-ratio contour diagram.

Table 13.7 Human vibration in buildings (ISO 2631: DADI). Weighting factors for acceptable building vibration (factors applied to basic level of Fig. 13.26).

Place	Time	Continuous or inter-mittent vibration, repeated impulsive shock	Impulsive shock excitation with not more than 3 occurrences per day
Hospital operating theatre and critical working areas	Day	1	1
	Night	1	1
Residential (minimum complaint level)	Day	$2^{(3)}$	16^5
	Night	1.41	1.41^5
Office	Day	$4^{(3)}$	128
	Night	4	128
Workshop	Day	$8^{(3)}$	128
	Night	8	128

Note 1: Curve 1 for the simplified method represents a rms velocity of 0.995×10^{-4} m/s above 8 Hz, an acceleration of 5×10^{-3} m/s^2 at 8 Hz and an acceleration of 3.6×10^{-3} m/s^2 below 2 Hz. The rms velocity above 8 Hz corresponds for sinusoidal motion to a peak velocity of 1.41×10^{-3} m/s.
Note 2: Levels of acceptable impulsive shock in hospital operating theatres and critical working areas could be the same as in residential areas, provided there is due agreement and warning.
Note 3: The figures given for continuous and intermittent vibration, together with repeated impulsive shock for residential, office and work shop areas, represent good environmental standards. If vibration levels are increased beyond these levels by a factor of up to two, there will be moderate complaint, and an increase by a factor of up to four will lead to major complaint.
Note 4: The levels for impulsive shock excitation occurring not more than three times per day in offices and work shop areas cannot be increased without some possibility of structural damage occurring, as well as a significant disruption of working activity.
Note 5: In BS 6472 these factors become 60–90 and 20 respectively, for intermittent and impulsive vibration, having several occurrences per day.

13.8 Mitigation

Measures to reduce the undesirable consequences of excessive ground vibration may involve actions at its source, on its pathway, or at the position of the structure or service affected.

Reduction of vibrations transmitted to the ground from machines requires provision of isolating elements between machine base and foundations. Provision must be made to dissipate the unwanted vibration energy, otherwise the reductions in ground-transmitted vibration are achieved at the expense of excessive oscillation on the machine itself, so leading to problems with services and process connections. Reduction in the actual out-of-balance force may sometimes as achieved and, in favourable circumstances where the troublesome motion has a fixed frequency, dynamic absorbers can help (Alloway and Grootenhuis, 1978). Machine isolation is a specialist area of technology that is well covered by standard texts (see, for example, Major, 1962). Where motion is not simply deterministic, optimum isolation and damping may prove difficult to design.

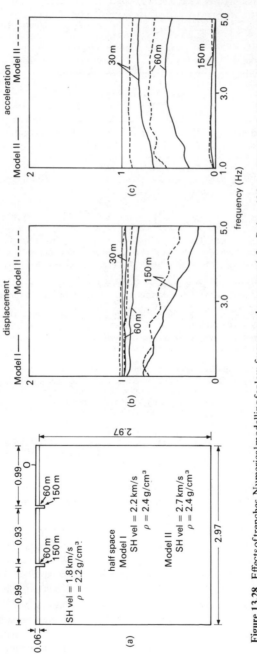

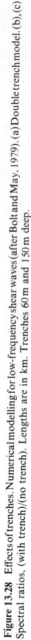

Figure 13.28 Effects of trenches. Numerical modelling for low-frequency shear waves (after Bolt and May, 1979). (a) Double trench model. (b), (c) Spectral ratios, (with trench)/(no trench). Lengths are in km. Trenches 60 m and 150 m deep.

The vibrations generated by traffic are reduced when the running surface is made smoother and speed is reduced.

Reduction of vibration from explosive sources is aided by good blasting practice, which is not inconsistent with economic advantage (Anderson, et al., 1982). Good public relations is advisable (Ashley, 1976). Millisecond delay blasting over short distances and longer delays for larger distances reduce the probability of constructive interference between vibrations from individual shot holes (Devine and Duvall, 1963); Langefors and Kihlström, 1978).

Piling vibrations may be reduced by changing the type of piling, its energy and coupling condition by means of a piling helmet, for example, or incorporation of an 'air cushion' as mentioned earlier (BSP ID 17 hammer). Pre-boring is often helpful if the piles are to terminate some distance into rock. Controlled in situ fragmentation by explosives may ease later driving and justify an explosion-induced vibration. Where piling vibrations could induce potentially troublesome pore-water pressure rises in saturated soils, steps can be taken to alter driving sequences to allow localized increments to dissipate.

Vibrations arising from impacts such as in demolitions can be significantly reduced by arranging for the object to fall on to a pile of rubble or even on a placed 'cushion' of suitable material such a colliery waste. Vibration levels may then be reduced by factors of two to four.

Reduction of vibration by such measures as trenches or barriers (Figs. 13.27, 13.28) has been much studied in the last 20 years (Barkan, 1960; Richart et al., 1970; Bolt and May, 1979). Although it has attractions where high-frequency energy (possibly as high-mode Rayleigh waves) are troublesome to sensitive equipment, the trenches to be effective must be at least one-third of a wavelength deep, which may preclude economic or even safe construction.

Modification of building-to-ground coupling by spring supports is now quite feasible and has been undertaken in several cases (Anon, 1979; BSI 1975).

References

Abbis, C.P. (1981) Shear wave measurement of the elasticity of the ground. Géotechnique 31 (1) 91–104.

Alloway, P.H. and Grootenhuis, P. (1978) The generation of and isolation of machinery vibration. In Proc. Conf. Control of Odours & Noise in the Process Industries, London, May 23 1978, Institution of Chemical Engineers, pp. 18–26.

Ambraseys, N.N. and Hendron, A.J. (1969) Dynamic behaviour of rock masses. In Rock Mechanics in Engineering Practice, eds. Stagg, K.G., Zienkiewicz, O.C., John Wiley, London, ch. 7, pp. 203–236.

Anderson, D.A., Winzer, S.R. and Ritter, A.P. (1982) Blast design for optimising fragmentation. In Proc. of Eighth Annual Convention on Explosives and Blasting Techniques, New Orleans, pp. 1–21.

Anon (1979) Isolation for London housing development. Rubber Developments 32 (7) 48–51.

Applied Technology Council (1978) Tentative Provisions for the Development of Seismic Regulations for Buildings. ATC Special Publication, 3.06.

Ashley, C. (1976) Blasting in urban areas. Tunnels and Tunnelling March 1967, 55–58, 63–67.

Attewell, P.B. and Ramana, Y.V. (1966) Wave attenuation and internal friction as functions of frequency in rocks. Geophysics 31 (6) 1049–1056.

Attewell, P.B. and Farmer, I.W. (1973) Attenuation of ground vibration from pile driving. Ground Engineering 6, 26–29.

Attewell, P.B. and Fry, R.H. (1982) The effects of explosive detonations and mechanical impacts on adjacent buried pipelines. In 'Europipe' '83' Conference, Basle, paper 16, pp. 123–128.

Attewell, P.B. (1983) Assessing the effects of explosive detonations and mechanical impact upon adjacent buried pipelines. *WRC Engineering Centre External Report* No. 98E, April 1983, p. 95.

Awojobi, A.O. and Sobayo, O.A. (1974) Ground vibration due to a seismic detonation in oil exploration. *Earthquake Eng. Struct. Dyn.* **3**, 171–181.

Awojobi, A.O. and Sobayo, O.A. (1977) Ground vibrations due to a seismic detonation of a buried source. *Earthquake Eng. Struct. Dyn.* **5**, 131–143.

Barkan, D.D. (1960) *Dynamics of Bases and Foundations*, McGraw–Hill Book Co., New York.

Bata, M. (1971) The effect on buildings of vibration caused by traffic. *Building Sciences* **6**, 221–246.

Beam, R. and Page, J. (1976) *Traffic-induced Vibration in the Vicinity of Road Tunnels*. Dept. of Environment. TRRL Report No. SR 2186, Crowthorne, Berks, England.

Bendat, J.S. and Piersol, A.G. (1971) *Random Data Measurement and Analysis Procedures*. Wiley–Interscience, New York.

Blackman, R.B. and Tukey, J.W. (1958) *The Measurement of Power Spectra*. Dover Publications, New York.

Blake, F.G. (1942) Spherical wave propagation in solid media. *T. Acoust. Soc. Am.* **34**, 211–215.

Bollinger, G.A. (1971) *Blast Vibration Analysis*. Southern Illinois University Press, and Fetter & Simons Inc., London.

Bolt, B.A. and May, T.W. (1979) The effectiveness of trenches and scarps in reducing seismic energy. In *Proc. 2nd U.S. Nat. Conf. Earth Eng.* Stanford, pp. 1104–1113.

Borcherdt, R.D. (1974) Rayleigh type surface wave on a linear viscoelastic half space. *J. Acoust. Soc. Am.* **55** (1) 13–15.

Brennan, B.J. Smylie, D.E. (1981) Linear viscoelasticity and dispersion in seismic wave propagation. *Dev. Geophysics Space Phys.* **19**, 223–246.

British Standards Institution (1975) Vibration isolation of structures by elastomeric bearings.

BS 6472: 1984 British Standards guide to evaluation of human exposure to vibration in buildings (1 Hz–80 Hz).

BS 4675: TPI 1976/ISO 2372–1974 Mechanical vibration in rotating and reciprocating machinery.

BSI (1974) Code of Practice on Foundations for Machinery, Part 1. Foundations for Reciprocating Machines. CP 2012 Part 1.

Broch, J.T. (1972) *Mechanical, Shock and Vibration Measurement*. Bruel and Kjaer, May 1972, Copenhagen.

Broch, J.T. (1980) *Mechanical Vibration and Shock Measurement*. (2nd edn) Bruel and Kjaer, Denmark.

Building Research Establishment (1983) Vibrations: building and human response. Digest 278, Oct. 1983.

Bycroft, G.M. (1956) Forced vibration of a rigid circular plate on a semi-infinite elastic half space and on an elastic stratum. *Phil. Trans. Roy. Soc. London* **248**, 327–368.

Bycroft, G.M. (1957) The magnification caused by partial resonance of the foundation of a ground vibration detector. *Trans. Am. Geoph. Union* **38** (5) 928–930.

Cherry, J.J. and Peterson, F.L. (1970) Numerical simulation of stress wave propagation from underground nuclear explosions. (IAEA PL 338/15). *Peaceful Nuclear Explosives*, IAEA, Vienna.

CIRIA (1977) *The rationalisation of safety and serviceability factors in structural codes*. Report 63, London.

Control of Pollution Act (1974) HMSO (London).

Crampin, S. (1980) *A review of wave motion in anisotropic and cracked media*. Inst. Geol. Sci. Global Seism. Unit. Report 132, October.

Crandell, F.J. (1949) Ground vibration due to blasting and its effect on structures. *Bos. Soc. C.E. Jnl.* **36** (2) 222–245.

Crockett, J.H.A. (1979) Piling vibrations and structural failure. In: *Recent Developments in Design and Construction of Piles*, Institution of Civil Engineers, London.

Den Hartog, J.P. (1956) *Mechanical Vibrations*. McGraw-Hill Book Co., New York.

Devine, J.F. and Duvall, W.I. (1963) Effect of charge weight on vibration levels for millisecond delayed quarry blasts. Earthquake Notes. N.E. Section, *Bull. Seismol. Soc. Am* **34**, 1202–1217.

Dieckmann, D.A. (1958) A study of the influence of vibration on man. *Ergonomics* **1** (4) 347–355.

DIN 4150 (1938) Vibrations in buildings—influence of constructions, Aug. 1983.

Docher, P. (1980) Compaction of loess by saturation and explosives. *Proc. Int. Conf. Compaction, Paris* 1, 313–317.

Duval, W.I., Atchison, T.C. and Fogelson, D.E. (1966) Empirical approach to problems in blasting research. *Proc. Symp. Rock Mech. Minnesota. Am. Inst. Min & Petr. Eng.* 1967, New York, Chapter 20.

Dvorak, A. (1962) Seismic effects of blasting on brick houses. *Geofysikalni Sbornik*, No. 169, pp. 189–2202 (in English).

Dvorak, A. (1970) Vibrations for tunnel blasting in cities. *Geofysikalni Sbornik* XVIII, No. 328, pp. 311–322 (in English).

Ewing, W.M., Jardetsky, W.S. and Press, F. (1957) *Elastic Waves in Layered Media*. McGraw Hill Book Co. New York.

Esparaza, E.D., Westine, P.S. and Wenzel, A.B. (1981) *Pipeline Response to Buried Explosive Detonations*. Final Report to the American Gas Association, by the South West Research Institute, Project PR-15-109 (SWRI Project No. 02–5567).

Fail, J.P., Grau, G. and Lavergne, M. (1961) Couplage des seismographes avec le sol. In 21*st Meeting European Association of Exploration Geophysicists, Trieste*, Dec. 1961, Vol. 10, No.2, pp. 128–147.

Fang, N.Y. and Kroemer, R.M. (1977) Soil structure interaction during blasting. In *Int. Symp. on Soil Structure Interaction, University of Roorkee, India*, Vol. 1, pp. 151–156.

Finn, Liam W.D. (1981) Liquefaction potential—developments since 1976. In *Int. Conf. on Recent Devel. in Geotechnical Earthquake Eng. and Soil. Dyn., Missouri*, Vol. 2, pp. 655–681.

Forsbladd, L. (1967) *Investigation of soil compaction by vibration*. Acta Polytechnica Scandinavia, Stockholm.

Franetti, G.E. (1977) Near-field elastic ground response spectra for multiple surface explosions. In *Proc. of 3rd Conf. on Explosives and Blasting Techniques, Pittsburgh*, pp. 49–117.

Fryba, L. (1962) *Vibration of Solids Under Moving Loads*. Noordhoff International Publishing Co., Groningen, Holland.

Gasch, R. (1967) The evaluation of dynamic stresses in structural parts of buildings. *VDI–Bericht* 113, pp. 77–81.

Graff, K.F. (1975) *Wave Motion in Elastic Solids*. Clarendon Press, Oxford.

Greenwood, D.A. and Kirsch, K. (1983) Specialist ground treatment by vibratory and dynamic methods. In *Int. Conf. on Advances in Piling and Ground Treatment for Foundations*, Inst. of Civil Engineers, London, pp. 17–45.

Heller, L.W. and Weiss, R.A. (1967) Ground transmission from surface sources. In *Int. Symp. on Wave Propagation and Dynamic Properties of Earth Materials. Univ. of New Mexico. Alberquerque*, pp. 71–83.

Hendron, A.J. and Auld, H.E. (1967) The effect of soil properties on the attenuation of airblast induced ground motions. In *Int. Symp. on Wave Propagation and Dynamic Properties of Earth Materials, Univ. of New Mexico, Albuquerque*, pp. 29–42.

Holmberg, R. and Persson, P.A. (1979) Design of tunnel perimeter blast hole patterns to prevent rock damage. In *Proc. Conf. 'Tunnelling' '79'*, ed. Jones, M.J. Institute of Mining and Metallurgy, London, pp. 280–283.

Holmberg, R., Lundborg, N. and Rundquist, G. (1981) *Ground Vibration Damage Criteria*. Swedish Council for Building Research (BFR), Report R85.

Horton, C.W. (1953) On the propagation of Rayleigh waves in the surface of a viscoelastic solid. *Geophysics* 18, 70–74.

Hwang, R.N. and Lysmer, J. (1981) Response of buried structures to travelling waves. *J. Geotechnical Engineering Division, ASCE*, 107, 183–200.

International Standards Orgaisation (1975) *Vibration and Shock Vocabulary*, pp. 750–2041.

International Standards Orgaisation (1978) Mechanical Vibration of Machines with Operating Speeds from 10–200 revs/s. No. 2372.

International Standards Organisation (1978) Guide for the evaluation of human exposure to whole body vibration. ISO 2631 2nd Edition.

International Standards Organisation (1983) Vibration and shock—methods for analysis and presentation of data. ISO/DIS 4865.

Ivanov, P. (1980) Consolidation of saturated soils by explosives. In *Proc. Int. Conf. on Compaction, Paris*, Vol. 1, pp. 334–337.

Iyer, W.M. and Hitchcock, T. (1976) Seismic noise survey in Long Valley, California. *J. Geophysical Research* **81** (5) 821–846.

Johnson, S.M. (1971) *Explosive Excavation Technology*, NCG Technical Paper No. 21, June 1971.

Kiel, L.D., Burgess, A.S. and Nielsen, A.N. (1977) Blast vibration monitoring of rock excavations. *Canadian Geotech. J.* **14**, 603–619.

Knopoff, L. (1956) The seismic pulse in materials possessing solid friction. *Bull. Seism. Soc. Amer.* **46**, 175–183.

Kolsky, H. (1963) *Stress Waves in Solids*, Dover Publications, New York.

Kriebel, T.S. (1980) Demolition of high rise flats by controlled explosives. *Building Technology and Management*, **18** (4) 3–7.

Lamb, H. (1904) On the propagation of tremors on the surface of an elastic solid. *Phil. Trans. Roy. Soc.* **203** (359A) 1–42.

Lande, G. and Rundqvist, G. (1981) *Measurement, recording and analysis of traffic generated vibrations.* BFR 770252–5 Uptec 81, p. 44, April 81.

Lande, G. (1974) Relation between traffic generated vibrations, their frequency, particle motion, displacement, velocity, and speed of the truck. Uppsala University No. 47.

Langefors, U. and Kihlström, B. (1978) *The Modern Techniques of Rock Blasting* (3rd edn.) J. Wiley & Sons, New York.

Larssen, D.B. (1982) Explosive energy coupling in geological materials. *Int. J. Rock Mech. Min. Sci. & Geomech. Abst.* **19**, 157–166.

Love, H. (1906) On wave patterns due to a travelling disturbance. *Phil. Mag.* **13**, 539–548.

Lyakov, G.M. (1961) Shock waves in the ground and the dilatancy of water saturated sand. *Zharucol Prikladnoy Mekhaniku Tekhnicheskoy Fiziki*, Moscow, **1**, 38–46.

Lysmer, J. (1978) Analytical procedure in soil dynamics. In *Proc. Geol. Eng. Dio., ASCE (Speciality Conference Earthquake Engineering and Soil Dynamics, Pasadena USA)*, Vol. 3, pp. 1207–1316.

Mallard, D.J. and Bastow, P. (1980) Some observations on the vibration caused by pile driving. In *Proc. Conf. on Recent Developments in Design and Construction of Piles*, ICE, London, pp. 261–284.

Major, A. (1962) *Vibration Analysis and Design of Foundations for Machines and Turbines.* (2nd edn.) Académiai Kiadó, Budapest.

Major, A. (1980) *Dynamics in Civil Engineering.* Académiai Kiadó, Budapest (Collets Holdings, London), Vol. 4.

Martin, D.J., Nelson, P.M. and Hill, R.C. (1979) Measurement and analyses of traffic induced vibration in buildings. *TRRL Supplementary Report* 402, Crowthorne, Berks.

Martin, D.J. (1980) Ground vibrations from impact pile driving during road construction. *TRRL Supplementary Report* 544, Crowthorne, Berks.

Medearis, K. (1976) *The development of rational damage criteria for low rise structures subjected to blasting vibrations.* A report to the National Crushed Stone Association, Washington, D.C., Aug. 1976.

Ménard, L. and Broise, Y. (1975) Theoretical and practical aspects of dynamic consolidation. *Géotechnique* **25**, (1) 3–18.

Miller, G.F. and Pursey, H. (1954) The field and radiation impedance of mechanical radiators on the free surface of a semi-infinite isotropic solid. *Proc. Roy. Soc.* **223A**, 521–541.

Miller, G.F. and Pursey, H. (1955) On the partition of energy between elastic waves in a semi-infinite solid. *Proc. Roy. Soc.* **233A**, 55–59.

Nelson, I., Baron, M. and Sandler, X. (1971) Mathematical models for geological materials for wave propagation studies. In *Shock Waves and Mechanical Properties*, Syracuse University Press, New York.

New, B.M. (1982) Vibration caused by underground construction. In *Proc. 'Tunnelling '82', Conf.*, ed. M.J. Jones, IMM, London, pp. 217–229.

New, B.M. (1982) Personal communication on vibration caused by trains.

Newland, D.E. (1975) *An Introduction to Random Vibrations and Spectral Analyses.* Longman, Harlow.

Newmark, M. and Rosenblueth, E. (1973) *Fundamentals of Earthquake Engineering.* Prentice–Hall Inc., Englewood Cliffs, N.J.

Novak, M. (1974) Dynamic stiffness and damping of piles. *Can. Geotech. J.* **11**, 574–598.

O'Brien, P.N.S. (1969) Some experiments concerning the primary pulse. *Geophysical Prospecting* **17**, 511–547.

Page, J. (1973) *Dynamic behaviour of two linked-twin-axle lorry suspension systems: a theoretical study*. TRRL Report CR 581, Crowthorne, Berks.

Palmer, D. (1983) *Ground-borne vibration from piling*. CIRIA RP 299 (in draft).

Peck, R.B. (1979) Liquefaction potential, science versus practice. *J. Geotechnical Engineering Division, ASCE*, **105** (G73) 393–398.

Pekeris, C.L. (1955a) The seismic buried pulse. *Proc. Nat. Acad. Sci.* **41**, 460–480.

Pekeris, C.L. (1955b) The seismic surface pulse. *Proc. Nat. Acad. Sci.* **41**, 629–639.

Pekeris, C.L. and Lifson, H. (1957) Motion of the surfaces of a uniform elastic half space produced by a buried pulse. *J. Acoustic Soc. America* **29** (11) 1232–1238.

Persson, P.A., Holmberg, R., Lande, G. and Larsson, B. (1980) Underground blasting in a city. In *Subsurface Space*, Pergamon Press, Oxford, Vol. 1, pp. 109–206.

Pfaffinger, D.D. (1983) Calculation of power spectra from response spectra. *J. Eng. Mech., ASCE* **109** (1) 351–372.

Pyke, R., Seed, H.B. and Chan, C.K. (1975) Settlement of sands under multidirectional shaking. *J. Geotech. Eng. Div., ASCE* **103** (GT4) 379–398.

Richart, F.E., Hall, J.R. and Woods, R.D. (1970) *Vibration of Soils and Foundations*. Prentice–Hall Inc., Englewood Cliffs, N.J.

Ricker, N. (1977) *Transient Wave Forms in Visco-Elastic Materials*. Elsevier, Amsterdam.

Rundqvist, G. (1981) *Criteria for acceptable traffic generated vibrations: I People in their homes; II Risk of damage to buildings*. BFR. 770259 5 UPTEC 81.44.12 April. Stockholm.

Sackmani, J.L. (1962) Uniform moving load on a layered half space. *Trans. ASCE* 1 (271) 823–837 (EM4).

Safar, M.H. (1978) On the minimisation of distortion caused by geophone-ground coupling. *Geophysical Prospecting* **26**, (3) 538–549.

Sarma, S.K. (1975) Seismic stability of earth dams and embankments. *Géotechnique* **25** (4) 743–761.

Sarma, S.K. (1981) Seismic displacement analysis of earth dams. *J. Geotechnical Eng. Div., ASCE* **107** (GT12) 1735–1739.

Scott, R.A. and Pearce, R.W. (1975) Soil compaction by impact. *Géotechnique* **25** (1) 19–30.

Seed, H.B. (1979) Soil liquefaction and cyclic mobility evaluation for level ground during earthquakes. *Proc. ASCE.* **105**, (GT4) 201–235.

Sharpe, J.A. (1942) The production of elastic waves by explosion pressures. Theory and field observations. *Geophysics* **7**, 144–154.

Sherif, M.A. and Ishibashi, I. (1982) A rational theory for predicting soil liquefaction, *Soil Dynamics and Earthquake Engineering* **1** (1) 20–29.

Sior, G. (1961) The damaging effects of vibration due to pile driving operations. *Bautechnik* **38** (6), 161–185 (BRS translation LC 1283).

Siskind, D.E., Stagg, M.G., Kopp, J.W. and Dowding, K.H. (1980a) *Structure response and damage produced by ground vibration from surface mine blasting*. US Bureau of Mines, R.I. 8507.

Sisking, D.E., Stachura, U.J., Stagg, M.G. and Koop, J.W. (1980b) *Structure response and damage produced by air blast from surface blasting*. US Bureau of Mines, R.I. 8485.

Skipp, B.O. and Tayton, J. (1971) Blasting vibrations—ground and structure response. In *Dynamic Waves in Civil Engineering*, Wiley-Interscience, pp. 182–212.

Skipp, B.O. and Marriot, M. (1973) The measurement and interpretation of ground vibration. In *Field Instrumentation in Geotechnical Engineering*, Part 1. Butterworths, pp. 396–410.

Skipp, B.O. and Buckley, J.S. (1977) Ground vibration from impact. In *9th Int. Conf. on Soil Mech. & Found. Eng., Tokyo*, Vol. 2, pp. 397–400.

Skipp, B.O. (1978) Ground vibration instrumentation—a general review. In *Proc. Conf. on Instrumentation for Ground Vibration and Earthquake*, ICE, London, pp. 11–34.

Splittgerber, H. (1969) Horizontal strains of buildings due to blasting. *VDI III* pp. 709–13, 1185–92.

Splittgerber, H. (1977) Einflusse die Starke von Erschutterungen bei Gewinnungsprengen (summary in English). *Schriftenreihe der Landesanstalt für Immissionsschutz* **42**, 86–105.

Steffans, R.J. (1974) *Structural vibrations and damage*. Building Research Establishment Report H.M.S.O., London.

Teichmann, G.A. and Westwater, R. (1957) Blasting and associated vibration. *Engineering*, London **183** (4753) 460–464.

Thoenen, J.B. and Windes, S.C. (1942) Seismic effect of quarry blasts *U.S. Bureau of Min. Bull.* 442.

Trulio, J.G. (1966) Theory and structure of the AFTON codes. *Report No. AFWL-TR* 66–19, Air

Force Weapons Laboratory, Kirtland, U.S.A.

Walker, S., Young, P.A. and Davey, P.M. (1982) Development of response spectra techniques for prediction of structural damage from open pit blasting vibration, *Trans. Inst. Min. Metall. (Section A, Mining Industry)* **91**, A53–A62.

Wall, J.F. (1967) Seismic induced architectural damage to masonary structures at Mercury, Nevada. *Bull. Seism. Soc. Am.* **57** (5) 991–1007.

Whiffin, A.C. and Leonard, D.R. (1971) A survey of traffic induced vibrations. Department of the Environment and the Department of Transport. *TRRL Report* LR 418, Crowthorne, Berkshire, England.

Winzer, S.R. (1978) *The firing times of MS delay blasting caps and their effect on blasting performance.* NSF DAR 77.05171, Martin Mariellor Labs, Baltimore, U.S.A., p. 36.

Wiss, J.F. (1981) Construction vibrations 'State of the Art'. *J. Geotechnical Engineering Div., ASCE,* **107**, 176–181.

Youd, T.L. and Perking, D.M. (1978) Mapping liquefaction induced ground failure potential. *J. Geotechnical Engineering Div., ASCE,* **104** (GT4) 433–446.

Zienkiewicz, O., Leung, K.H., Hinton, F. and Chang, C.T. (1980) Earth dam analysis of earthquake—numerical solution and constitutive relations for non-linear damage. In *Proc. Conf. on Design of Dams for Earthquake,* ICE, pp. 141–156.

Index